Food Science and Technology Series

FOOD PROCESSING AND ENGINEERING TOPICS

Food Science and Technology Series

Food Science and Technology: New Research
Lorenzo V. Greco and Marco N. Bruno (Editors)
2008. ISBN: 978-1-60456-715-1

The Price of Food
Meredith N. Fisher (Editor)
2009. ISBN: 978-1-60692-440-2

Food Processing and Engineering Topics
Maria Elena Sosa-Morales and Jorge F. Vélez-Ruiz (Editors)
2009. ISBN: 978-1-60741-788-0

Food Science and Technology Series

FOOD PROCESSING AND ENGINEERING TOPICS

MARIA ELENA SOSA-MORALES

AND

JORGE F. VÉLEZ-RUIZ

EDITORS

Nova Science Publishers, Inc.

New York

For permission to use material from this book please contact us:
Telephone 631-231-7269; Fax 631-231-8175
Web Site: http://www.novapublishers.com

LIBRARY OF CONGRESS CATALOGING-IN-PUBLICATION DATA

Sosa-Morales, Maria Elena.
Food processing and engineering topics / Maria Elena Sosa-Morales and Jorge F. Velez-Ruiz.
p. cm.
Includes index.
ISBN 978-1-60741-788-0 (hardcover)
1. Food--Testing. 2. Food texture. 3. Rheology. 4. Food--Quality. I. Vélez-Ruiz, Jorge Fernando, 1955- II. Title.
TP372.5.S66 2009
664'.02--dc22

2009012381

Published by Nova Science Publishers, Inc. † New York

CONTENTS

PREFACE

'Food Processing and Engineering Topics' offers a combination of theoretical support, practical examples, process applications, and recent findings on diverse aspects of food science and engineering, such as rheology, heat transfer, evaporation, osmotic dehydration, air drying, ultrasound and deep-fat frying. Topics upon selected fluids, powders, cheese, concentrated foods, and frozen dough are also included. Presenting an interesting, complete and current vision of important food processing and food engineering, food products and food technologies, the manuscript is a useful tool for teaching, processing and researching. The book could be used as a textbook by students, finding in it some academic themes such as: rheological applications an its relation with moment transport and flow, measure of textural attributes for cheese, particle size distributions for food powders; also, the fundamentals of heat transfer focused to explain the convective heat transfer evaluation, the heat transfer complications due to the fouling formation, and the evaporation of food liquids; mass transfer principles and applications on osmotic concentration, air drying, and frying; and finally some innovative and practical applications of ultrasound, baking and frying will complete the panorama. Industrial people could use this work as a tool for specific food items or problems, like rheology of some liquid foods, particle distributions of food powders, measurement of cheese texture, approaches for analysis of fouling of heat transfer exchangers, effect of evaporation on food properties; furthermore, they will find recent information and applications of osmotic and air dehydration, combined treatments on fried foods, ultrasound and baking in food processing. Researchers may compare their results with some data presented in tables and graphics included in each chapter.

Chapter 1 - Fruit and vegetable fluids represent a very extensive and important sector of the food industry. Most of them are manufactured by industrial processes that change their original structure generating a fluid in which the dispersed material contributes significantly to the rheological behavior. Therefore the specific processing, solid concentration, solid particle and temperature affect the flow response. Vegetal fluids, such as apple, mango, orange, peach and pear juices on one side; avocado, pea, potato, and tomato purees, among others behave as Newtonian or non-Newtonian fluid as a function of particular characteristics and process treatment. Physical properties of vegetal fluids are needed for evaluation and design of equipment and processes, being the rheological properties one of the most studied and required.

Chapter 2 - Stress and strain are the main parameters involved in the measurement of flow behavior of fluids. Fluids can be studied by subjecting them to a constant shear rate to measure shear stress. Ideally, this can be accomplished placing a fluid between two parallel plates using a predetermined gap. The lower plate is fixed and the upper plate moves at a selected velocity. Simple shear flow is also called viscometric flow. It includes axial flow in a tube, rotational flow between concentric cylinders, rotational flow between a cone and a plate, and torsional flow between parallel plates. Several mathematical functions can describe the relationship between shear stress and shear rate data. Newtonian and non-Newtonian behavior can be observed when testing different types of fluid or semi-fluid foods. Newtonian fluids have a linear relation between shear rate and shear stress. Non-Newtonian fluids, on the other hand, can behave in different ways, which will depend on the composition and structure of the food system. Several models to describe the flow behavior of fluid foods are explained in this chapter. Examples of selected liquid and semi-liquid foods will be given along with their flow behaviour. Also, the rheological properties assessed using various rheological models and the temperature effect will be discussed for four selected food products products.

Chapter 3 - Particle size distribution (*PSD*) is one of the most important characteristics of food powders, the distribution depends on the kind of food powder or mixture of different food powders. There are several factors that may modify the *PSD* in a food powder system such as processing, handling, storing, and transportation. Changes in *PSD* may bring up alteration on their appearance, physical properties, functionality and nutritional characteristics caused by attrition, segregation orand agglomeration. Several methods are used to assess *PSD* (microscopy observations, resistive pulse technique, optical methods, sieve analysis, sedimentation-photometry, and centrifuging method). AlsoFurther, various distribution models can be used to fit *PSD* (normal function, log-normal distribution, Rosin-Rammler or (Weibull) distribution, Gates-Gaudin-Schuhmann distribution, modified Gaudin-Meloy function, and modified Beta function). However, information in literature, regarding *PSD* models, is limited. Reported dData in literature are only presented in a table or in graphical representation, but no fitting of particle size using distribution models are available. Reported particle size data and experimental data of selected foods (freeze-dried coffee, spray-dried coffee, wheat flour, granulated sugar and starch) were plotted and modeled using selected distribution functions.

Chapter 4 - Cheese as dairy product, is one of the first, most popular and universal elaborated food item. Cheese represents a product in which the milk components are preserved. Cheesemaking practice has changed from an incipient and experimental art to an industrial technology within strong scientific support. Since Rrheology studies the deformation of solids and the flow of fluids, when the instrumental characterization is done on cheese, is essentially a quantitative expression of its physical behavior, particularly it is a measurement of its mechanical properties. Cheese as many solid foods hasis a viscoelastic in nature.

In 1995, more than 2000 types of cheese already existed (Gunasekaran and Ak, 2003), although only 500 have been recognized by the International Dairy Federation (Fox, 2003), and the variety still is augmenting every time, due to the presence of milk from different mammalian species, the diversity of manufacturing methods, and the consumption amplitude by human population.

Chapter 5 - Food process operations frequently involve heating or cooling stages between fluids and solid phases, being the heat transfer by conduction and convection the dominant mechanisms, that in many of the processes both are presenthappen simultaneously.

The increasing interest of the Food Industry for manufacturing high quality, nutritious and safe food products, as well as modeling of thermal processes and design of food equipment, have resulted in a need for a precise evaluation of the convective heat transfer coefficient. It is necessary to understand the contribution of this convective heat transfer coefficient to the overall heat transfer processes involving liquid foods and solid surfaces, as well as solid foods within Newtonian and non-Newtonian fluids, for an accurate evaluation and supply of the heat transfer process and equipment.

Chapter 6 - Plate and tubular heat exchangers are widely used in food industry, whatever is the production of the food item. They are more utilized in to handle of liquid foods, such as beer, concentrates, juices, milk fluids, nectars, soups, wine, and yogurt, just to mention a few. Food industry faces a common problem of fouling formation inside of heat exchanger equipments; it is a frequent concern in which the information is scarce.

Fouling may be defined as an accumulation or deposition of undesirable material on a heat transfer surface. It decreases the heat flowtransport as well as fluid velocity flow, by reducing the heat transfer efficiency and increasing the pressure drop. In the food industry, fouling deposits can also act as sites for bacteria growing, causing product safety concerns. Fouling in the food industry is therefore an important economic factor.

It is well known that surfaces of the plate and tube heat exchangers do not remain clean after some time of operation. Scaling or deposits formeded over the interior and/or exterior surfaces, may be analyzed as an "static" resistance, but the phenomenon is really a gradual build up of layers of damaged major food components, salts, minerals, dirty impurities, among others. Those formed deposits can severely affect the heat transfer equipment and process, particularly affecting the overall heat transfer coefficient. Two pictures of fouled surfaces may be appreciated in figure 1. When fouling becomes substantial, cleaning needs to be carried out, sometimes taking an important time that stops the normal production.

Chapter 7 - Many liquid foods are concentrated to decrease water content implying some considerations, such as enhancing of shelf life, inducing consistency and flavor changes, reduction of volume and weight, and diminishing of energy consumption for drying. The concentration of liquid foods by evaporation is usually utilized for liquid foods such as milk, fruit juices, vegetable purees, soups and syrups; whereas other concentration processes are applied to a limited extent (Vélez-Ruiz, 2008). Products like cheese whey, fermentation wastes, molasse residues, sea water among others, also can be concentrated by evaporation (APV, 2003).

Food evaporation is the unit operation in which a liquid food is heated by means of high quality steam, reaching the boiling point of the solution. T, then the generated vapor, in equilibrium with the liquid food is separated from the rest of the solution. C; consequently, the remnant food increases its solids content and a preservative effect is gotten due to a reduction inof the water activity. lowering.

Chapter 8 - The world market for fried foods is continuously growing and it represents great quantities of money. Important products/snacks such as chicken nuggets, French fries, fried chicken, meat balls and beef burgers, regional items, toasted tortillas, among others, have a very high demand and consumption. These and other food products include a frying treatment as main part of the manufacturing process, in which oil or fat is incorporated to the product. On the other side, consumer preferences for low and free fat food items is a present driven force to modify them in the food industry.

In recent years, some research efforts have been focused in how to reduce oil uptake of fried items. Thus, some treatments and innovations such as dehydration, edible films, thermal treatments, vacuum application, and vapor injection in combination with immersion frying have been tested, in order to get fried foods of high quality with lower fat content.

The chapter covers the fundamentals and characteristics of the frying process, basic modeling, physical properties of fried items, and some obtained modifications of fried foods when one of the aforementioned treatments is applied. Experimental and reported data also have been included.

Chapter 9 - Food preservation and supplying are the main goals for the food industry. Food preservation involves a diversity of knowledge oriented to maintain nutritional, biochemical, physical, and other properties of food items. In order to preserve those food properties at the best conditions, the characteristics of the food process operations should be correctly identified to get maximum benefits.

Among the many transformation processes used for making foods, such as refrigeration, freezing, heating, concentration, frying, and dehydration, among others, they are classified in several groups depending of the main preservation agent. Each one of them offers characteristics and advantages that may be combined to get a better food product. Furthermore, the combination of processes has been and should be oriented to influence positively the properties of the final product.

Most of the studies about the combination of methods, also known as hurdle technology, have covered microbiological, chemical, and quality aspects and a minor quantity of works have focused on engineering and physical aspects. Nowadays, foods based on combined preservation methods are prevalent in developing and industrialized countries. These foods, just to mention some, may be grouped as chilled foods, fermented foods, functional foods, heated foods, intermediate and high moisture foods, healthful foods, and packaged foods depending of the main preservation factor or final purpose; all of them are the result of different and combined preservation processes. This technology of hurdle combination has allowed the preservation or developing of new products, such as: fruits and vegetables in Latin America, meats in China, probiotic food items in Europe, and dairy products in India.

Although the combination of food preservation methods is very extensive and covers a diversity of aspects, the aim of this chapter is to cover the most important aspects of two mass transfer process operations, osmotic dehydration and air drying and the combination of both technologies to improve the food properties of the final food products.

Chapter 10 - Among the emerging technologies used as alternatives to traditional thermal processing, ultrasound has demonstrated to be a good option to provide safe food with high quality characteristics.

Applications of ultrasound in food processing can be classified in regards of the applied intensity. Low-intensity ultrasound (high frequency, 2 - 10 MHz) does not cause chemical or physical changes in the material where the waves pass through, being a non-destructive technique which is commonly used for diagnostic tests. In contrast, high-intensity ultrasound (low frequency, 20-100 kHz) can cause physical damage and promote certain chemical reactions. The advantages provided by high-intensity ultrasound had led to investigate numerous applications in food and chemical engineering: emulsion homogenization, cell rupture, microbial destruction, enzyme inactivation, enhanced filtration and extraction, among others.

This chapter will present high-intensity ultrasound research/ advances in food processing for microbial destruction (bacteria, yeasts and molds) non-thermal juice processing and homogenization and pasteurization of milk for yoghurt processing. Microbial destruction studies include mold inhibition in model systems subjected to different stress factors (a_w, pH, synthetic or natural antimicrobials), and prediction of microbial survival. Predictive modeling in fruit juices (orange and grapefruit) was utilized to validate ultrasound treatments. For dairy processing, studies concerning the application of ultrasound as a pre-treatment of pasteurization will be discussed, as well as milk homogenization by means of ultrasound for yoghurt preparation.

Chapter 11 - Bakery is one of the oldest food industries. In Mexico, the sales of bakery products in 2000 were estimated over US$1,700 millions, and for 2003 they reached up US$2,300 millions (INEGI, 2008). As it can see, bakery industry keeps a big growing up trend, not only in Mexico, also around the world. Research and development haves been conducted to increase the diversity of breads types, formulations and presentations, even preservation technologies for the bakery product as well. Because bread is a perishable product (mainly conceived as softness loss), some resources have been employed to avoid its deteriorate, such as use of additives, temperature reduction for dough or pre-cooked products and packaging in modified atmospheres.

Breads taste is much better when they are just baked, because thedue to aroma and texture are optimal. Frozen dough may be an alternative to reduce losses and costs; only the required dough pieces are placed into the oven and thus, the consumer can get just baked products at any time. To get this advantage, the design of process for frozen dough and potential use of adequate additives to improve the final bread characteristics must be investigated.

Pabisan, S.A. is a Mexican local industry devoted to the manufacturing of frozen bread, which is commercialized in the rest of the country. Several studies have been conducted to get products of good quality such as biscuits, sponge cake bread and Danish bread. In this Chapter, conditions for processing, storage and tested additives blends for this kind of sweet breads are presented. The effect on density, texture, color and sensory perception of the final product are discussed. Some engineering aspects, such as dough freezing point and description of freezing process are included as well.

Chapter 12 - Doughnuts, also called cake donuts or simply donuts, are a kind of bread characterized by a golden-brown exterior color, a crisp crust and an inner core that resembles a baked product more than a fried food (Shih *et al.*, 2001). Doughnuts are the most famous fried bakery product. Their market represented US$3-4 billion business in the US in the last years (Shi and Daigle, 2002).

Based on the origin of volume increase, two types of doughnuts are distinguished: doughnuts prepared with baking powder (sodium carbonate) and doughnuts leavened through a fermentation process carried out by yeasts (*Saccharomyces cerevisiae*). Fermentation may be done in two stages, the first stage involves around 90 min, when only a part of flour, water and the total of yeast are mixed. The resultant fermented dough is called “sponge”. Then, the rest of ingredients (flour, egg, butter, egg, water or milk) are mixed to form the dough and the sponge will be added lately. The final dough is sheeted, and cut in rings, the typical doughnuts shape. Dough rings are submitted to the second fermentation stage for 30 min, where the increasing of volume is high. This procedure is known as the sponge-cake method

(Pyler, 1988). When yeast-leavened doughnuts are prepared, the recommended fermentation time is 90 min and the frying temperature is around 190°C (Amendola, 1993).

Being doughnuts a fried product, they absorb substantial amounts of oil/fat during the frying process. The excessive consumption of fat and oils has been related with health concerns, such as obesity, cardiac diseases and digestive troubles. Then, the current trend of consumers is related to reduce the oil content in foods. Food technologists have the challenge to answer that concern and to develop doughnuts without excess of oil content and keeping the desirable properties of the traditional food (Shih *et al.*, 2001).

This chapter is devoted to present the characteristics of the traditional fried doughnuts and the recent advances in the development of formulations to improve nutrition factors of this food, such as to increase the protein and fiber content, reduce the caloric content and, overall, diminish the oil content in this appreciated fried productfood.

ABOUT THE EDITORS

Maria Elena Sosa-Morales is an associate professor at Chemical and Food Engineering Department in the Universidad de las Américas, Puebla (Mexico). She has been advisor of more than 40 theses, and has co-authored several chapters in books and articles. She received her B.Sc. in Food Engineering from Universidad de Guanajuato in Mexico (1997), and her M.S. in Food Science from Universidad de las Americas, Puebla (1999). She has served in Food Industry during several residences. Her research interests are focused to physical properties of foods, frying, baking and recently, microwave applications for foods, area in which developed doctoral studies. She received her PhD from Instituto Tecnológico de Veracruz, Mexico in 2009.

Jorge F. Vélez Ruiz was born in Puebla city, Mexico. He earned a B.Sc. degree in Biochemical Engineering (1977) and his M.Sc. in Food Science (1981) from the ITESM (Instituto Tecnológico y de Estudios Superiores de Monterrey), Mexico. After one year of work (1977-1978) in the food industry and thirteen years of teaching and researching (1980-1993), he traveled to USA to get the Ph.D. in Food Engineering from Washington State University, Pullman WA in 1996. He has visited WSU (Summer 1999 and Summer 2006) as researching residence, and IATA (Instituto de Agroquímica y Tecnología de Alimentos at Valencia, Spain; 2003-2004) as sabatical year. He has directed more than 100 theses of B.Sc, M.Sc. and Ph.D., is author and coauthor of several book chapters and scientific papers. He is part of the National System of Researchers in Mexico, since 1990. His research activities are focused on physical properties of foods; food engineering topics; rheology of foods; dairy products and milk processing; heat and mass transfer through evaporation, dehydration, freezing and frying; application of new technologies; combination of food process operations; and product development. Since 1980 he has been with the Chemical and Food Engineering Department at Universidad de las Américas, Puebla, Mexico.

CONTRIBUTORS

Gustavo V. Barbosa-Cánovas
Biological Systems Engineering Department
Washington State University
Pullman WA, 99164 USA.

Hugo González-Loo
Pabisan, S.A.
Cerrada Amatlán 1115, Col. Mayorazgo
Puebla, Pue., 72450 Mexico

José A. Guerrero-Beltrán
Departamento de Ingeniería Química y Alimentos
Universidad de las Américas, Puebla
Ex-Hacienda Sta. Catarina Mártir
Cholula, Pue., 72820 Mexico

Maria Teresa Jiménez-Munguía
Departamento de Ingeniería Química y Alimentos
Universidad de las Américas, Puebla
Ex-Hacienda Sta. Catarina Mártir
Cholula, Pue., 72820 Mexico

Aurelio López-Malo
Departamento de Ingeniería Química y Alimentos
Universidad de las Américas, Puebla
Ex-Hacienda Sta. Catarina Mártir
Cholula, Pue., 72820 Mexico

Enrique Palou
Centro para la Educación de las Ciencias, Ingenierías y Tecnologías
Departamento de Ingeniería Química y Alimentos
Universidad de las Américas, Puebla
Ex-Hacienda Sta. Catarina Mártir
Cholula, Pue., 72820 Mexico

María Elena Sosa-Morales
Departamento de Ingeniería Química y Alimentos
Universidad de las Américas, Puebla
Ex-Hacienda Sta. Catarina Mártir
Cholula, Pue., 72820 Mexico

Jorge F. Vélez-Ruiz
Departamento de Ingeniería Química y Alimentos
Universidad de las Américas, Puebla
Ex-Hacienda Sta. Catarina Mártir
Cholula, Pue., 72820 Mexico

Jorge Welti-Chanes
Departamento de Biotecnología e Ingeniería en Alimentos
Instituto Tecnológico y de Estudios Superiores de Monterrey
Av. Eugenio Garza Sada 2501 Sur, Col. Tecnológico,
Monterrey, N. L. 64849 Mexico

PART I: TOPICS OF MOMENTUM TRANSPORT AND RHEOLOGY OF FOODS

In: Food Processing and Engineering Topics
Editors: M. E. Sosa-Morales and J. F. Vélez-Ruiz
ISBN: 978-1-60741-788-0

Chapter 1

RHEOLOGICAL PROPERTIES OF VEGETAL FLUIDS

Jorge F. Vélez-Ruiz
Chemical and Food Engineering Department
Universidad de las Américas, Puebla (Mexico)

1. INTRODUCTION

Fruit and vegetable fluids represent a very extensive and important sector of the food industry. Most of them are manufactured by industrial processes that change their original structure generating a fluid in which the dispersed material contributes significantly to the rheological behavior. Therefore the specific processing, solid concentration, solid particle and temperature affect the flow response. Vegetal fluids, such as apple, mango, orange, peach and pear juices on one side; avocado, pea, potato, and tomato purees, among others behave as Newtonian or non-Newtonian fluid as a function of particular characteristics and process treatment. Physical properties of vegetal fluids are needed for evaluation and design of equipment and processes, being the rheological properties one of the most studied and required.

Precisely, rheology is defined as the science of flow of fluids and deformation of solids, it has been gaining more and more acknowledgement in the field of food processing and engineering. Rheology and its instrumental determinations have been utilized in several aspects of the food science knowledge, such as (Steffe, 1992; Vélez-Ruiz, 1996; Rao, 1999; Vélez-Ruiz, 2002).

1. Process engineering applications involved in equipment and operation design and calculations,
2. In development of new products and reformulations of existing food items,
3. In quality assurance and control of intermediate and final products,
4. In understanding of food structure and structural changes by processing,
5. In correlation of sensory assessment with rheological parameters,

6 In characterization and testing of shelf life.

Rheology may be utilized and applied widely in many areas of the industry, not only for foodstuff, also for biological, inorganic and complex materials. Consequently, basic rheological information of materials is important not only to engineers, but also to processors, food scientists and people who are related with the food industry. The evolution of rheology is related to important contributions of outstanding scientists: Newton, Kelvin, Maxwell, Einstein, Flory, de Gennes, and Chu, among others (Rao, 1999) and many studies have been carried out on foods in the last three decades that have contributed to the knowledge of the food science.

One important part of the food industrial production is that related with vegetal fluids in which many studies and researches have been conducted; therefore an overall vision of fruit and vegetable fluids and their rheological behavior is presented.

1.1. Rheology of Food Materials

In according to Mohsenin (1986), a general classification of food materials covers two extremes. On one side, the elastic ideal behavior of solids, in which deformation is the physical key concept; and on the other side, the ideal behavior of fluids being flow the important property. The solids may be divided in two subgroups: elastic (Hookean and non-Hookean), or inelastic (viscoelastic and viscoplastic); whereas the liquids also may be grouped in two: viscous (Newtonians and non-Newtonian) or plastic (Bingham and non-Bingham). For this classification of food materials, the response to forces, stresses, time, or deformations need to be known; these parameters are frequently obtained with different types of instruments, graphically represented by rheograms, and quantitative fitted by rheological models.

Food items as fluids, semisolids or solid materials in nature, encompass a diversity of composition, origin and structure exhibiting some of the aforementioned responses. Foods as raw materials, are normally transported and processed in liquid form due to easiness for handling, and only a minor part of the manufacturing is related with solid items; certainly many food products are in solid or semisolid state.

Liquids and semisolids may exhibit Newtonian, plastic, rheopectic, shear thinning, shear thickening, thixotropic, as well as viscoelastic behavior in which flow properties like viscosity, flowability, pourability, flow index, consistency coefficient, yield stress, storage modulus, and stress relaxation, among others are indispensable. They may be objectively determined by steady shear tests. Similarly, solid and semisolid foods may show creep compliance, relaxation profile, plastic, viscoelastic or inelastic behavior, that are characterized by strain or stress as a function of time, a phase angle and different moduli by dynamic determinations. Furthermore, solids characterization of texture and measuring of textural attributes are frequently needed and utilized with practical purpose.

Determination of rheological properties of foods, from the simplest to the most complex, is necessary at different stages of the food transformation, storage and consumption. These needs include since the transportation from the ground or farm as far as the final product, passing by the quality attributes, and finalizing with those food items ready to be eaten by an exigent consumer. To get the rheological properties of foods, is necessary to know those fundamental physical measurements that are obtained with rheometers, texturemeters and viscometers. Additionally, rheometry in conjunction with mathematical modeling permit attainment of flow equations. They are applied into food processes involving momentum and heat transfer phenomena, and with less frequency into mass transfer operations.

1.2. Instrumentation in Rheological Characterization

Any instrument utilized to characterize the physical response of a given food item is able of measuring forces, stresses, torques, deformations, or deformation rates as a function of time. Fundamental properties of vegetal fluids are independent of the measuring instrument in which they are characterized, therefore different instruments will generate the same final results; but some of the instruments facilitate too much the rheological determinations. And even though, there is a wide gamma of instruments, including the empirical and fundamentals device systems; only a few characteristics of fundamental instruments are presented, without minimize the importance and usefulness of the numerous empirical instruments.

Common and commercial instruments, capable of providing fundamental determinations may be divided in two types: rheometers and viscometers used to characterize the physical response of fluids and semisolid, whereas that texture analyzers are employed to test mainly the nature of solid and semisolid food items. Fluid testing instruments may be grouped in two categories taking the basic supporting geometry: tube type and rotational type. In the first group the basic determinations of pressure drop and volumetric flow are converted into shear stress and shear rate, these viscometers are simpler and may be fabricated on site. Whereas in rotational instruments the torque is transformed to shear stress, and the rotational velocity is utilized to compute the shear rate taking into account the geometry dimensions; most of the rotational variants are marketed with a computer software unit by specialized commercial brands. Costs vary tremendously from the inexpensive glass capillary and tube viscometers to very expensive stress control rheometers.

The main geometries utilized in measuring instruments, under steady shear and dynamic conditions, are: concentric cylinder, cone and plate, parallel plate, agitators with different arrangements, capillary at atmospheric and high pressure, piston and cylinder in extrusion, tubes, others are minority used.

Rheometers

These instruments allow continuous determinations, commonly at a selected shear rate and less frequently at a given shear stress. The time of testing may be short or long, allowing flow characterization of both behaviors: time independent and time dependency. They can be used in dynamic mode, in which a sinusoidal varying stress or strain is applied for characterization of vegetal fluids from the viscoelastic viewpoint. Rheometers have become widespread nowadays.

Due to the proper structure of vegetal products, less popular in research applications are those rheometers able to measure shear and normal stresses, simultaneously. Commercial rheometers have incorporated constant improvements and technological advances: electric motors, transducers, higher speeds and larger memories, digital processors and microprocessors (Barnes *et al.*, 1999).

Additionally, it has been recognized a significantly growth in the utilization and analysis of complex geometries as part of the rotational rheometry, like vane and helical, to overcome measurement limitations found in conventional rheometers. Some of the studied new fixtures include different types of mixers or agitation impellers, special designs as the vane and other geometries. Vane methodology has been recognized as a reliable and easy method for rheological characterization of concentrated suspensions (Martínez-Padilla and Rivera-Vargas, 2006).

Viscometers

They are single instruments in which the basic geometry is a couple of concentric coaxial cylinders (cup and bob), are capable of providing readings, normally torque and rotational velocity that can be converted to shear stress and shear rate, respectively. These instruments are unable of dynamic performances, thus they are cheaper. Some instruments for rheological characterization of vegetal fluids and their basic characteristics are included in table 1.

Table 1. Important characteristics of rheological instruments*

Instrument/Geometry	Costs (US$)	Commercial brands	Characteristics and determinations	Reference
Capillar viscometers (D: 0.1 – 3.0 mm)	Inexpensive (0.5 – 1.5) x 10^3. Piston and pressure increase costs	Wescan, Schott, Cannon-Fenske Ostwald	Viscosity for Newtonian fluids Cleanness of the capillary is very important.	Steffe, 1996 Rao, 1999; Vélez-Ruiz, 2009a
Tube viscometers (D: 6 – 25 mm)	Inexpensive (3.0 – 8.0) x 10^3	Self manufactured with tubes, fittings and controls	Newtonian and non Newtonians High shear rates. Constructed “on site”	Vélez-Ruiz, 2009a
Viscometers	Medium (3.0 – 15) x 10^3	Brookfield, Ofite, Haake	For laboratory and plant applications, in line for processing	Vélez-Ruiz, 2008a
Rotational rheometers (general)	Expensive (25 -100) x 10^3	Haake, Contraves, Paar Physica, TA Instruments	Controlled shear rate and stress. Flow and dynamic properties. Fixtures diversity. Computarized	Barnes *et al.*, 1999; Anton Paar, 2004; Vélez-Ruiz, 2009a
Mixer (helical ribbon screw impeller, vane)	Designed and built up at place Medium	Broolfield, Haake The impeller is adapted to a rotational device	Fluids with large particles. Mininizing of slippage. Temperature control	Steffe, 1996 Vélez-Ruiz, 2009a

Table 1. (Continued)

Instrument/Geometry	Costs (US$)	Commercial brands	Characteristics and determinations	Reference
Controlled strain/shear rate	Expensive (60 – 120) x 10^3	Contravest, Haake, Paar Physica, Weissenberg, TA Instrs., Rheometrics,	Controlled shear rate, oscillatory testing, normal forces. Peltier plate system	Holdsworth, 1993; TA Instruments, 2005
Controlled stress	Expensive (80 – 150) x 10^3	Carri-Med, Bohlin, Physica, Haake, TA Instruments	Controlled stress, oscillatory characterization, magnetic bearing	Holdsworth, 1993; TA Instruments, 2005

* For exactitude of technical details and costs, the correspondent company should be consulted.

Texturemeters

Are instruments that allow the characterization of texture and textural identification of solid foods of any type through mechanical loading. These instruments display the viscoelastic characterization through creep and stress relaxation tests. Of course, some measurements on vegetal fluids may be carried out at these instruments, such as adhesivity, compression, and penetration, allowing the determination of various modulli, but normally, they are not able of flow characterization.

For flow determinations by rheometers and viscometers some general assumptions are considered: i) the liquid is incompressible, ii) the regime of the flow is laminar, iii) the steady flow implies only a velocity component as a function of the radius, neglecting axial and radial flows, iv) there is not slippage phenomenon, v) final and edge effects are neglected, and vi) the experiment is conducted at isothermal condition.

The choice of instrument to be utilized depends on the nature of the vegetal product and the properties that need to be characterized. As fluid materials, and based on the objective determinations and influence of time, vegetal liquids may classified in two groups as independent of time and dependent of time. Although there are more studies for the first class, those are divided in Newtonians and non-Newtonians; time dependency is quite important for some structural foods, this second group is divided into thixotropic and rheopectic behaviors. The rheograms correspondent to both behaviors is schematized in figure 1. Viscoelastic characterizations of vegetal fluids are scarce yet.

1.3. Vegetal Fluid Foods

Vegetal fluids are a mixture of dispersed solids in a continuous liquid phase, they are complex materials from structural and rheological point of view. The constituents of plant fluids may be macromolecules, particles, droplets, air bubbles, or combination thereof. The solid particles, besides of the size as an influent factor; involve different shapes, interparticle forces, different arrangements, colloidal interactions, among other aspects that influence the rheological nature of the liquid in a distinct manner. On the other hand, the continuous phase is an aqueous solution of sugar, organic acids, pectic derivates and salts.

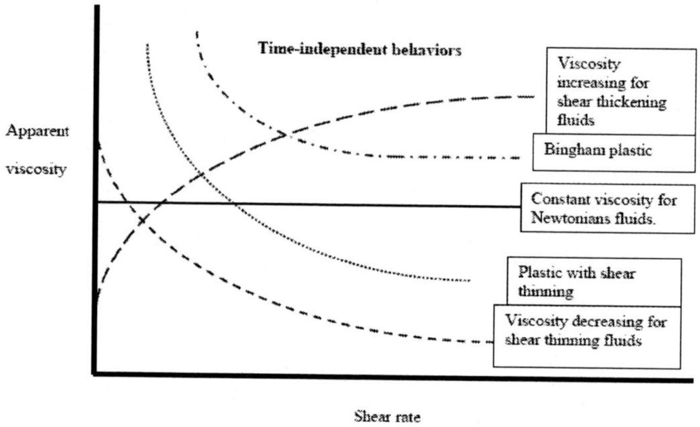

Figure 1a. Rheograms for non-Newtonian fluids (time-independent).

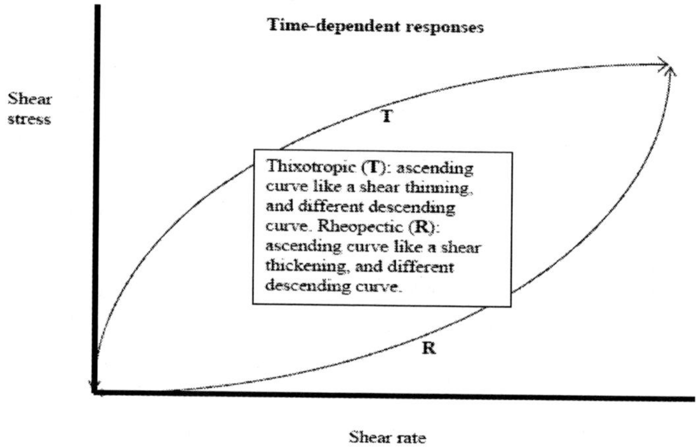

Figure 1b. Rheograms for non-Newtonian fluids (time-dependent).

Fruits and vegetable are parts of plants. Fruits may be regarded as the reproductive organs, they normally have a high content of sugars and acids, together with aromatic components. Vegetables on the other side, include bulbs, leaves, roots, stems and tubers and tend to have lower acid and sugar contents. The distinction between a fruit and a vegetable is often unclear, both are preserved as juices (Jones and Beckett, 1995). Inhomogeneity of shape and size on a microscopic scale is probably the rule rather than the exception in this type of fluids, with some exceptions, such as clarified juices. Some particularities of vegetal fluids are briefly commented next.

Juices and Purees

From the physicochemical view point, these fluids are dispersions of solid particles (pulp) in an aqueous phase (serum). Juices may be prepared from fresh fruits and vegetables by comminution or extraction of pulp, separation of coarse particles or pressing of the comminuted. In several studies, the serum has been characterized by separate, as part of the dispersion (Jones and Beckett, 1995). Before the clarification and filtration processes, that are very common in manufacturing of many juices and some purees; the flow nature of fruit juices and vegetable purees has been reported as non Newtonian.

Concentrates

Concentrates are commercial vegetable fluids in which part of the water is removed, they are also dispersions in which the suspended solids are relatively immobile, and they do not settle to a significant magnitude (Rao, 1987)

Jams

As a mixture of fresh crushed fruit, cold or canned fruit, sugar, acids and pectin, it contains low water content. In addition to particular ingredients and low moisture, the cooking process affects the consistency of the mixture, due to evaporation of water and thermal modification of the ingredients (Jones and Beckett, 1995). Jam is a fruit product with high consistency and semisolid appearance developed during the manufacturing process.

Baby Foods

Baby foods are structured vegetal suspensions. The yield stress of these products is very important in handling, processing, transportation, storage and consumption. It must be thick enough to stay on the spoon and not so thick that an infant could have swallowing troubles (Alonso and Zapico, 1996).

Sauces

Sauces are multiphase dispersed systems constituted mainly by solid rigid particles, solid deformable particles, liquid deformable particles such as oil droplets and a continuous phase formed by water or a solution. Normally, the solution contains polysaccharides or gums, proteins, salts, organic acids and other preservation components (Martínez-Padilla and Rivera-Vargas, 2006).

In general, the non Newtonian response of many vegetal fluids may be attributed to the presence of components of high molecular weight, surface forces, concentration of particles in solution or dispersed in the liquid phase. In these types of foods, both the particle size and particle shape are not uniform, contributing to their complex flow behavior. In it, time dependency may be caused by continuous structural changes due to the applied forces. The commercialization of fruit and vegetable juices has expanded remarkably in the last decades.

1.4. Rheological Behavior and Food Process Operations

Most of the food process operations like transportation through pipes, pumping, agitation, mixing, centrifugation, filtration, cooling and heating within heat exchangers, pasteurization, evaporation, distillation, extraction, membrane separation, drying, high pressure and electric pulse treatments, just to mention some, work with materials in a liquid phase. Flow properties of vegetal fluids, such as the flow index, yield stress or consistency coefficient have been related to different process variables, trying to predict engineering parameters for process performance and equipment design. Correct design of unit operations is essential for optimum processing.

Flow equations and their parameters have been employed in engineering aspects of food process operations, for derivation of relationships or correlations related with velocity profiles and correction factor of kinetic energy, volumetric rates, pressure drops, friction factors in cylindrical and not cylindrical conductions, pumping needs, filtration rates, etc. (Vélez-Ruiz, 2009a). Furthermore, they are valuable for heat transfer aspects, such as convective coefficients in processes for temperature increasing and decreasing, heating and cooling rates, overall coefficient, thermal efficiency, evaporation rates, fouling of heat exchangers (Vélez-Ruiz, 2008c). And finally, they are also important in the mass transfer phenomena and separation unit operations flow properties have been involved less frequently. Dynamic relationships have been related to phase changes, thermal transformations, development of new products, process transformations, and sensory evaluations of foods (Vélez-Ruiz, 2002).

The calculations of head loss through fittings, tubes and valves are highly important to design pipe systems for transportation of vegetal fluids, but there are few data to predict pressure losses in these type of liquids. Only few aspects of this topic are commented next. A pipe transport parameter that has been studied by some investigators, although with different results is the friction, either expressed as factor or as a coefficient. This magnitude is needed for evaluation of energy loss due to friction as part of the pumping requirements.

Some equations that have been proposed for fluid handling are presented below:

Steffe *et al.*, 1984, obtained the friction factor for applesauce as function of the flow regime:

$$f = \frac{14.14}{GRe^{1.048}} \qquad (1)$$

$$GRe = \frac{8 v^{2-n} D^n \rho}{2^n K \left(\frac{3n+1}{n} \right)^n} \qquad (2)$$

where: f is the Fanning friction factor (dimensionless), GRe is the generalized Reynolds number (dimensionless), v is the average velocity (m/s), n is the flow behavior index (dimensionless), D is the internal diameter (m), ρ is the fluid density (kg/m^3), and K is the consistency coefficient (Pa s^n).

The same authors (Steffe *et al.*, 1984) developed three empirical relationships for the friction loss coefficient (K_f) for non Newtonian fluids, also as a function of the generalized Reynolds number.

$$K_f = \frac{29.4}{GRe^{0.504}} \text{ for a tee} \qquad (3)$$

$$K_f = \frac{191}{GRe^{0..896}} \text{ for a } 90^\circ \text{ bend} \qquad (4)$$

$$K_f = \frac{30.3}{G\,Re^{0..492}} \text{ for a three-way plug valve} \quad (5)$$

García and Steffe (1987) found significant differences when the friction factor was evaluated with equations based on fluids without a yield stress, to predict friction factors for fluids having a yield stress.

A Telis-Romero *et al.* (1999) proposed an equation to evaluate the friction factor, that lately was corroborated experimentally, for cooling and heating of orange juice in laminar flow through circular tubes:

$$f = \frac{16}{\phi\, Re_M} \quad (6)$$

$$\phi = \frac{4n}{(3n+1)}\left(\frac{\sigma_0}{\sigma_w}\right)\left(\frac{1-\frac{\sigma_0}{\sigma_w}}{\frac{\sigma_0}{\sigma_w}}\right)^{\frac{1}{n}}\left[1-\frac{\frac{\sigma_0}{\sigma_w}}{(2n+1)}\left(\frac{1+n+2n\frac{\sigma_0}{\sigma_w}\left(1+n\frac{\sigma_0}{\sigma_w}\right)}{n+1}\right)\right] \quad (7)$$

$$Re_M = \frac{\rho v D}{\sigma_0\left(\frac{K}{\sigma_0}\right)^{\frac{1}{n}}} \quad (8)$$

where: ϕ is a fitting factor for non-Newtonian fluids (dimensionless), Re_M is the modified Reynolds number (dimensionless), n is the flow behavior index (dimensionless), σ_0 is the yield stress (Pa), and σ_w is the shear stress at the wall (Pa).

In addition to the generalized Reynolds number, other relationships are frequently involved in momentum and heat transfer phenomena. Particularly, when the heat transfer convective coefficient needs to be estimated as a function of the fluid movement, by empirical relations (Vélez-Ruiz, 2003; Vélez-Ruiz, 2009c):

Nu = f (GRe, GPr, GGr, among others)

$$G\,Pr = \frac{Cp}{k}\left[\frac{K\left(\frac{3n+1}{n}\right)^{n} 2^{n-1}}{4v^{1-n}D^{n-1}}\right] \quad (9)$$

$$GGr = \frac{g\beta\rho^2 D^3 \Delta T}{\left[\frac{K\left(\frac{3n+1}{n}\right)^n}{4v^{1-n}R^{n-1}}\right]^2} \qquad (10)$$

where: Nu (= hD/k) is the Nusselt number (dimensionless), h is the convective coefficient ($W/m^2{}^\circ C$), k is the liquid thermal conductivity (W/m°C), GPr is the generalized Prandtl number; Cp is the liquid specific heat (kJ/kg°C); GGr is the generalized Grashoff number, g is the gravity acceleration (m/s^2), β is the thermal expansion coefficient (1/°C), ΔT is the temperature difference (°C), and R is the internal radius (m).

Rodriguez-Luna *et al.* (1987) studied the rates of heat transfer in mango and papaya nectars, passing through horizontal tubes under laminar flow, in which they used a mixed equation for describing experimental heat transfer coefficients:

$$Nu = 1.75\left[Gz + 0.0083\left(Gr_w\ Pr_w\right)^{0.75}\right]^{1/3}\left(\frac{3n+1}{4n}\right)^{1/3}\left(\frac{K_b}{K_w}\right)^{0.14} \qquad (11)$$

where: Gz (= wCp/kL) is the Graetz number (dimensionless), w is the mass flow rate (kg/s), L is the heat transfer length (m), Gr_w (= $\beta\Delta TD^3\rho^2 g/\eta$) is the Grashoff number at wall temperature (dimensionless), η is the fluid viscosity, Pr_w (= Cpη/k)) is the Prandtl number at wall temperature (dimensionless), K_b is the consistency coefficient at bulk temperature (Pa s^n), K_w is the consistency coefficient at wall temperature (Pa s^n).

In relation to thermal processing of non-Newtonian fluids, Moyano *et al.* (1997) obtained an experimental equation to compute the residence time of non-Newtonian fruit pulp, based on the generalized Reynolds number.

$$t_{min} = t_{mean}\left(\frac{0.698}{G\,Re^{0.26}}\right) \qquad (12)$$

where: t_{min} is the minimum residence time (min), and t_{mean} is mean residence time (min).

In a recent work of Cabral *et al.* (2007), on fluidized bed drying of guava puree, the pressure loss and the minimal velocity of fluidization were experimentally related to the apparent viscosity of the puree.

Modeling, simulation and optimization of unit operations handling vegetal fluids are difficult because rheological properties exhibit substantial changes as a function of several factors, particularly temperature and water content, such is the case of juices, purees and concentrates. Furthermore, physical properties, particularly the flow and dynamic properties, as well as the overall industrial process, are not only convenient, they are indispensable for the adequate design of individual process equipment, and for the accurate evaluation of engineering parameters. Contributing in this way to improve energy efficiency, environment respect and product quality

2. Rheological Modeling

A rheological model or constitutive equation, as a mathematical relationship that describe rheological or flow information, provides a convenient and accurate way of quantifying the measured data. In addition to the mathematical description, the rheological model includes properties that may be correlated to process variables, such composition, concentration, formulation, pressure, shear rate, temperature, time, and chemical factors, among others.

In the analysis of flow behavior of liquid foods there exist numerous rheological models; such models that have found extensive application in vegetal fluids responses are presented, they are divided in those that fit the independent time nature, and those that have been used for time dependent behavior.

2.1. Time Independent Fitting

The more utilized flow models include the Newton equation (NE), Power Law (PL) or Ostwald-De Waele equation, the Bingham plastic (BP) model, and the Hershel-Bulkley (HB) relationship, and sometimes the Casson (CA) equation, that are presented below, for non-time dependency behavior of vegetal fluids:

$$\sigma = \eta \dot{\gamma} \qquad \text{NE equation} \qquad (13)$$

$$\sigma = K_P \dot{\gamma}^{n_P} \qquad \text{PL equation} \qquad (14)$$

$$\sigma = \sigma_0 + \eta_P \dot{\gamma} \qquad \text{BP equation} \qquad (15)$$

$$\sigma = \sigma_0 + K_H \dot{\gamma}^{n_H} \qquad \text{HB equation} \qquad (16)$$

$$\sigma^{0.5} = \sigma_0^{0.5} + C_c \dot{\gamma}^{0.5} \qquad \text{CA equation} \qquad (17)$$

where: σ is the shear stress (Pa), η is the viscosity (Pa·s), γ is the deformation or shear rate (s^{-1}), K is the consistency coefficient (Pa·s^n), K_P for Power law and K_H for Herschel and Bulkley model, n is the flow behavior index, or simply, the flow index (dimensionless), n_P for Power law and n_H for Herschel and Bulkley model, σ_0 is the yield stress (Pa), η_P is the plastic viscosity (Pa·s), and C_C is a Casson parameter $(Pa \cdot s)^{0.5}$.

The Newton relationship allows the flow characterization of Newtonian vegetal fluids, such those plant juices and purees obtained after clarification, enzymatic treatment or filtration. Clarified apple, grape, orange and peach juices are examples.

The Ostwald De Waelle or Power Law (PL) model has been extensively used in vegetal fluids because it gives a very good description of flow nature in the shear rate range that is easily measured by most of instruments. It is the most utilized model for studies of processing of fluids, such as handling, transportation, heating, cooling and evaporation. And though, this model is very popular and useful, it has an empirical nature with two flow parameters.

If, in addition to the flow behavior index and consistency coefficient, the yield stress of the vegetal fluid is quantified, then the rheological fitting is known as Herchel and Bulkley model. Although the existence of the yield stress has been questioned, it represents a very important concept for foods, and in particular for some vegetal fluids, it has utility in consumer acceptability, coating, agitation and transport.

Bingham and Casson models have been applied to vegetal fluids with much less extension. Some vegetal fluids, although showing linear relation of shear stress and rate, exhibit a yield stress and therefore, they are characterized as plastics. In according with Holdsworth (1993), Casson model is limited to fluids with linear chains in their structure.

Occasionally, models such the Cross (CR) and Carreau equations (CA), among others have been used. Alvarez *et al.* (2006a) applied these two models, for the rheological characterization of five jams (apricot, fruits, peach, prune, raspberry and strawberry).

$$\eta = \eta_\infty + \frac{\eta_0 - \eta_\infty}{1 + \left(\frac{\gamma}{\gamma_b}\right)^n} \quad \text{CR equation} \quad (18)$$

$$\eta = \eta_\infty + \frac{\eta_0 - \eta_\infty}{\left[1 + \left(\frac{\gamma}{\gamma_b}\right)^2\right]^n} \quad \text{CA equation} \quad (19)$$

where: η_∞ is the viscosity at infinite time (Pa s), η_0 is the initial viscosity (Pa-s), and γ_b is a time parameter (1/s).

More recently, Alvarez *et al.* (2006b) utilized the Yasuda model (YA) as a modification of Carreau equation, and Nindo *et al.* (2007) used the Sisko model (SI) for high shear rates, finding this model very good to fit the flow of blueberry puree.

$$\eta = \eta_\infty + \frac{\eta_0 - \eta_\infty}{\left[1 + (\lambda\gamma)^2\right]^{\frac{1-n}{2}}} \quad \text{YA equation} \quad (20)$$

$$\eta_a = \eta_\infty + K_S \dot{\gamma}^{n-1} \quad \text{SI equation} \quad (21)$$

where: η is the viscosity (Pa s), λ is a time parameter (s), η_∞ is the viscosity at infinite time (Pa s), η_0 is the initial viscosity (Pa s), n is the flow index (dimensionless), η_a (= σ/γ) is the apparent viscosity (Pa s), and K_S is is the consistency coefficient for the Sisko model (Pa s^n).

Experimental data of time independent modeling with some of the above models, obtained for vegetal fluids: fruits or sweet products and vegetables or salty products, are shown in tables 2 and 3, including some additional details.

Table 2. Flow properties and parameters of fruit or "sweet" vegetal products

Reference	Fluid	Studied variables	Best model	K (Pa s^n)	n	Other parameters/ Instrument
Durán & Costell, 1982	Apricot puree	Variety	Herschel & Bulkley	0.82-1.65	0.59-0.79	σ_0: 10-57 Pa; Rheomat15
Vitali & Rao, 1984	Orange juice	Serum viscosity, pulp content and T (-19-30° C)	Power law	0.05-12.7 0.43-58.9	0.81-1.0 0.67-0.82	Serum of 4 varieties; Haake RV2 Juices of 2 varieties as a f (T), Ea: 11.2-13.6 kcal/gmol
Durán & Costell, 1985	Apricot puree	Homogenization	Herschel & Bulkley	1.62-2.61 1.95-2.82	0.60-0.72 0.60-0.69	σ_0: 9.0-15; (not treatment) σ_0: 12.6-18 Pa; (with treatment), Rheomat15
Qiu and Rao, 1988	Apple sauce	Pulp content and serum viscosity	Bingham	-	-	σ_0: 45.8-87.6 Pa; Concentric cylinders geometry. Haake RV2
Ibarz *et al.*, 1992a	Peach juices	T (5-60°C) and C (40-69° Bx)	Newton	-	-	η: 0.0039-2.087 Pa s; Ea: 4.63-13.3 kcal/gmol; Rotovisco RV12
Ibarz *et al.*, 1992b	Blackcurrant Juices	T(5-60°C) and C (35-65°Bx)	Newton	-	-	η: 0.0024-0.499 Pa s; Ea: 4.65-10.6 kcal/gmol; Rotovisco RV12
Alonso *et al.*, 1995	Infant food	Type of food	Herschel & Bulkley	1.89	0.72	σ_0: 8.03 Pa; Haake RV12
Alonso and Zapico, 1996	Fruit baby food	Storage and T (5-25° C)	Herschel & Bulkley	1.44-2.74	<1	σ_0: 5.8-8.0 Pa; Haake RV12
Ibarz *et al.*, 1996	Sloe fruit juices	C(25-58.5° Bx) and T (15-65° C)	Bingham	2.5-772 mPa s	1.0	σ_0: 1.8-11.2 Pa; Rotovisco RV12
Villarán *et al.*, 1997	Apricot, rose hips and bilberry jams	Fruit and T (5-65° C)	Power law	0.65-31 0.76-15 2.72-99	0.33-0.65 0.39-0.62 0.36-0.69	Apricot, Bilberry, Ea: 5.35-11.43 kcal/gmol Rose hips; Mettler RM180

Table 2. (Continued).

Reference	Fluid	Studied variables	Best model	K (Pa s^n)	n	Other parameters/ Instrument
Giner *et al.*, 1996	Clarified cherry juices	C (22-74° Bx) and T (5-70° C)	Newton	-	-	η: 0.0009-0.2986; Ea: 3.44-14.6 kcal/gmol; Rotovisco RV12
Guerrero & Alzamora, 1997	Banana puree	pH, T (10-55° C), and glucose (21-51° Bx)	Herschel & Bulkley	0.30-6.7	0.41-0.63	σ_0: 4.0-32.5 Pa; Rotovisco RV12
Bhattacharya and Rastogi, 1998	Mango pulp	Enzyme, treatment time and T (25-60°C)	Herschel & Bulkley	1.05-3.5	0.38-0.49	σ_0: 2.9-3.9 Pa; Haake RT10
Guerrero & Alzamora, 1998	Mango puree Papaya puree Peach puree	pH, T (10-55° C), and glucose addition (12-52°Bx)	Herschel & Bulkley	0.30-3.4 0.50-2.5 0.06-3.3	0.43-0.78 0.32-0.67 0.48-0.58	σ_0: 4.4-14.3; Rotovisco RV12 σ_0: 3.8-28.5; σ_0: 2.3-5.2 Pa
Grigelmo-Miguel *et al.*, 1999	Peach dietary fiber suspension	C (5-20%) and T (15-65° C)	Power law	0.05-170	0.24-0.83	Haake Rotovisco RV12
Grigelmo-Miguel & Martín-Belloso, 1999	Strawberry jam	C (< 55° Bx) and fiber level (1-4%)	Power law	1.4-11	0.44-0.58	Haake Rotovisco RV12
Cantú-Lozano *et al.*, 2000	Apple	C (15-28%) and particle size		2.0-70	0.15-0.70	σ_0: 8-60 Pa, - Vane in Haake RV2
Singh and Eipeson, 2000	Mango juice (clarified)	C (15-66° Bx) and T (15-85°C)	Newton	-	-	η: 0.0009-0.2986 Pa s; Ea: 1.64-8.44 kcal/gmol; Haake VT 550
Quispe & Cruz, 2001	Camu-camu juice	T (10-50°C)	Power law	1.0-1.7	0.22-0.28	Brookfield RVT Ea: 258 kcal/gmol

Table 2. (Continued).

Reference	Fluid	Studied variables	Best model	K (Pa s^n)	n	Other parameters/ Instrument
Suárez *et al.*, 2003	Juices and concentrates	Fruit and C (8-40%)	Power law, Herschel & Bulkley	2.2-4.4 0.6-2.1 0.4-3.0	0.50-0.51 0.50-0.51 0.50-0.53	melon (8-25%), papaya (10-35%), watermelon (8-40%) σ_0: 3.0-6.0 Pa; Brookfield DV I
Alvarez *et al.*, 2006a	Jams	Fruit and T (20-40° C)	Cross Herschel & Bulkley	-	0.52-0.83	η_0: 9.82-42.8 Pa s; γ_b: 0.21-3.06 s^{-1}; Haake VT 550
Maceiras *et al.*, 2007	Purees and jams	Fruit and T (20-40° C)	Herschel & Bulkley	5.3-6.8 3.1-18 4.3-6.6 23-59 2.0-4.3. 8.8-15 5.3-5.8 6.6-37	0.36-0.39 0.33-0.53 0.31-0.32 0.22-0.30 0.36-0.38 0.32-0.35 0.26-0.27 0.28-0.45	σ_0: 10.2-12.7 Pa, peach puree σ_0: 1.8-6.1, peach jam σ_0: 1.8-3.8, prune puree σ_0: 1.1-7.8, prune jam σ_0: 2.0-3.5, raspberry puree σ_0: 1.7-12.6, raspberry jam σ_0: 2.8-5.1, strawberry puree σ_0: 1.5-3.9 Pa, strawberry jam Haake VT 550
Cabral *et al.*, 2007	Guava pulp	T (38-72° C) and C (25-36°Bx)	Power law	3.3-135	0.21-0.64	Rheotest 2.1
Nindo *et al.*, 2007	Blueberry puree	C (10-25%) and T (10-100 s-1)	Sisko			Ea: 11.4-17.1 (kJ/mol); AR 2000 (TA Instr.)

K: consistency coefficient (Pa s^n); n: flow behavior index (dimensionless); σ_0: yield stress (Pa); η: viscosity (Pa s); η: plastic viscosity (Pa s); η_0: initial viscosity (Pa s); γ_b: time parameter (1/s); E_a: activation energy; C: concentration (°Bx or % w/w); Temperature (°C).

Table 3. Flow properties and parameters of vegetable or "salty" vegetal products

Reference	Fluid	Studied variables	Best model	K (Pa s^n)	n	Other parameters	Instrument
Rao *et al.*, 1981	Tomato concentrates	Cultivar, T (15-55° C) and flow models	Herschel & Bulkley	1.7-195	0.27-0.66	Ea: 2.3 kcal/gmol	Haake RV2
Rao & Cooley, 1983	Tomato concentrates	Cultivar, C (6-36%) and T (15-55° C)	Herschel & Bulkley	-	0.52-0.54	Ea: 1.96-15.1 kcal/gmol	Haake RV2
Tanglertpaibul & Rao, 1987	Tomato concentrate	Serum concentration and pulp content	---	---	---	η_a: 0.10-0.88 Pa s	Haake RV2
Bhamidipaty & Singh, 1990	Tomato sauce	Particles, T (70-95° C) and C (7 and 14° Bx)	Power law	0.04-0.60	0.20-0.86	Ea: 2.9-4.8 kJ/mol	Tube geometry
Alonso *et al.*, 1995	Infant food	Type of food	Herschel & Bulkley	2.74	0.73	σ_0: 28.6 Pa	Haake RV12
Correia & Mittal, 1999	Tomato paste	Shear rate and type of rheometer	Herschel & Bulkley	18-44	0.40-0.55	σ_0: 139-180 Pa	Rotovisco RV3
Dogan *et al.*, 2002	Tomato suspension	Volumetric flow	Herschel & Bulkley	0.6-0.9	0.20-0.25	σ_0: 0.61-0.84 Pa	In line viscometer
Cepeda & Gómez, 2002	Pimento puree	Pulp content and T (5-40°C)	Power law	11-27 3.7-17	0.25-0.29 0.27-0.38	Ea: 5.9-13 kJ/mol	Mettler RM180
Montoya Lázaro *et al.*, 2003	Carrot, green tomato and tomato purees	Vegetable type and C (8-29%)	Power law	- -	0.49-0.63 0.80-0.89 0.41-0.80	carrot (10-29%), green tomato (8-13%), tomato (8-13%)	Brookfield DVI

Table 3. (Continued).

Reference	Fluid	Studied variables	Best model	K (Pa s^n)	n	Other parameters	Instrument
Canet *et al.*, 2005	Potato puree	Two purees and T (25-65°C)	Herschel & Bulkley	22-367	0.17-0.55	σ_0: 5.7-388 Pa	Bohlin CVR50
Martínez-Padilla & Rivera-Vargas, 2006	Mexican sauces	Sauce type and rheometer type	Power law	2.9-19	0.24-0.45	6 – 500 s^{-1} 10-100 s^{-1}	Haake RT20 Vane fixture
Alvarez *et al.*, 2006b	Sauces with tomato	T (20-40°C)	Power law	17-46	0.19-0.37	18-445 s^{-1}	Haake VT550

K: consistency coefficient (Pa s^n), n: flow behavior index (dimensionless), σ_0: yield stress (Pa), Ea: activation energy; η_a: apparent viscosity (Pa s), T: temperature, C: concentration (°Bx or % w/w).

Table 4. Time dependent parameters of vegetable structured products

Reference	Fluid	H: a (x10-3)	P	FS: σ_{max}-σ_e	k ($x10^{-3}$)	W: A	B	Instrument
Durán & Costell, 1985	Apricot puree	6.60-11.2 3.10-6.20	0.13-0.172 0.20-0.23	without treatment with homogenization		22.2-41.6 54.0-72.6	1.36-3.70 5.32-9.25	Rheomat15
Alonso *et al.*, 1995.	Infant foods	19-27	1.42-1.93 Pa	-	-	73-120	4.3-16.7	Haake RV12
Alonso & Zapico, 1996	Baby foods	Storage and T (5-25°C)		- -	- -	62-73 120-154	4.3-5.3 (fruits), Haake RV12 11-16.7 (vegetables)	
Villarán *et al.*, 1997	Apricot, rose hips and bilberry jams	Fruit type and T		σ_0: 50-94, σ_0: 15-609, σ_0: 90-419 Pa, σ_e: 47-75 σ_e: 10-523 σ_e: 74-379	2.3-16 8.1-34 8.2-11	- - -	Apricot Bilberry Rose hips Apricot Bilberry Rose hips	Mettler RM180
Choi & Yoo, 2004	Baby food Ketchup Mustard	0.354-0.514 0.355-0.455 0.459-0.536	1.05-1.29 1.17-1.51 1.30-1.51	17.2-22.0 12.2-34.1 25.2-44.4	0.865-1.19 0.818-1.05 1.060-1.23	69.5-82.4 75.4-117 82.2-164	3.68-5.81 2.97-1.51 5.40-8.81	5– 35 s-1 Haake VT550
Alíns, 2008	Mandarine concentrate at low temperatures (-12 to 6 ° C)	0.700-10.00	0.26-3.00	0.70-20.0	1.000-9.80	16.4-136	0.18-4.66	Haake RS80

a in Pa s^{-1}; P in Pa; σ_{max} in Pa, σ_e in Pa, k in s^{-1}; A in Pa; B in Pa s^{-1}; T: temperature.

2.2. Time Dependency Modeling

Whereas for time dependency exhibited by some vegetal fluids, in which structural breakdown has been measured. Two typical models have been applied to those fluids, the Weltman (WE) and Hahn (HA) relationships (Alíns, 2008; Vélez-Ruiz, 2009a). Morever, the Figoni and Shoemaker (FS) model has been incorporated in some fluids to characterize time dependent flow properties (Choi and Yoo, 2004; Alíns, 2008); also a modification of the FS was utilized by Villarán *et al.* (1997) for the rheological characterization of fruit jams. And in few cases the tixotropic nature obtained between the up and down curves of a rheogram is measured as a function of the area between curves.

$$\sigma = A - B \ln t \qquad \text{WE model} \qquad (22)$$

$$\log(\sigma - \sigma_e) = P - at \qquad \text{HA model} \qquad (23)$$

$$\sigma = \sigma_e + (\sigma_{max} - \sigma_e) \exp kt \qquad \text{FS model} \qquad (24)$$

$$\sigma = \sigma_e + (\sigma_0 - \sigma_e) \exp kt \qquad \text{FS modified model} \qquad (25)$$

where: σ is the shear stress (Pa), at time of shearing (t, in s), σ_e is the shear stress at equilibrium (Pa), σ_{max} is the maximum shear stress (Pa), A and B are the Weltman parameters, A is the maximum viscosity (Pa), B is the coefficient of thixotropic breakdown (Pa/s), P and a are the Hahn parameters, P is related to the maximum tangential tension, a is a function of the structural rupture (s^{-1}), k is the Figoni and Shoemaker parameter (s^{-1}), and σ_0 is the initial shear stress (Pa).

Few data of time dependent fittings obtained for vegetal fluids, with some of the above models are included in table 4.

2.3. Effect of Temperature and Concentration

It must be highlighted that very often is needed to handle rheological models, since much more information may be obtained from the rheological property (viscosity, apparent viscosity, flow index, consistency coefficient, yield stress, storage modulus, among others) as a function of the important experimental variable or variables. Temperature and concentration, as two important variables affecting the rehology of vegetal fluids, have been related to rheological parameters in different forms (Rao, 1987; Holdsworth, 1993; Vélez-Ruiz and Barbosa-Cánovas, 1997; Rao, 1999; Krokida *et al.*, 2001; Ibarz and Barbosa-Cánovas, 2003; Alvarez *et al.*, 2006a; Nindo *et al.*, 2007; Vélez-Ruiz, 2009b). Several approaches have been employed, such as a linear relationship to fit the flow index dependency with concentration, a power relation for consistency coefficient as a function of the concentration, and a combined exponential-potential functionality of consistency coefficient against concentration (X, C) and temperature (T).

$$n = mX + b \tag{26}$$

$$n = n_0 - mX \tag{27}$$

$$K = B\exp(CX) \tag{28}$$

$$K = B\exp(C/T) + D^X \tag{29}$$

Furthermore, the effect of temperature on consistency (viscosity, apparent viscosity, or consistency coefficient) of vegetal foods with non Newtonian response has usually been fitted by an Arrhenius-type relation, that allows the evaluation of the activation energy.

$$\eta = A_0 \exp(E_a/RT) \tag{30a}$$

$$\eta_a = A_0 \exp(E_a/RT) \tag{30b}$$

$$K = A_0 \exp(E_a/RT) \tag{30c}$$

where: η is the viscosity (Pa s), η_a the apparent viscosity (Pa s), A_0 is an empirical constant (Pa s or Pa s^n), E_a is the activation energy for flow (kJ/mol), R is the universal constant for gases (kJ/mol K)), and T is the absolute temperature (K).

With respect to viscoelastic characterization of vegetal fluids, there are few studies. Yoo and Rao (1996) carried out creep and dynamic tests to correlate the concentration of tomato juice with viscoelastic response. They found that magnitudes of the instantaneous modulus ($G_0 = 1/J_0$) for tomato concentrates were of the same order as the storage modulus (G') at 1 rad/s. And particularly, the viscoelastic properties for the concentrate 21°Bx were related to particle size.

For transient viscoelastic flow of vegetal fluids there are also few studies, but several models have been developed for this type of response (Rao, 1999).

More details and fundamentals of rheology, instruments, models, discussion and applications on rheological nature of foods may be found in texts (Mohsenin, 1986; Steffe, 1996; Rao, 1999; Ibarz and Barbosa-Cánovas, 2003)

Although there are various published reports about rheological data, including energy of activation (tables 2 and 3), yield stress and different effects have not been reported plenty. Therefore, there is a lack of rheological data for vegetal fluids and specific equations applicable to them needs to be established.

3. Experimental Work

Investigation of flow, structural and viscoelastic properties of vegetal fluids requires different experimental techniques and features. Steady shear methodology implies that the food item is subjected to a net force, stress or shear stress force to flow. In dynamic experimentation, the sample is altered by a very small oscillatory perturbation resulting in no net deformation.

A deeper analysis of the rheology and flow of liquid foods requires knowledge of important and practical physical concepts, such is the case of tensors and vectors, as well as basic principles of transport phenomena, such as the equation of mass conservation, momentum theorem, and conservation of mechanical energy. In a plain and practical way, those concepts and laws have been incorporated to the dynamic of fluids, to the rheological characterization and modeling, as well as to the unit operations fundamentals.

Due to the complex chemical and physical structure of vegetal fluids, the viscosity of the liquids can not be predicted by theoretical methods, such molecular dynamics and semi empirical models that are applied to pure fluids, consequently the viscosity and other flow parameters of these fluids may be instrumentally measured (Krokida *et al.*, 2001).

Rheological or flow properties are directly linked to consistency of juices, purees, and concentrates. Most vegetal fluids show shear thinning behavior that, in according with Rao (1999), is an indication of structural breakdown as a consequence of hydrodynamic forces created during shear.

3.1. Characteristics of Measurement

Experimental setups for rheological measurements have been predominantly based on rheometers, different variables of study, and selecting the specific fixture as a function of the particular vegetal fluid or the objective of the investigation.

Concentric cylinder has been preferred (Durán and Costell, 1985; Ibarz *et al.*, 1992a; Alonso *et al.*, 1995; Ibarz *et al.*, 1996; Bhattacharya and Rastogi, 1998; Grigelmo-Miguel and Martín-Belloso, 1999; Choi and Yoo, 2004; Canet *et al.*, 2005; Alvarez *et al.*, 2006a; Alvarez *et al.*, 2006b) because it avoids separation of the pulp (Nindo *et al.*, 2007). Other experiments have been carried out with the controlled stress mode when yield stress needs to be quantified with precision (Canet *et al.*, 2005; Martínez-Padilla and Rivera-Vargas, 2006; Nindo et al, 2007). The mixer, vane and helical methodologies, as non conventional geometries, have been used as an alternative to cylindrical geometry for rheological characterization of vegetal fluids containing coarse particles (Qiu and Rao, 1988; Cantú-Lozano *et al.*, 2000; Martínez-Padilla and Rivera-Vargas, 2006).

Works utilizing rotational viscometers (Rao *et al.*, 1981; Tanglertpaibul and Rao, 1987; Ibarz *et al.*, 1992a; Ibarz *et al.*, 1992b; Guerrero and Alzamora, 1997; Singh and Eipeson, 2000; Suárez *et al.*, 2003), viscometers with tube geometries (Vitali and Rao, 1984; Bhamidipaty and Singh, 1990; Correia and Mittal, 1999) and in line instruments (Dogan *et al.*, 2002) are scarce, in comparison with those investigations in which the concentric geometries are utilized.

Most of the studies on vegetal fluids are completed with the fluid obtained from the fresh fruit or vegetable; Rao *et al.*, 1981, and Rao and Cooley, 1983 with tomato concentrates; Qiu and Rao, 1988 with apple sauce; Giner *et al.*, 1996 with cherry juice; Guerrero and Alzamora, 1998 with mango, papaya, and peach purees; Singh and Eipeson, 2000 with mango juice; Dogan *et al.*, 2002 with tomato suspensions; Suárez *et al.*, 2003 with papaya, melon and watermelon purees; Canet *et al.*, 2005 with prepared potato puree; Cabral *et al.*, 2007 with guava pulp; Rodriguez-Martinez *et al.*, 2007 with prickly pear concentrate; even for room, high temperature and/or for concentrated samples (Vitali and Rao, 1984; Ibarz *et al.*, 1992a; Singh and Eipeson, 2000; Suárez *et al.*, 2003 ; Nindo et al, 2007).

Sometimes the vegetal fluid has been directly supplied by industries (Ibarz *et al.*, 1992a with peach juice; Nindo et al, 2007 used blueberry puree), or well obtained from supermarkets (Durán and Costell, 1985 with apricot puree; Ibarz *et al.*, 1992b with blackcurrant juice; Villarán *et al.*, 1997 with apricot, bilberry and rose hip jams; Bhattacharya and Rastogi, 1998 with mango pulp; Canet *et al.*, 2005 with potato puree; Martínez-Padilla and Rivera-Vargas, 2006 with Mexican sauces; Alvarez *et al.*, 2006a and b, utilizing five groups of sauces, and five different jams, respectively). And only few studies have prepared the fluid from the dehydrated plant, in which died pimento was used (Cepeda and Gómez, 2002).

There exist specific methods for yield stress determination, where the steady shear methodologies are preferred. Determination of yield stress from dynamic tests is extremely tedious, is not always applicable and is more uncertain (Canet *et al.*, 2005)

The fluid sample should be temperature equilibrated for a reasonable time (5-30 min); and when the vegetal fluid is being heated and /or concentrated, to much care should be done in order to avoid water losses. However, the interaction of other factors, such as changes in particle to particle proximity with concentration and temperature, must no be overlooked.

Computational rheology (Crochet and Walter, 1993) has been proposed as a new science to face problems and situations of complex fluids and flows that experimentally are difficult of handling. Its applications and utility has been growing continuously, but with reticence in the food field. This approach is based on numerical simulation of flows which are supported on governing fundamental equations.

Vegetal fluids have been rheologically modeled by following different approaches in which the physicochemical characteristics play an important role; mainly, there are equations to fit the effect of solids and temperature. A good number of researches have been conducted to know the influence of different ingredients, processing, solids content and temperature on their flow properties (tables 2 and 3). Where data are available, the role of the formulation of the food fluid on the rheological behavior has been pointed out.

3.2. Research Trends

As it was briefly mentioned at the beginning, rheological modeling has a broad field of application. Rheology may be utilized and applied widely in many areas of the food industry, from the raw items characterization, passing by processing, and finalizing in consumers influence.

Thus, the investigation in food rheology has been focused on: quality assurance and control of intermediate and final products; understanding of food structure and structural changes by processing, engineering applications involved in equipment and operation design and calculations; development of new products and reformulations of existing food items; correlation of sensory assessment with rheological parameters; and on characterization and testing of shelf life, just to mention the most important.

4. Examples of Application

4.1. Illustrative Example 1

Rheological data from a rotational viscometer, for a prickly pear (*Opuntia ficus-indica)* juice at 25°C and 42°Bx, are shown next (Rodriguez-Martinez *et al.*, 2007). By applying of the Power law equation, determine the flow properties.

γ (1/s):	3.38	6.77	10.15	13.54	16.92	20.30	23.69	27.07	30.46	33.84
σ (Pa):	0.45	0.94	1.12	1.47	1.68	1.86	1.98	2.13	2.36	2.53

Solution: Taking the Power law equation and doing some modifications to get linearity of the relationship (Eq. 14):

$$\sigma = K\gamma^n$$
$$\log\sigma = n\log\gamma + \log K \tag{31}$$

Therefore, doing the correspondent calculations, the slope, the intercept and correlation coefficient of a linear regression, may be obtained:

. γ (1/s)	. σ (Pa)	log γ	log σ		
3.38	0.45	0.5289	-0.3468		
6.77	0.84	0.8306	-0.0757	correlation coeff.	0.9951
10.15	1.12	1.0065	0.0492		
13.54	1.47	1.1316	0.1673	slope	0.7303
16.92	1.68	1.2284	0.2253		
20.30	1.86	1.3075	0.2695	intercept	-0.6967
23.69	1.98	1.3746	0.2967		
27.07	2.13	1.4325	0.3284		
30.46	2.36	1.4837	0.3729		
33.84	2.53	1.5294	0.4031		

And from this information, $R^2 = 0.995$; being the flow properties: n = 0.73; $K = 10^{-0.69668} = 0.2011$ Pa s^n.

Conclusion: with these flow properties, the nature of this vegetal fluid is a *shear thinning or pseudoplastic type.*

Comment: For other cases, the fluid exhibites a yield stress, then the flow behavior should be fitted by other rheological models, such the Herschel and Bulkley, the Bingham or the Mizrahi and Berk, just to mention three of them.

4.2. Illustrative Example 2

A tube viscometer with 1.384 cm diameter and 1.22 m length was used to obtain the following data (pressure drop vs. volumetric flow) for a concentrated tomato puree (Vélez-Ruiz, 2009a).

ΔP (Pa):	(1.45	1.56	2.96	3.10	3.31	3.51	4.20	4.69	5.10) x 10^5
V (mL/s)	8.80	10.1	33.77	40.31	50.89	53.36	67.83	104.4	107.5

i) Determine the flow characteristics by fitting data to Power law (PL) equation
ii) Determine the flow characteristics by fitting data to Herchel-Bulkley (HB) model.

Solution: First, to be able for fitting to PL, the approach developed by Heldman (1975), as a linear expression for shear thinning and shear thickening responses, may be utilized:

$$\log\left(\frac{\Delta P}{2L}\right) = n\log\dot{V} + \left[\log K - n\log\pi - n\log\left(\frac{n}{3n+1}\right) - (3n+1)\log R\right] \quad (32)$$

ΔP (Pa)	V (mL/s)	ΔP/2L (Pa/m)	V (m^3/s)	log (ΔP/2L)	log V		
145000	8.80	59426.229	8.8E-06	4.7740	-5.0555		
156000	10.10	63934.426	1E-05	4.8057	-4.9957		
296000	33.77	121311.475	3.4E-05	5.0839	-4.4715	Corr. coeff.	0.9969
310000	40.31	127049.180	4E-05	5.1040	-4.3946		
331000	50.89	135655.738	5.1E-05	5.1324	-4.2934	slope	0.4905
351000	53.36	143852.459	5.3E-05	5.1579	-4.2728		
420000	67.83	172131.148	6.8E-05	5.2359	-4.1686	intercept	7.2580
469000	104.4	192213.115	0.0001	5.2838	-3.9813		
510000	107.5	209016.393	0.00011	5.3202	-3.9686		

The flow behavior index is 0.49, and to get the consistency coefficient, the intercept value must be substituted in the second part of the equation 32:

$$7.26 = \log K - 0.49\log(3.1416) - 0.49\log\left(\frac{0.49}{3x0.49+1}\right) - (3x0.49+1)\log(0.00692)$$

$$\log K = 7.26 + 0.49(0.492) - 0.49(0.703) - 2.472(2.16) = 1.816$$

Therefore, the consistency coefficient is 65.464 Pas^n.

Secondly, for the HB model; it is assumed that the minimum pressure required to cause flow in the horizontal tube is known (ΔP = 45000 Pa). And by evaluation of the shear stress and the pseudo shear rate ($\gamma = 32V/\pi D^3$; Rao, 1999), the approach to obtain the flow properties, would be the following:

For the yield stress.

$$\sigma_0 = \frac{\Delta P_{min} R}{2L} = \frac{45000\left(\frac{N}{m^2}\right)x0.00692(m)}{2(1.22m)} = 127.62Pa\ ,$$

consequently some modifications must be carried out to evaluate n and K for the Herschel and Bulkley model (Eq. 16).

$$\sigma = \sigma_0 + K_H \dot{\gamma}^{n_H}$$

That may be linearized to:

$$\log\sigma = \log\sigma_0 + n_H \log K_H\ ;$$ H was added to n and K

then the next table show the main calculations:

ΔP (Pa)	V (m3/s)	σ (Pa)	σ - σ₀ (Pa)	γ (1/s)	logγ	Log (σ-σ₀)	
		(ΔPR/2L)		(32V/πD³)			
145000	0.0000088	411.2295	283.609508	33.8248507	1.5292359	2.45272	Corr. coeff.:
156000	0.0000101	442.4262	314.80623	38.8217036	1.5890746	2.49804	0.9960
296000	0.0000338	839.4754	711.85541	129.802865	2.1132843	2.85239	
310000	0.0000403	879.1803	751.560328	154.940879	2.190166	2.87596	Slope:
331000	0.0000509	938.7377	811.117705	195.607574	2.2913857	2.90908	0.6017
351000	0.0000534	995.4590	867.839016	205.101595	2.311969	2.93844	
420000	0.0000678	1191.1475	1063.52754	260.720412	2.416175	3.02675	Intercept:
469000	0.0001041	1330.1148	1202.49475	401.285729	2.6034537	3.08008	1.5471
510000	0.0001075	1446.3934	1318.77344	413.201301	2.6161617	3.12017	

Conclusion: with these flow properties (*$n = 0.73$, $K = 65.464\ Pas^n$ for the LP model, and $\sigma_0 = 127.6\ Pa$, $n = 0.60$, and $K = 35.34\ Pa\ s^n$* for the HB model, the nature of this vegetal fluid *is a shear thinning or a plastic* with shear thinning behavior.

Comment: To decide which one of the two applied models does a better fit of experimental data (even both had a high correlation coefficient); a test of goodness (root square mean error, percentage mea error, for instance) should be completed, and then to decide for that model showing a lower magnitude with the selected test of goodness.

4.3. Illustrative Example 3

Some flow properties have been reported for vegetal fluids. Establish the type or relationship among the flow property and the experimental variable for three different cases.

i) Apparent viscosity of a cherry juice at 22°Bx as a function of temperature (Giner *et al.*, 1996):

η (mPa s):	4.31	3.79	2.94	2.32	2.0	1.7	1.46	1.36
T (°C):	5	10	20	30	40	50	60	70

ii) Flow behavior index of an apple dispersion as a function of concentration (Cantú-Lozano *et al.*, 2000)

n (dimensionless):	0.69	0.36	0.27	0.25	0.23	0.2
Concentration (% pulp content):	16.0	19.5	20.5	21.4	23.0	24.7

iii) Consistency coefficient of a strawberry jam as a function of temperature (Álvarez *et al.*, 2006a)

K (Pa s^n):	36.92	21.45	15.59	10.59	6.62
T (°C):	20	25	30	35	40

***Solution*:**

i) Taking the given data to fit to the equation: $\eta = \eta_0 \exp(Ea/RT)$ (Eq. 30a), in which the converted and needed data are tabulated next:

T (°C)	η (mPa s)	T (K)	1/T	ln η		
5	4.31	278	0.003597	1.46094		
10	3.79	283	0.003534	1.33237	Corr. coeff.	0.9950
20	2.94	293	0.003413	1.07841		
30	2.32	303	0.003300	0.84157	slope	1729.76
40	2	313	0.003195	0.69315		
50	1.7	323	0.003096	0.53063	intercept	-4.8054
60	1.46	333	0.003003	0.37844		
70	1.36	343	0.002915	0.30749		

From the slope and with the proper value of R, Ea = 14382 kJ/kmol (3.44 kcal/gmol, which agrees with that reported by the authors, being of 3.4 kcal/mol). Thus, the model becomes: $\eta = -4.805 \exp(14382/RT)$.

ii) For the second case, three empirical fittings are tested: a linear model ($n = a + bC$), an exponential relation ($n = a \exp^{bC}$), and a power law ($n = a\, C^b$); the results of three fittings are shown below.

Solids (%)	n (-)		ln n		log C	log n		
16.0	0.69		-0.37106		1.20412	-0.16115		
19.5	0.36		-1.02165		1.29004	-0.4437		
20.5	0.27		-1.30933		1.31175	-0.56864		
21.4	0.25		-1.38629		1.33041	-0.60206		
23.0	0.23		-1.46968		1.36173	-0.63827		
24.7	0.20		-1.60944		1.39270	-0.69897		
linear fitting:			exponential fitting:			power fitting:		
	Corr. coeff	-0.9180		Corr. coeff	-0.9611		Corr. coeff	-0.9771
	slope	-0.0558		slope	-0.1433		slope	-2.9100
	intercept	1.4977		intercept	1.7942		intercept	3.3083

Taking the correlation coefficient as a criterion of decision, and as can be observed, the best fitting is reached up with the power fitting, then the exponential trend, and being the less good the linear relationship. Thus, the flow behavior could be predicted as $n = 2033.6C^{-2.91}$, obtaining approximate values to those reported by the authors.

iii) For the third case, something similar to the first example may be developed. Then applying the Arrhenius equation ($R^2 = 0.997$, slope = 7607.696, intercept =-22.3749), giving as predictive equation: K = -22.38 exp(7607.7/T), with a Ea = 63254.2 kJ/kmol. K was fitted by a linear relationship as a function of temperature by the authors.

iv) Yield stress of a blueberry puree as a function of concentration (Nindo *et al.*, 2007)

σ_0 (Pa):	1.32	4.53	18.03	25.0
Concentration (°Brix):	10	15	20	25

This last data can be also fitted by the three practical relations mentioned in the case 2; being the yield stress of blueberry puree best fitted by a power relationship ($R^2 = 0.989$), followed by the linear ($R^2 = 0.976$) and the exponential ($R^2 = 0.974$). Therefore the yield stress could be predicted as $\sigma_0 = 5.452 x 10^{-4} C^{3.38}$.

Conclusion: all the flow properties may be fitted to different types of relationships, apparent viscosity and consistency coefficient were fitted by the Arrhenius equation, whereas the flow index and yield stress were best fitted by a power law relation.

Comment: Normally, flow index is mainly affected by concentration; while viscosity, apparent viscosity, consistency coefficient and yield stress are affected by both, concentration and temperature. Concentration augments the flow property; in contrary, temperature has a decreasing effect on the fluids flow .

5. Final Remarks

Solving rheological problems of vegetal fluids involves a critical analysis of the specific fluid and a very careful decision of the appropriate instrument and the right technique to apply in finding a good solution. Thus, the objective of this chapter was to present basic information, a gamma of experimental works, and a varied list of references to help students, industrial people and researchers.

Vegetal fluids, as important items of the food industry and commercialization, exhibit a rheological behavior that may be characterized by flow and dynamic tests. When solids have been removed from the dispersion, the fluid becomes Newtonian or close to it, and when the solids are not removed, a non Newtonian or viscoelastic behavior is expected.

The theoretical relationships between the rheological properties of vegetal fluids and their mechanical response when different testing methods are applied are well understood and a wide set of experiments have been carried out. Consequently there exists a good quantity of specialized literature, as it was presented in this chapter. The practical relevance of rheology in food science and food engineering is therefore self evident and, typically experimental. There exists an interesting and challenging new approach, the computational rheology that offers some options in labor saving, and surely will play an outstanding role in the future addressing complex flow phenomena.

The advances in software and complementary techniques (X rays, nuclear magnetic resonance, electronic microscopy, ultrasound, among others) will facilitate and complete the rheological characterization of vegetal fluids, and foods in general.

However, there are aspects and studies that need to be completed and understood in relation to vegetal fluids, including complex flows and wall slippage during the experimental characterization of some. It will have more utilization of the squeezing flow methodology, and utilization of vane fixture and other impellers as generalized devices of measuring, in addition to others advances and goals.

References

Alíns V. 2008. Rheological characterization of mandarin concentrate. Bachelor Thesis. Universidad de Lleida/Universidad de las Américas, Puebla (In Spanish).

Alonso M.L. and Zapico J. 1996. Effect of storage on the rheological behaviour of baby foods. *Journal of Texture Studies,* 27: 361-369.

Alonso M.L., Larrodé O. and Zapico J. 1995. Rheological behaviour of infant foods. *Journal of Texture Studies,* 26: 193-202.

Anton Paar. 2004. Commercial brochure. Germany Gmbh.

Alvarez E., Cancela M.A. and Maceiras R. 2006a. Effect of temperature on rheological properties of different jams. *International Journal of Food Properties,* 9 (1): 135-146.

Alvarez E., Cancela M.A. and Maceiras R. 2006b. Comparison of rheological behavior of salad sauces. *International Journal of Food Properties,* 9 (4): 907-915.

Barnes H.A., Schimansky H. and Bell D. 1999. 30 Years of progress in viscometers and rheometers. *Applied Rheology,* 9 (2): 69-76.

Bhamidipaty S. and Singh R.K. 1990. Flow behavior of tomato sauce with or without particulates in tube flow. *Journal of Food Process Engineering,* 12: 275-293.

Bhattacharya S. and Rastogi N.K. 1998. Rheological properties of enzyme-treated mango pulp. *Journal of Food Engineering,* 36 (3): 249-262.

Cabral R.A.F., Telis-Romero J., Telis V.R.N., Gabas A.L. and Finzer J.R.D. 2007. Effect of apparent viscosity on fluidized bed drying process parameters of guava pulp. *Journal of Food Engineering,* 80 (4): 1096-1106.

Canet W., Alvarez M.D., Fernández C. and Luna P. 2005. Comparison of methods for measuring of yield stress in potato puree: effect of temperature and freezing. *Journal of Food Engineering,* 68: 143-153.

Cantú-Lozano D., Rao M.A. and Gasparetto C.A. 2000. Rheological properties of noncohesive apple dispersion with helical and vane impellers: effect of concentration and particle size. *Journal of Food Process Engineering,* 23: 373-385.

Cepeda E. and Gómez R. 2002. Rheological characteristics of pimento puree: Theoretical interpretation. *Journal of Food Science,* 67 (7): 2734-2738.

Choi Y.H. and Yoo B. 2004. Characterization of time-dependent flow properties of food suspensions. *International Journal of Food Science and Technology,* 39: 801-805.

Correia L.R. and Mittal G.S. 1999. Food rheological studies using coaxial rotational and capillary extrusion rheometers. *International Journal of Food Properties,* 2 (2): 139-150.

Crochet M.J. and Walter K. 1993. Computational rheology: a new science. *Endeavour, New Series*, 17 (2): 64-77.

Dogan N., McCarthy M.J. and Powell R.L. 2002. In line measurement of rheological parameters and modeling of apparent wall slip in diced tomato suspensions using ultrasonics. *Journal of Food Science,* 67 (6): 2233-2240.

Durán L. and Costell E. 1982. Rheology of apricot puree: characterization of flow. *Journal of Texture Studies,* 13: 43-58.

Durán L. and Costell E. 1985. Influence of homogenization on the rheological behaviour of apricot puree. *Acta Alimentaria,* 14 (2): 201-210.

García E.J. and Steffe J.F. 1987. Comparison of friction factor equations for non-Newtonian fluids in pipe flow. *Journal of Food Process Engineering*, 9: 93-120.

Giner J., Ibarz A., Garza S. and Xhian-Quan S. 1996. Rheology of clarified cherry juices. *Journal of Food Engineering,* 30: 147-154.

Grigelmo-Miguel N. Ibarz-Ribas A. and Martín-Belloso O. 1999. Rheology of peach dietary fibre suspensions. *Journal of Food Engineering,* 39: 91-99.

Grigelmo-Miguel N., and Martín-Belloso O. 1999. Influence of fruit dietary fibre addition on physical and sensorial properties of strawberry jams. *Journal of Food Engineering,* 41: 13-21.

Guerrero S.N. and Alzamora S.M. 1997. Effect of pH, temperature and glucose addition on flow behavior of fruit purees: I. Banana puree. *Journal of Food Engineering,* 33: 239-256.

Guerrero S.N. and Alzamora S.M. 1998. Effect of pH, temperature and glucose addition on flow behavior of fruit purees: II. Peach, papaya and mango purees. *Journal of Food Engineering,* 37: 77-101.

Heldman D. 1975. Food Process Engineering. The AVI Publishing Co. Inc., Westport, CO, USA. pp. 25-44.

Holdsworth S.D. 1993. Rheological models used for the prediction of the flow properties of food products: A literature review. *Transactions of Institute for Chemical Engineers,* 71 (C): 139-179.

Ibarz A. and Barbosa Cánovas G.V. 2003. Unit Operations in Food Engineeering. CRC Press, Boca Raton, FL, USA. pp. 625-670.

Ibarz A., Gonzalez C., Esplugas S. and Vicente M. 1992a. Rheology of clarified fruit juices. I: Peach juices. *Journal of Food Engineering,* 15: 49-61.

Ibarz A., Pagán J. and Miguelsanz R. 1992b. Rheology of clarified fruit juices. II: Blackcurrant juices. *Journal of Food Engineering,* 15: 63-73.

Ibarz A., Garvin A. and Costa J. 1996. Rheological behavior of sloe (*Prunus spinosa*) fruit juices. *Journal of Food Engineering,* 27: 423-430.

Jones H.F. and Beckett S.T. 1995. Fruits and Vegetables. In Physico-Chemical Aspects of Food Processing (Ed. S.T. Beckett). Blackie Academic & Professional. Cambridge, GB. Pp. 292-314.

Krokida M.K., Maroulis Z.B. and Saravacos G.D. 2001. Rheological properties of fluid fruit and vegetable puree products: Compilation of literature data. *International Journal of Food Properties,* 4 (2): 179-200.

Maceiras R., Alvarez E. and Cancela M.A. 2007. Rheological properties of fruit purees: Effect of cooking. *Journal of Food Engineering,* 80: 763-769.

Martínez-Padilla L.P. and Rivera-Vargas C. 2006. Flow behavior of Mexican sauces using a vane-in-a-large cup rheometer. *Journal of Food Engineering,* 72: 189-196.

Mohsenin N. N. 1986 Physical Properties of Plant and Animal Materials. Gordon and Breach Science Publishing. USA.

Montoya Lázaro M., Sosa Morales M.E. and Vélez Ruiz J.F. 2003. Effect of the solids content on the flow properties of three vegetable fluids (PF05). *IV Congreso Iberoamericano de Ingeniería de Alimentos (CIBIA IV)*, October 5-8, Valparaíso, Chile. (In Spanish).

Moyano P.C., Esquerre C.A. and Osorio F.A. 1997. Temperature and generalized-Reynolds number effects on minimum residence time of fruit pulp. In *Engineering and Food at ICEF 7,* (R. Jowitt, Ed.), Food Rheology Section: E56-E59.

Nindo C.I., Tang J., Powers J.R. and.Takhar P.S. 2007. Rheological properties of blueberry puree for processing applications. *Food Science and Technology* (LWT), 40: 292-299.

Qiu C.G. and Rao M.A. 1988. Role of pulp content and particle size in yield stress of apple sauce. *Journal of Food Science, 53* (4): 1165-1170.

Quispe-Ramírez A.A. and Cruz-Romero M.C. 2001. Rheological behavior of camu-camu (*Myrciaria dubia*) juice as affected by temperature. *Proceedings of the Eighth International Congress on Engineering and Food (ICEF 8)*. Technomic Pub. Co. Inc. Vol. I: 483-486.

Rao M.A. 1987. Predicting the flow properties of food suspensions of plant origin. *Food Technology,* 3: 85-88.

Rao M.A. 1999. Rheology of Fluid and Semisolid Foods. Aspen Publishers, Inc. Gaithersburg, MR, USA.

Rao M.A. and Cooley H.J. 1983. Applicability of flow models with yield for tomato concentrates. *Journal of Food Process Engineering,* 6: 159-173.

Rao M.A., Bourne M.C. and Cooley H.J. 1981. Flow properties of tomato concentrates. *Journal of Texture Studies,* 12: 521-538.

Rodriguez-Luna G., Segura-Jauregui J.S., Torres J. and Brito E. (1987). Heat transfer to non-Newtonian fluid foods under laminar flow conditions in horizontal tubes. *Journal of Food Science,* 52 (4): 975-979.

Rodríguez Martinez V., Sosa Morales M.E. y Vélez-Ruiz J.F. 2007. Physicochemical and flow characterization of prickly pear concentrates. *VI Congreso Ibero Americano de Ingeniería de Alimentos (CIBIA VI),* November 5-8, Ambato, Ecuador. (In Spanish).

Singh N.I. and Eipeson W.E. 2000. Rheological behavior of clarified mango juice concentrates. *Journal of Texture Studies,* 31: 287-295.

Steffe J.F. 1992. Rheological Methods in Food Process Engineering. Freeman Press. East Lansing, MI, USA.

Steffe J.F.; Mohamed I.O. and Ford E.W. 1984. Pressure drop across valves and fittings for pseudoplastic fluids in laminar flow. *Transactions of the ASAE*: 616-619.

Suárez D., Macedo C. and Vélez J. 2003. Rheological behavior of three non-processed fruit fluids (papaya, melon and watermelon) and their concentrates. *Proceedings of the 3rd International Symposium on Food Rheology and Structure.* 429-430

TA Instruments. 2005. Commercial brochure. New Castle, DE, USA.

Tanglertpaibul T. and Rao M.A. 1987. Flow properties of tomato concentrates: effect of serum viscosity and pulp content. *Journal of Food Science,* 52 (2): 318-321.

Telis-Romero J., Telis V.R.N. and Yamashita F. (1999). Friction factors and rheological properties of orange juice. *Journal of Food Engineering,* 40: 101-106.

Vélez-Ruiz, J.F. 1996. Rheological Properties of Concentrated Milk. Ph.D. Thesis. Washington State University.

Vélez-Ruiz, J.F. 2002. Relevance of Rheological Properties in Food Processing Engineering. In *Engineering and Food for the 21st Century.* (J. Welti-Chanes, G.V. Barbosa-Cánovas and J.M Aguilera; Eds.). CRC Press. Nueva York, NY. USA. pp: 185-197.

Vélez Ruiz-J.F. 2003. Introductory aspects of momentum transfer phenomena. In *Transport Phenomena in Food Processing.* (J Welti, J. Vélez and G. Barbosa; Eds.). CRC Press. Boca Raton, USA. 45-65.

Vélez-Ruiz J.F. 2009a. Notes of Food Engineering I. Universidad de las Américas, Puebla. Unpublished document. Puebla, Mexico.

Vélez-Ruiz J.F. 2009b. Evaporation of Liquid Foods. In *Food Processing and Engineering Topics* (M.E. Sosa-Morales and J.F. Vélez-Ruiz, Eds.). Ed. Nova Publishers. New York, NY, USA.

Vélez-Ruiz J.F. 2009c. Evaluation of Convective Heat Transfer Coefficient Between Fluids and Particles. In *Food Processing and Engineering Topics* (M.E. Sosa-Morales and J.F. Vélez-Ruiz, Eds.). Ed. Nova Publishers. New York, NY, USA.

Vélez-Ruiz J.F. and Barbosa-Cánovas G.V. 1997. Effects of concentration and temperature on the rheology of concentrated milk. *Transactions of the ASAE,* 40 (4): 1113-1118.

Villarán M.C., Cepeda E., Llorens F.J. and Ibarz A. 1997. Influence of temperature on rheological behavior of jams of apricot, bilberry and rose hip. *Food Science and Technology International,* 3 (1): 13-19.

Vitali A.A. and Rao M.A. 1984. Flow properties of low-pulp concentrated orange juice: Serum viscosity and effect of pulp content. *Journal of Food Science,* 49: 876-881.

Yoo B. and Rao M.A. 1996. Creep and dynamic rheological behavior of tomato concentrates: effect of concentration and finisher screen size. *Journal of Texture Studies,* 27 (4): 451-459.

In: Food Processing and Engineering Topics
Editors: M. E. Sosa-Morales and J. F. Vélez-Ruiz
ISBN: 978-1-60741-788-0

Chapter 2

RHEOLOGICAL MODELS AND CHARACTERISTICS FOR SELECTED FOODS

***José A. Guerrero-Beltrán*[1], *Jorge Welti-Chanes*[2] and *Gustavo V. Barbosa-Cánovas*[3]**

[1] Departamento de Ingeniería Química y Alimentos
Universidad de las Américas, Puebla, Mexico)
[2] Departamento de Biotecnología e Ingeniería en Alimentos
Instituto Tecnológico y de Estudios Superiores de Monterrey, Mexico
[3] Biological Systems Engineering Dept.
Washington State University (USA)

1. INTRODUCTION

Stress and strain are the main parameters involved in the measurement of flow behavior of fluids. Fluids can be studied by subjecting them to a constant shear rate to measure shear stress. Ideally, this can be accomplished placing a fluid between two parallel plates using a predetermined gap. The lower plate is fixed and the upper plate moves at a selected velocity. Simple shear flow is also called viscometric flow. It includes axial flow in a tube, rotational flow between concentric cylinders, rotational flow between a cone and a plate, and torsional flow between parallel plates. Several mathematical functions can describe the relationship between shear stress and shear rate data. Newtonian and non-Newtonian behavior can be observed when testing different types of fluid or semi-fluid foods. Newtonian fluids have a linear relation between shear rate and shear stress. Non-Newtonian fluids, on the other hand, can behave in different ways, which will depend on the composition and structure of the food system. Several models to describe the flow behavior of fluid foods are explained in this chapter. Examples of selected liquid and semi-liquid foods will be given along with their flow behaviour. Also, the rheological properties assessed using various rheological models and the temperature effect will be discussed for four selected food products.

2. Types of Fluids

Fluid foods are classified taking into account several properties. They can be classified in terms of their composition, the technique used to measure their flow characteristics, or their rheological properties.

2.1. Classification According to Their Composition

Noncomplex Foods

In liquid and semi-liquid foods, composition may be diverse. In fruits and vegetable products such as juices, purées and nectars the main constituents are water, simple carbohydrates (glucose, fructose, and, or sucrose), organics acids (such as citric, malic, and tartaric) along with minor components (minerals and vitamins) (Haard, 1985). In milk and liquid yogurt, on the other hand, the main components are water, sugar (lactose), proteins (casein and whey proteins) and fat (Creamer and MacGibbson, 1996; Swaisgood, 1985). In honey and syrups the major components are water, sugar, high fructose syrup, and maltodextrinas (White, 1985); just to mention some examples of liquid or semi-liquid foods. Therefore, these types of products in general are Newtonian or non-Newtonian time independent materials. Thus, simple liquids, true solutions, low molecular weight solvents, diluted dispersions of macromolecules, solutions of polymer that do not interact, and food pastes containing few amounts of solids can exhibit Newtonian behaviour (Ibarz and Barbosa-Cánovas, 2002). For instance, Genovese and Lozano (2000) pointed out that cloudy apple juice (up to 50°Bx) behaves as a Newtonian fluid. Mango pulp behaves as a pseudoplastic material that show yield stress and thixotropic properties (Bhattacharya and Rastogi, 1998; Bhattacharya, 1999). Nindo *et al.* (2005) reported that blackberry and raspberry juices behave as Newtonian fluids in a wide total soluble solids range (10-65%) and temperature (20-60°C). A number of rheological properties for selected food products have been published by Barbosa-Cánovas and Peleg (1982), Rao (1986), Hayes (1987), Mohsenin (1996), and Bourne (2002), just to mention some examples.

Complex Foods

In many processed foods, besides of their usual compounds already mentioned above, a number of high molecular components are used to produce or to give some textural characteristics to foods. These compounds are in general biopolymers that come from plants (hydrocolloids) such as gums (arabic, tragacanth, xanthan, agar, carrageenan), pectin and starches (Whistler and Daniel, 1985), from microorganisms (dextran, xanthan) (Whistler and Daniel, 1985), or from animal products such as muscle tissue proteins (contractile and sarcoplasmic proteins, gelatin) (Hultin, 1985), egg proteins (albumin) (Powrie and Nakai, 1985), and milk proteins (casein and whey proteins) (Swaisgood, 1985). Depending on the concentration of the biopolymer, the rheological properties can be in the liquid or semi-liquid range or give some viscoelastic properties or even produce textural characteristics that have to be measured using texturemeters. Nevertheless, the use of biopolymers at low concentration can give some consistency to liquid or semi-liquid foods. Much research has been conducted regarding these types of biopolymers. Starch-in-water pastes, for instance, fall in the non-

Newtonian time independent food materials (Núñez-Santiago *et al.*, 2003); hydrocolloids are used as thickening agents to give textural characteristics to fluid foods and can behave as Newtonian or Non-Newtonian fluids depending on type and concentration of the hydrocolloid (Marcotte *et al.*, 2001).

2.2. Classification According to the Measurement Instrument

The common instruments used to measure fundamental rheological properties in liquid and semi-liquid foods can be classified in two types: the rotational type and the tube type. From these, the parallel plate, cone and plate, concentric cylinder and the mixer geometry (rheometers) are in the rotational type. On the other hand, the glass capillary, high pressure capillary and pipe geometry are in the tube type classification (Barbosa-Cánovas *et al.*, 1996; Steffe, 1996). The rotational approaches are commonly used to investigate the time dependency behaviour because in the tube approaches the fluid passes only one time in the system (Steffe, 1996). The information obtained with each measurement system may be different and it can be used to define the rheological properties of each tested product.

2.3. Classification According to Their Flow Behavior

Liquid and semi-liquid foods may present different flow behaviour and thus the rheological properties can be described using different rheological models. The viscous behaviour of fluids is classified in two main groups: Newtonian (time independent) and non-Newtonian fluids. The non-Newtonian fluids are classified in time-independent (those which rheological properties can be described using the power low, Bingham, and Herschel-Bulkley equations, among other models) and time-dependent (rheopectic and thixotropic) fluids (Steffe, 1996). Newtonian fluids show a linear behavior when shear stress vs. shear rate (rheogram) is plotted. The intercept is cero and the slope is the absolute viscosity (η). However, fluids other than Newtonians do not show this straight line behavior. Consequently, the shear stress vs. shear rate representation is different. The non-Newtonian time-independent fluids can show shear thickening (dilatant) or shear thinning (pseudoplastic) behaviour or present a straight line with yield stress (Bingham plastics). The rheogram of the dilatant behavior fluids grew up showing upward curves and the trend of the pseudoplastic behaviour grew up in convex shape curves (Mohsenin, 1996; Steffe, 1996; Ibarz and Barbosa-Cánovas, 2002). Therefore, the flow behavior index (n) describes the rheogram curvature. Salad dressings are examples of pseudoplastic materials and chocolate syrups and starch suspensions are examples of dilatant fluids (Bourne, 2002).

2.4. Classification According to Their Time-Dependency

In time-dependency the answer of the components in the material to stress is time-dependent due to its structure. For viscous materials the answer to stress is not instantaneous. The thixotropic and rheopectic materials show decreased and increased shear stress (and apparent viscosity), respectively, as a function of time at constant shear rate (Mohsenin, 1996; Steffe, 1996; Carreau *et al.*, 2002). Therefore, the structure of the material may change during flow. Gels from starch behave as thixotropic materials (Bourne, 2002). Peanut butter exhibits yield stress, shear thinning, and thixotropic characteristics (Carreau *et al.*, 2002). However, it is difficult to find rheopectic behaviour in food materials (Bourne, 2002).

3. Rheology

3.1. Physical Forces

When a force is applied to a given material, the response is the flow of that material; however, some resistance to flow or deformation appears which is called stress. Therefore, shear stress (σ) is the stress produced when molecules slip along one another on a parallel plane. Shear rate ($\dot{\gamma}$), on the other hand, is the velocity gradient or rapidity at which molecules slip one with another. This shear rate, when a force is applied, is given by:

$$\dot{\gamma} = -\frac{dv}{dy} \tag{1}$$

where v is the velocity (m/s) and y is the distance (m) (Toledo, 1999; Ibarz and Barbosa-Cánovas, 2002;). The shear stress and shear rate data representation (rheogram) for Newtonian and non-Newtonian fluids can be plotted. Therefore, data can be described using different rheological models to obtain their particular flow behavior and rheological properties (figure 1).

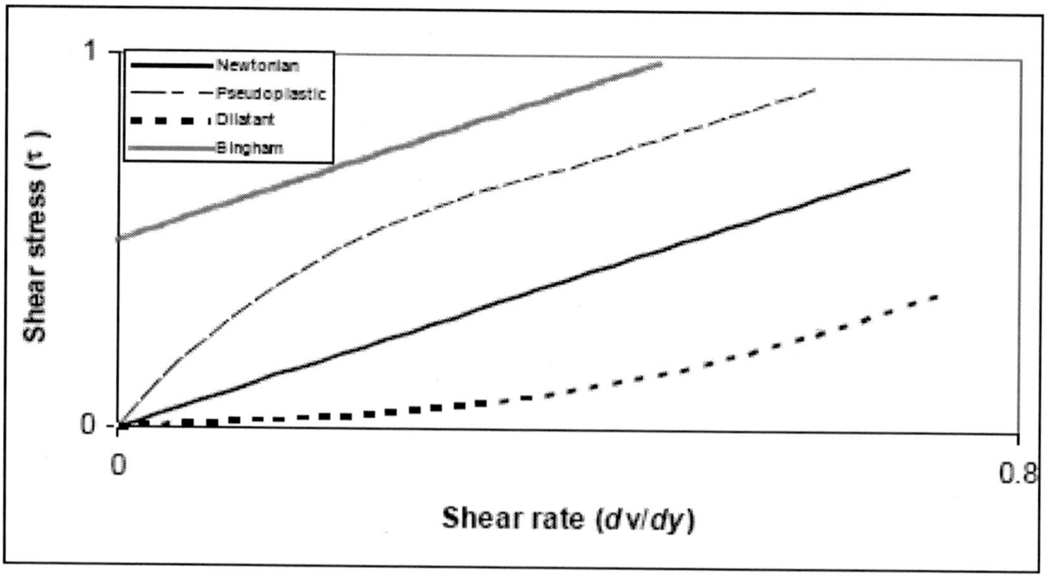

Figure 1. Types of fluids according to their shear stress and shear rate behaviour.

3.2. Rheology Definition

Rheology is the science that studies the deformation and flow of materials. It studies the way in which materials act in response to a stress or strain. All materials possess rheological properties. In the food area, rheology can be defined as the material science of foods (Steffe, 1996; Ibarz and Barbosa-Cánovas, 2002). It is clearly understood that food rheological properties are of paramount importance in the food industry to design processes such as food concentration or flow of fluid foods through pipes or tubes (Nindo *et al.*, 2005). Therefore, all fluid foods have to be rheologically characterized to know their flow behaviour when a unit operation is performed in a particular food industry.

3.3. Viscosity

Viscosity is a physical property that describes the resistance of a fluid to flow. Viscosity for Newtonian and Bingham plastics corresponds to the absolute viscosity (η) and plastic viscosity (η_{Pl}), respectively. The apparent viscosity (η_{app}), on the other hand, in non-Newtonian fluids is related to a specific shear rate ($\dot{\gamma}$). The apparent viscosity decreases as shear rate increases in pseudoplastic materials and it increases as shear rate increases in dilatant materials (Mohsenin, 1996; Steffe, 1996; Bourne, 2002). In some cases, the viscosity of diluted biopolymer solutions can be reported as relative viscosity ($\eta_{rel} = \eta_{solution}/\eta_{solvent}$), specific viscosity ($\eta_{sp} = \eta_{rel} - 1$) or reduced viscosity ($\eta_{red} = \eta_{sp}$/C), in which C is the mass concentration of the solution (g/100 mL, Steffe, 1996). The consistency coefficient (*K*), on the other hand, that describes the textural characteristics of non-Newtonian fluids corresponds to the absolute viscosity and plastic viscosity for Newtonian and Bingham plastic materials, respectively, in both cases *n* is equal to 1.0.

3.4. Yield Stress

Yield stress (σ_o) is defined as the minimum shear stress required by fluids to initiate their flow (Kovalenko and Briggs, 2002). This value is zero in a number of fluid foods. Yield stress in some ways describes textural properties and structural changes and it is useful to determine the operation conditions during processing and handling (Kovalenko and Briggs, 2002). Yield stress, for practical application is in fact significant since it is necessary to determine the power needed for moving fluids. Also, yield stress imparts stability to foods in low-stress situations. Even though yield stress is generally recognized to be important in the food industry, some researchers state that yield stress does not really exist; however, Cheng (1986) pointed out that "the magnitude of the yield stress obtained depends on the time interval that is used to determine whether flow is established or has ceased." Bourne (2002) has reported a list of yield stress values for food products. One of the rheological models to determine rheological properties, such as *K*, *n*, y σ_o, is the Hershel-Bulkley relationship.

3.5. Rheological Parameters

Table 1 lists the limits or intervals for rheological parameters (n, K and σ_o) of fluids. A number of actual values for fluids can be found in food literature elsewhere.

Table 1. Flow behaviour index (n), consistency coefficient (K), and yield stress (σ_o) values for Newtonian and non-Newtonian fluids

	σ_o	n	K
Newtonian	0	1	$K = \eta$
Non-Newtonian:			
Bingham plastic	> 0	1	$K = \eta_{pl}$
Dilatant (shear thickening)	0	$n > 1$	> 0
Dilatant with yield stress	> 0	$n > 1$	> 0
Pseudoplastic (shear thinning)	0	$n < 1$	> 0
Pseudoplastic with yield stress	> 0	$n < 1$	> 0

(Barbosa-Cánovas *et al.*, 1996; Steffe, 1996; Bourne, 2002)

4. Composition and Rheological Behavior of Selected Foods

Honey, Molasses and Syrups

Pure honey bee is a semi-liquid food product. Bees take nectar from the flowers surrounding the place where they build the apiary. Therefore, bees elaborate honey with nectar from different types of flowers, but in general honey is composed by water, glucose, fructose, sucrose, and few quantities of maltose, enzymes, minerals, vitamins and color (White, 1985; Barhate *et al.*, 2003). Molasses come from the sugar production, they contain sugar, water, inorganic matter, organic substances (organic acids, lipids), and high molecular weight macromolecules (starch, cellulose and hemicellulose (Toğrul and Arslan, 2004). Broadfoot and Miller (1990) pointed out that pure molasses, and molasses containing starch (Kaur *et al.*, 2002) or ethanol (Toğrul and Arslan, 2004) behaved as pseudoplastic materials. Syrups, on the other hand, are made with corn syrup, high fructose corn syrup, nonfat milk, fructose, corn starch-modified, color, and natural and artificial flavors. Rao (2007) states that corn syrups (77.4°Bx; 20°C) behave as Newtonian fluids since the flow behavior index is equal to 1.0. Nevertheless, the main components in sweeteners are sugars and oligosaccharides from deferent sources and molecular weights.

Ketchup (or Catsup or Catchup) and Other Tomato Products

Ketchup is made by concentrating tomato pulp until reaching around 28% of total solids and it is added with high fructose corn syrup, distilled vinegar, corn syrup, onion powder, garlic powder and natural flavors. The consistency of ketchup is due to the soluble and insoluble solids content, including pectins (Farrell, 1985). Tomato paste (concentrate) is obtained by concentrating tomato pulp until reaching around 28 – 33% of total solids content

(Farrell, 1985). Tomato purée on the other hand, is prepared concentrating tomato pulp until reaching a total solids content of around 10%; it is also added with sugar, and salt. Rheological properties of tomato based products are commonly measured using the Adams and Bostwick consistometers and Brookfield viscometer (Cullen *et al.*, 2001). Tehrani and Ghandi (2006) evaluated the consistency of tomato concentrates at 25°C using the Bostwick instrument. Rheological properties (n, K, σ_o) data are reported for tomato products in Rao (2007).

Mustard and Salad Dressings

Mustard is prepared with distilled vinegar, water, mustard seeds, salt, paprika, spices, natural flavor and garlic powder; however, sugar, wine, starches and gums may also be added (Farrell, 1985). The rheological characteristics will depend on the type of product: semi-liquid (83.1±0.1% of moisture content) or paste (77.7±0.4% of moisture content). Salad dressings are semi-liquids with high vegetable oil content (other than low fat salad dressings); however, their composition will depend on type of dressing. Besides of vegetable oil (soy, corn, girasol, olive, among other oily seeds) water, and vinegar (Farrell, 1985), salad dressings can also content a wide variety of other components. Examples of these components are mustard, garlic powder, onion powder, high fructose corn starch, whey powder, egg yolk, corn starch-modified, hidrocoloides (xanthan gum for instance), lemon juice, calcium ethylenediaminetetraacetic acid (EDTA), spices, and natural and artificial flavors.

Mayonnaise (Traditional or Low Fat Products)

Mayonnaise is an oil-in water emulsion (Powrie and Nakai, 1985; Blonk, 2002) prepared with vegetable oil (soy, corn, girasol, among other oily seeds), whole egg, egg yolk (emulsifying ingredient containing phospholipids, lipoproteins, and proteins), and distilled vinegar (Powrie and Nakai, 1985; Maruyama and Sakashita, 2006; Rao, 2007). It also contents few amounts of water, salt, spices, lemon juice, calcium disodium EDTA, and oleoresin paprika, as well as, gums (guar and xanthan), propylene glycol, and carboxymethylcellulose (Rao, 2007).

5. Rheology Models for Liquid and Semi-Liquid Foods

5.1. Newtonian Fluids

The viscosity of the Newton fluids is defined as (Barbosa-Cánovas *et al.*, 1996; Bourne, 2002; Ibarz and Barbosa-Cánovas, 2002):

$$\eta = \frac{\sigma}{\dot{\gamma}} \tag{2}$$

where η is the absolute viscosity (Newtonian viscosity) (Pa s), σ is shear stress (Pa), and $\dot{\gamma}$ is shear rate (1/s). Reports regarding absolute viscosity can be found elsewhere; some of these reports are by Rha (1975) and Hayes (1987). As stated before, water, fruit and vegetable clarified juices and simple liquids and diluted solutions can show Newtonian behaviour (Ibarz and Barbosa-Cánovas, 2002). As an example, the cloudy apple juice (up to 50°Bx) behaves as Newtonian fluid (Genovese and Lozano, 2000) and many liquid foods may behave as Newtonian fluids.

5.2. Non-Newtonian Fluids

For non-Newtonian fluids the viscosity is not longer constant, but is dependent on the shear rate and temperature. Therefore, the shear rate-dependent viscosity is called apparent viscosity (η_{app}) (Steffe, 1996). The three main categories of non-Newtonian materials are time-independent, time-dependent, and viscoelastic materials. Even though the Power Law or Ostwald-de Waele model and the Herschel-Bulkley model characterize a wide variety of non-Newtonian materials (Núñez-Santiago *et al.*, 2003). There are an important number of rheological models that have been proposed to fit experimental data and determine their rheological properties. Thus, the apparent viscosity and rheological properties for some fluids can be computed using these uncomplicated models:

a) Power law or Ostwald-de Waele model (Steffe, 1996; Yu and Gunasekaran, 2001)

$$\eta_{app} = f(\dot{\gamma}) = \frac{K\dot{\gamma}^n}{\dot{\gamma}} = K\dot{\gamma}^{(n-1)} \tag{3}$$

The Power Law model has been applied to determine the rheological properties of many types of foods. Some examples are given here. Molasses, containing ethanol (1-5%), showed a flow behaviour index in the range 0.756-0.97. The consistency coefficient decreased and the n parameter increased as the amount o ethanol increased (Toğrul and Arslan, 2004). Kaur *et al.*, (2002), on the other hand, pointed out that the consistency coefficient increased and the flow behaviour index decreased for molasses as the starch concentration increased from 0.3 to 1.9% (w/w). Marcotte *et al.* (2001) assessed the rheological properties (at temperatures up to 80°C) of the hydrocolloids pectin (2.5%) and starch (4%) using the Power Law model; they reported good correlation coefficient ($R^2 > 0.99$).

b) Herschel-Bulkley (H-B) model (De Kee *et al.*, 1983; Steffe, 1996; Toledo, 1999)

$$\eta_{app} = f(\dot{\gamma}) = \frac{K(\dot{\gamma})^n + \sigma_o}{\dot{\gamma}} = K\dot{\gamma}^{(n-1)} + \frac{\sigma_o}{\dot{\gamma}} \tag{4}$$

Marcotte *et al.* (2001) evaluated the rheological properties (σ_o, K and n) of the hydrocolloids xanthan (2%) and carrageenan (1.5%) at temperatures up to 80°C using this model with high correlation coefficients ($R^2 > 0.99$).

c) Bingham plastic equation (Bingham, 1922)

$$\eta_{Pl} = \frac{\sigma - \sigma_o}{\dot{\gamma}} \tag{5}$$

where η_{pl} is the plastic viscosity (Pa·s) (Steffe, 1996; Toledo, 1999). The Bingham model fitted the rheological data of pizza sauce very well since high correlation coefficients ($R^2 >$ 0.99) were observed by Cullen *et al.* (2001). Bhattacharya (1999) observed a correlation coefficient of 0.924 when using this model to fit mango pulp rheological data.

d) Casson model (Casson, 1959; Stroshine, 2000)

$$\sigma^{0.5} = \sigma_o^{\ 0.5} + K\dot{\gamma}^{0.5} \tag{6}$$

or (Yoo and Rao, 1995)

$$\sigma^{0.5} = k_{oc} + k_c\dot{\gamma}^{o.5} \tag{7}$$

or (Rao *et al.*, 1984)

$$\sigma^{0.5} = \sigma_o^{0.5} + k\dot{\gamma}^n \tag{8}$$

where σ_o is yield stress (Pa), which is equal to $(k_{oc})^2$ and η_c is the Casson viscosity (Pa·s), which is equal to $(k_c)^2$. The Casson model has been applied to fit rheological data of mango purée to determine yield stress (Bhattacharya, 1999) and also to fit rheological data of pizza sauce ($R^2 > 0.99$) (Cullen *et al.*, 2001).

e) Ree-Eyring equation (Barbosa-Cánovas *et al.*, 1996)

$$\eta = \eta_\alpha + (\eta_o - \eta_\alpha)\frac{\arcsin h(\beta\dot{\gamma})}{\beta\dot{\gamma}} \tag{9}$$

where η_α is the viscosity at very large shear rate ($\dot{\gamma} \rightarrow \alpha$), η_o is the viscosity at zero shear rate ($\dot{\gamma} \rightarrow 0$), and β is the molecular relaxation time (s).

This model has been used to fit guar gum and locust bean gum solutions data with reasonable results (Doublier and Launay, 1981).

f) Cross equation (Cross, 1965; Rao, 2007)

The Cross equation has been used to fit rheological data of dispersed systems of polysaccharide solutions and other types of biopolymers such as protein gels:

$$\frac{(\eta_o - \eta_\alpha)}{\eta - \eta_\alpha} = 1 + (t\dot{\gamma})^m \tag{10}$$

where t is the relaxation time (s) and m is a non-dimensional exponent. When both, $\eta << \eta_o$ and $\eta >> \eta_\alpha$ the Cross equation predicts a power behavior where $(1 - m)$ come close to the flow behaviour index (n). The Cross equation predicted the viscosity of stirred yogurt at low shear stresses (Geraghty and Butler, 1999).

g) Skriver *et al.* model (Skriver *et al.*, 1993 ; Bhattacharya, 1999)

This model is named as the three-parameter model which is similar to the Michaelis-Menten equation, containing an additional constant (σ_o) that would show a relationship between shear rate/shear stress data:

$$\sigma = \frac{k_1\dot{\gamma}}{k_2 + \dot{\gamma}} + \sigma_o \tag{11}$$

where k_1 is a constant in the Michaelis-Menten equation (Pa) and k_2 is the shear rate at which the shear stress is equal to (σ_o + $k_1/2$) (1/s). Bhattacharya (1999) reported values of 104.9±1.5 Pa and 230.6±9.5 s^{-1} for k_1 and k_2, respectively, for mango pulp.

h) The exponential model (De Kee *et al.*, 1983)

$$\eta(\dot{\gamma}) = \frac{\sigma_o}{\dot{\gamma}} + \sum_{p=1}^{\alpha} \eta_p e^{-t_p\dot{\gamma}} \tag{12}$$

The series can be written as:

$$\sum_{p=1}^{\alpha} \eta_p e^{-t_p\dot{\gamma}} = \sum_{p=1}^{M} \eta_p e^{-t_p\dot{\gamma}} + \eta_\alpha \tag{13}$$

where η_α is the limiting viscosity at high rates of shear (Pa·s) and η_p (Pa·s) and t_p (s) are constants. This model has been applied to determine rheological properties of mayonnaise, honey, yogurt, tomato juice, and salad dressings with adequate fittings. De Kee *et al.* (1983) reported values of 1.424 Pa·s, 3.962x10^{-3} s, 0.3420 Pa·s, 3.062x10^{-4} s, 0.1019 Pa·s, and 103.6 Pa for η_1, t_1, η_α, t_2, η_e, and σ_o for mayonnaise, respectively.

5.3. Temperature Dependency of Viscosity and Rheological Properties

The more useful mathematical model used to describe the effect of temperature on viscosity of Newtonians fluids and rheological parameters of non-Newtonian fluids (particularly the K parameter) is the Arrhenius relationship representation:

$$\eta_{app} = \eta_{Ar} \exp\left[\frac{E_a}{RT}\right] \tag{14}$$

or

$$K = K_o \exp\left[\frac{E_a}{RT}\right] \tag{15}$$

where η_{Ar} is a constant (Pa·s), E_a is the activation energy (kJ/mol), T is the absolute temperature (K), and R is the ideal gas constant (kJ/mol K) (Rao *et al.*, 1984; Stroshine, 2000, Rao, 2007).

6. Rheological Evaluation for Selected Foods

Selected condiments, sauces and sweeteners were studied to determine some physicochemical and rheological characteristics. Samples were dried using an oven at 75°C to determine moisture content. Total soluble solids, on the other hand, were assessed using an Atago refractometer. Rheological data were obtained using a Paar Physica US 200 Rheometer (Searle type) using a plate and plate geometry with a gap of 0.8 mm. The equipment was operated in the control shear rate mode (CSR) (30 points were taken). Various rheological models were used to fit the experimental shear stress and shear rate data of pure honey, ketchup, mustard, and mayonnaise obtained at three temperatures (20, 30, and 40°C) in order to obtain the rheological parameters. The temperature dependency of viscosity or consistency coefficient was obtained using the Arrhenius type model.

6.1. Food Characteristics

Moisture and total soluble solids content in foods may have a very important influence in their rheological characteristics. The experimental moisture content and total soluble solids of some food materials (condiments, sauces, and sweeteners) are presented in table 2. Ingredients included in total soluble solids were already described.

Table 2. Moisture and total soluble solids (TSS) for selected foods

Product	Moisture content (% w.b.)	TSS (%)
Condiments and sauces:		
Salad dressing (Italian style)	34.0 ± 1.2	18.3 ± 0.3
Salad dressing (Ranch)	48.1 ± 0.2	37.2 ± 0.1
Ketchup	65.5 ± 0.2	33.3 ± 0.1
Mayonnaise	16.2 ± 0.3	84.3 ± 0.1
Mustard (semi-liquid)	83.1 ± 0.0	17.6 ± 0.4
Mustard (semi-solid)	77.7 ± 0.4	21.3 ± 0.1
Tomato purée	90.0 ± 0.1	10.2 ± 0.1
Sweeteners:		
Honey	15.6 ± 0.4	82.5 ± 0.7
Syrup caramel	27.7 ± 0.3	66.3 ± 0.1

6.2. Rheological Characteristics

Rheological Data Analysis

Table 3 lists the experimental rheological data (average of three trials) for ketchup.

Table 3. Rheological data for ketchup evaluated at three temperatures

Temperature (°C)					
20		30		40	
γ (1/s)	σ (Pa)	γ (1/s)	σ (Pa)	γ (1/s)	σ (Pa)
0.2	18.5	0.1	17.0	0.2	17.6
10.4	49.2	10.4	53.7	10.4	47.8
20.8	51.0	20.8	55.3	20.8	46.4
31.1	54.4	31.1	58.4	31.1	47.9
41.5	58.7	41.5	61.1	41.5	50.9
51.8	62.3	51.8	64.0	51.8	54.0
62.1	64.9	62.1	67.2	62.1	57.4
72.5	67.6	72.5	70.0	72.5	59.8
82.8	69.5	82.8	72.4	82.8	62.7
93.2	71.6	93.2	74.6	93.2	65.1
104.0	73.4	104.0	76.4	104.0	67.1
114.0	74.9	114.0	78.2	114.0	68.8
124.0	76.3	124.0	79.8	124.0	70.5
135.0	77.8	135.0	81.3	135.0	71.9
145.0	79.1	145.0	82.8	145.0	72.8
155.0	80.3	155.0	84.2	155.0	74.3
166.0	81.2	166.0	85.4	166.0	75.4
176.0	82.2	176.0	86.5	176.0	76.4
186.0	83.3	186.0	87.4	186.0	77.2
197.0	83.9	197.0	88.4	197.0	77.8
207.0	84.7	207.0	89.2	207.0	78.7

Table 3. (Continued).

Temperature (°C)					
20		30		40	
γ (1/s)	σ (Pa)	γ (1/s)	σ (Pa)	γ (1/s)	σ (Pa)
217.0	85.4	217.0	90.0	217.0	79.4
228.0	86.1	228.0	90.6	228.0	80.1
238.0	86.6	238.0	91.2	238.0	80.7
248.0	87.2	248.0	91.8	248.0	81.4
259.0	87.8	259.0	92.4	259.0	82.1
269.0	88.3	269.0	93.0	269.0	82.6
279.0	88.8	279.0	93.5	279.0	83.1
290.0	89.5	290.0	94.2	290.0	83.5
300.0	90.0	300.0	94.7	300.0	84.0

Rheological data for ketchup were plotted (figure 2) and rheological models were used to obtain the rheological properties (n, K, and σ_o).

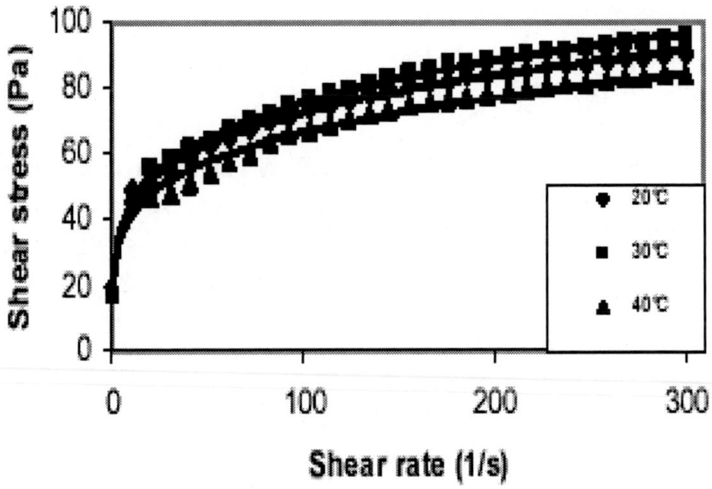

Figure 2. Experimental data and Ostwald-de Weale modeling for rheological data of ketchup.

a) Ostwald-de Weale model

$$\sigma = K\dot{\gamma}^n$$

Shear stress and shear rate were plotted in a linear fashion:

$$\log\sigma = n\log\dot{\gamma} + \log K$$

to obtain the consistency coefficient ($Pa{\cdot}s^n$) and the flow behaviour index (dimensionless) at three temperatures. An Excel program was used to fit all rheological data. The constants are included in the model:

$$\sigma_{20°C} = 27.3\dot{\gamma}^{0.21}$$

$$\sigma_{30°C} = 27.9\dot{\gamma}^{0.21}$$

$$\sigma_{40°C} = 25.3\dot{\gamma}^{0.21}$$

All rheological parameters for ketchup, in addition to the correlation coefficient, are listed in table 4. Also, rheological values for honey, mustard and mayonnaise are listed in table 4 which were obtained in similar way.

b) Herschel- Bulkley model

$$\sigma = K\dot{\gamma}^n + \sigma_o$$

Rheological data were plotted in the linear way:

$$\log(\sigma - \sigma_o) = n \log \gamma + \log K$$

to obtain the consistency coefficient ($Pa{\cdot}s^n$), the flow behaviour index (dimensionless), and yield stress (Pa) at three temperatures. A solver program from Excel was used to fit all rheological data. The constants are listed in table 5. Also, rheological values for honey, mustard and mayonnaise are listed in table 5.

c) Casson model

The square root of the shear rate and shear stress were plotted to obtain the yield stress for ketchup:

$$\sigma^{0.5} = K'\dot{\gamma}^{0.5} + \sigma_o^{0.5}$$

The following equations depict the constants for ketchup after a linear regression obtained using the Excel program:

$$\sigma_{20°C}^{0.5} = 0.28\dot{\gamma}^{0.5} + 5.292$$

$$\sigma_{30°C}^{0.5} = 0.30\dot{\gamma}^{0.5} + 5.215$$

$$\sigma_{40°C}^{0.5} = 0.27\dot{\gamma}^{0.5} + 5.06$$

All, constants and the correlation coefficient for ketchup, mustard and mayonnaise are listed in table 6.

d) Bingham model

Shear rate and shear stress data were plotted to obtain the plastic viscosity and yield stress as fallow:

$$\sigma = \eta_{Pl}\dot{\gamma} + \sigma_o$$

The equations below show the constants for ketchup after the linear regression using the Excel program:

$$\sigma_{20°C} = 0.26\dot{\gamma} + 31.8$$
$$\sigma_{30°C} = 0.29\dot{\gamma} + 29.7$$
$$\sigma_{40°C} = 0.24\dot{\gamma} + 29.5$$

All data and the correlation coefficients for ketchup, mustard and mayonnaise are listed in table 7.

e) Activation energy

The ln *K* values, from the Ostwald-de Weale model, vs. 1/T (absolute temperature) for ketchup were plotted as follow:

$$\ln K = \frac{E_a}{R}\frac{1}{T} + const.$$

The regression procedure was performed using the Excel program. Activation energies for honey (using viscosities or consistency coefficients), ketchup, mustard, and mayonnaise are listed in table 8.

Rheological Properties Discussion

Honey

Newton and Ostwald-de Waele models were used to fit shear stress rate and shear rate experimental data for honey (figure 3). Viscosity values of 37.1 ± 1.4 ($R^2 = 0.997$), 10.3 ± 0.5 ($R^2 = 0.995$), and 3.4 ± 0.2 ($R^2 = 0.986$) Pa·s at 20, 30 and 40°C were obtained, respectively, when applying the Newton model. The Ostwald-de Weale model, on the other hand, provided better correlation coefficients (table 4) when modeling rheological data of honey, however *n* values near to 1 are an indication of a Newtonian behavior of the samples of honey. Dickie and Kokini (1983) found values of 15.39 Pa·s^n and 0.989 for *K* and *n* parameters, respectively, at 30.5°C. Yu and Gunasekaran (2001) confirmed that honey behaved as a true Newtonian fluid when tested a 30 and 35°C. Witczak and Juszczak (2006) pointed out that heather honeys behaved as non-Newtonian fluids showing some yield stress and thixotropy. De Kee (1983) reported a viscosity of 9.17 (Pa·s) for honey at 25°C. Tavakolipour and

Ghandi (2006) pointed out that Iranian honey varieties behaved as Newtonian fluids. Acacia and chestnut honeys from Croatia also presented Newtonian behaviour (Tiban *et al.*, 2006). Much research has been performed to elucidate the rheological properties of honey from different sources and places. Thus, Mohsenin (1996) has concluded that the flow behaviour of honey depends entirely on the plant source from which honey is obtained.

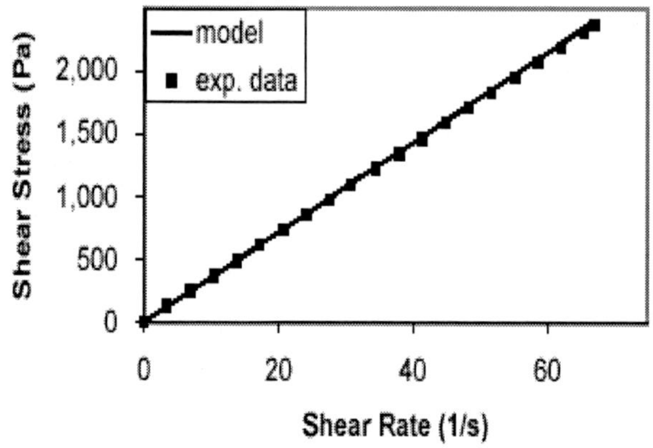

Figure 3. Newtonian model for honey at 20°C.

Table 4. Ostwald-de Waele model parameters for honey, ketchup, mustard, and mayonnaise

Product	Temp. (°C)	K (Pa.s^n)	n	R^2
Honey	20	44.1 ± 4.3	0.95 ± 0.03	0.999 ± 0.001
	30	15.0 ± 2.8	0.92 ± 0.03	0.999 ± 0.000
	40	5.1 ± 0.5	0.94 ± 0.01	0.997 ± 0.003
Ketchup	20	27.3 ± 2.7	0.21 ± 0.02	0.996 ± 0.002
	30	27.9 ± 0.9	0.21 ± 0.01	0.995 ± 0.001
	40	25.3 ± 3.0	0.21 ± 0.03	0.988 ± 0.008
Mustard	20	32.0 ± 1.8	0.24 ± 0.02	0.990 ± 0.009
	30	24.8 ± 3.3	0.27 ± 0.02	0.988 ± 0.002
	40	23.8 ± 1.4	0.25 ± 0.03	0.982 ± 0.015
Mayonnaise	20	133.7 ± 6.3	0.22 ± 0.06	0.974 ± 0.011
	30	115.4 ± 5.2	0.25 ± 0.02	0.270 ± 0.256
	40	96.2 ± 2.8	0.21 ± 0.01	0.615 ± 0.038

Ketchup

The ketchup rheological data are illustrated in figure 2. It was observed that the flow behavior of ketchup (pseudoplastic) was adequately described by the Ostwald-de Waele (table 4) and Herschel–Bulkley (H-B) (table 5) models in view of the fact that correlation coefficients were higher than 0.988 and 0.991, respectively. The consistency coefficient decreased as temperature increased. *K* values computed with the Ostwald-de Waele model showed less variability than those computed with the H-B model. Similar flow behavior index values (n = 0.21) were observed for the Power Law and Herschel- Bulkley models. The

Casson (R^2 < 0.907) (table 6) and Bingham (table 7) (R^2 < 0.69) models did not fit satisfactory the ketchup rheological experimental data.

Yield stresses were obtained in the range of 25.6 – 28.0 and 29.5 – 31.8 Pa with the Casson and Bingham models, respectively. Steffe (1996) reported yield stress data ranging from 15 to 30 Pa; such data were calculated using different rheological models and reported by different authors. De Kee *et al.* (1983) reported a yield stress value of 48 Pa for tomato juice. Yu and Gunasekaran (2001) reported values of 30.36 and 27.38 Pa·s^n obtained at 30 and 35°C, respectively. Regarding the *n* parameter, Yu and Gunasekaran (2001) reported values of 0.15 and 0.13 at same temperatures, respectively. Thus, different types of models can be applied to determine the more accurate rheological properties of ketchup.

Table 5. Herschel-Bulkley model parameters for ketchup, mustard, and mayonnaise

Product	Temp. (°C)	K (Pa.s^n)	n	R^2
Ketchup	20	30.1 ± 7.0	0.20 ± 0.02	0.997 ± 0.000
	30	36.2 ± 3.1	0.18 ± 0.01	0.996 ± 0.001
	40	19.3 ± 4.2	0.25 ± 0.02	0.991 ± 0.004
Mustard	20	39.0 ± 4.1	0.21 ± 0.01	0.993 ± 0.006
	30	18.7 ± 3.3	0.32 ± 0.03	0.985 ± 0.001
	40	12.6 ± 7.7	0.42 ± 0.14	0.981 ± 0.014
Mayonnaise	20	627.7 ± 357.2	0.09 ± 0.07	0.987 ± 0.010
	30	834.7 ± 74.5	0.04 ± 0.02	0.757 ± 0.091
	40	1745.4 ± 350.5	0.02 ± 0.01	0.970 ± 0.009

Mustard

Figure 4 illustrates the rheological data for mustard. Mustard showed a pseudoplastic flow behavior. The Ostwald-de Waele (table 4) and Herschel-Bulkley (table 5) models highly described the mustard flow behavior data. The correlation coefficients for both, the Ostwald-de Waele and Herschel-Bulkley models, were higher than 0.982 and 0.981, respectively. The consistency coefficient values decreased as temperature increased ranging from 24.8 to 32.0 and 12.6 to 31.9 Pa·s^n for the Ostwald de Waele and Herschel-Bulkley models, respectively. An *n* value of 0.25 ± 0.02 was computed with the Power Law model; however, higher variability was observed for the *n* value obtained with the Herschel-Bulkley model. Dickie and Kokini (1983) reported values of 35.05 Pa·s^n and 0.196 for *K* and *n* parameters, respectively, obtained at 30.5°C. The Casson (R^2 < 0.925) and Bingham (R^2 < 0.799) models did not fit the experimental data for mustard adequately. Yield stresses computed from the Casson and Bingham models ranged from 20.4 to 28.3 and from 24.8 to 33.3 Pa, respectively. Yield stress values computed with the Casson model showed less variability than data computed from the Herschel-Bulkley model. Steffe (1996) reported yield stress data from 52 to 78 Pa obtained by different measuring methods.

Table 6. Casson model parameters for ketchup, mustard, and mayonnaise

Product	Temp. (°C)	K' ($s^{0.5}$)	σ_o (Pa)	R^2
Ketchup	20	0.28 ± 0.02	28.0 ± 3.1	0.877 ± 0.029
	30	0.30 ± 0.01	27.2 ± 2.1	0.850 ± 0.018
	40	0.27 ± 0.03	25.6 ± 3.7	0.907 ± 0.030
Mustard	20	0.51 ± 0.03	28.3 ± 2.3	0.884 ± 0.041
	30	0.52 ± 0.03	20.5 ± 2.1	0.890 ± 0.033
	40	0.52 ± 0.17	20.4 ± 6.4	0.925 ± 0.019
Mayonnaise	20	0.49 ± 0.08	140.1 ± 13.5	0.730 ± 0.000
	30	0.58 ± 0.05	97.7 ± 2.5	0.723 ± 0.064
	40	0.54 ± 0.02	101.1 ± 7.8	0.660 ± 0.040

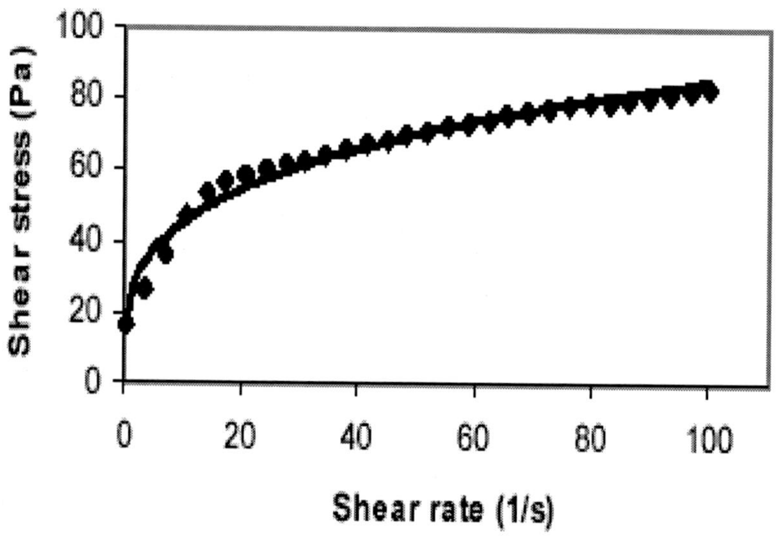

Figure 4. Power Law model for mustard at 30°C.

Table 7. Bingham model parameters for ketchup, mustard, and mayonnaise

Product	Temp. (°C)	η_{pl} (Pa.s)	σ_o (Pa)	R^2
Ketchup	20	0.26 ± 0.02	31.8 ± 4.2	0.583 ± 0.119
	30	0.29 ± 0.01	29.7 ± 2.9	0.455 ± 0.090
	40	0.24 ± 0.03	29.5 ± 5.2	0.690 ± 0.111
Mustard	20	0.82 ± 0.05	33.3 ± 3.1	0.649 ± 0.116
	30	0.76 ± 0.05	25.5 ± 3.4	0.708 ± 0.098
	40	0.72 ± 0.36	24.8 ± 6.3	0.799 ± 0.102
Mayonnaise	20	0.97 ±0.19	148.3 ± 17.4	< 0.5
	30	1.13 ±0.10	97.7 ± 10.8	< 0.5
	40	1.03 ± 0.07	98.5 ± 9.4	< 0.5

Mayonnaise

Figure 5 depicts the flow behavior data for mayonnaise. The flow behavior of mayonnaise was different as temperature increased. However, it can be observed that slippage occurred at 30 and 40°C. This means that increasing temperature; the emulsion changed its flow characteristics. Therefore, the rheological behavior also changed dramatically. The flow behavior data obtained at 20°C were adequately fitted with the Ostwald-de Waele (R^2 = 0.974, table 4) and Herschel-Bulkley models (R^2 = 0.987, table 5) up to 140 s^{-1}. Dickie and Kokini (1983) reported *K* and *n* values of 100.13 $Pa \cdot s^n$ and 0.131, respectively, measured at 30.5°C. Maruyama and Sakashita (2006) fit rheological data of mayonnaise adequately using the Power Law model to compute *K* and *n*. Yu and Gunasekaran (2001) reported *K* values of 123.92 and 80.35 $Pa \cdot s^n$ for mayonnaise at 30 and 35°C, respectively. Concerning to the flow behaviour index, they reported values of 0.09 and 0.17 for mayonnaise at 30 and 35°C, respectively. Mayonnaise flow behavior data could not be adjusted with the Casson and Bingham models since correlation coefficients, shown in tables 6 and 7, respectively, are not satisfactory. On the other hand, no good correlation was obtained at 30 and 40 °C with no one of the tested models. Yield stress data using the Casson (97.7 – 140.1 Pa) and the Bingham (97.7 – 148.3 Pa) models were computed. Yoo *et al.* (1995) reported yield stress data values ranging from 163.8 to 204.4 Pa obtained with the Vane method at 25°C. De Kee *et al.* (1983) reported a yield stress value of 134.8 Pa for mayonnaise obtained at 25°C.

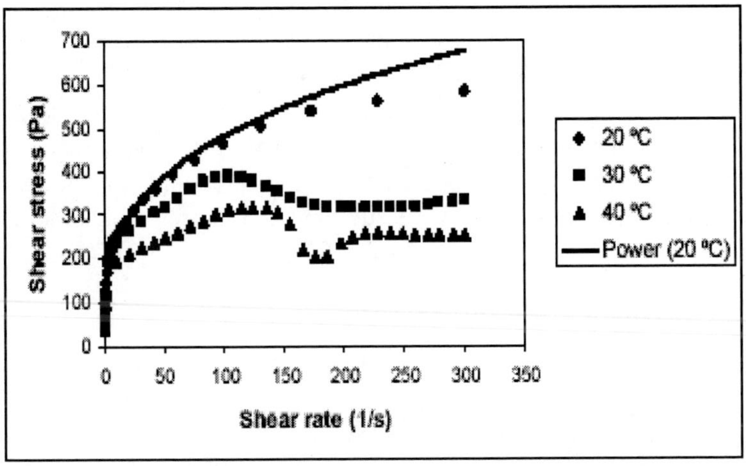

Figure 5. Rheological flow behavior for mayonnaise at different temperatures.

6.3. Activation Energy

Table 8 illustrates the activation energy for selected food products using the rheological parameters *K* or η. The E_a values were computed with the Ostwald-de Waele data since the E_a data obtained with the Herschel-Bulkley model did not fit a straight line (R^2 < 0.700) adequately. Tiban *et al.* (2006) reported activation energies of 500.1 and 490.7 kcal/g mole for acacia and chestnut honeys, respectively. The activation energy was lower for Ketchup and mustard than that for honey and mayonnaise, indicating that food composition may play an important role in the rheological characteristics which are temperature dependant.

Table 8. Activation energy using *K* and η parameters for selected food products

Product	Rheological parameter	Model	Ea (kcal/g mole)	R^2
Honey	η	Newton	309.3	0.906
Ketchup	K	Power Law	357.5	0.915
	K	Power Law	65.5	0.897
Mustard	K	Power Law	66.4	0.801
Mayonnaise	K	Power Law	324.2	0.933

7. Final Remarks

Information regarding rheological behavior of some liquid and semi-liquids food products was pointed out along with information reported elsewhere. Most of the information was related with condiments and sauce or sweeteners. Also, composition data are provided for the tested products along with similar products. All tested food products behaved as pseudoplastic non-Newtonian fluids. The Ostwald-de Waele and Herschel-Bulkley models can be applied to fit the experimental data of ketchup and mustard to obtain *K* and *n* values at different temperatures. However, it was not possible to obtain yield stress data when using the Herschel-Bulkley model. Even though Casson and Bingham models in general did not fit experimental data, yield stress values could be computed for ketchup and mustard. Rheological data for mayonnaise at 20°C was obtained, however, the product consistency changed dramatically as temperature increased showing slippage. Activation energy values for the tested food products using *K* data from the Ostwald-de Waele model were obtained. Thus, the evaluation of rheological properties and their modeling can be useful for quality control in the food industry and to improve equipments required to handle these liquid and semi-liquid foods.

References

Barbosa-Cánovas G. V. and Peleg M. 1982. Flow properties of liquid and semi-liquid foods. *Revista de Tecnología de Alimentos*. (Mexico), 17(2): 4- 28. (In Spanish).

Barbosa–Cánovas G.V., Kokini J.L., Ma L. and Ibarz A. 1996. The rheology of semi-liquid foods. *Advances in Food and Nutrition Research.* 39: 1-69.

Barhate R.S., Subramanian R., Nandini K.E. and Umesh H.H. 2003. Processing of honey using polymeric microfiltration and ultrafiltration membranes. *Journal of Food Engineering.* 60: 49-54.

Bhattacharya S. 1999. Yield stress and time-dependent rheological properties of mango pulp. *Journal of Food Science,* 64(6): 1029-1033.

Bhattacharya S. and Rastogi N.K. 1998. Rheological properties of enzyme-treated mango pulp. *Journal of Food Engineering,* 36: 249-262.

Bingham J.L. 1922. Fluidity and Plasticity. McGraw-Hill. New York, NY. USA.

Blonk J.C.G. 2002. Viewing food microstructure. In *Engineering and Food for the 21st Century*. (J. Welti-Chanes, G.V. Barbosa-Cánovas and Aguilera, J.M., Eds.). CRC Press. Boca Raton, FL, USA. pp.513-529.

Broadfoot R. and Miller K.F. 1990. Rheological studies of massecuites and molasses. *International Sugar Journal,* 92: 107-112.

Bourne M. 2002. Food Texture and Viscosity: Concept and Measurement. 2nd ed. Academic Press. China.

Carreau P.J., Cotton F., Citerne G.P. and Moan M. 2002. Rheological properties of concentrated suspensions: Applications to foodstuffs. In *Engineering and Food for the 21st Century.* (J. Welti-Chanes, G.V. Barbosa-Cánovas and Aguilera, J.M., Eds.). CRC Press. Boca Raton, FL, USA. pp. 327-359.

Casson N. 1959. A flow equation for pigmented-oil suspensions of the printing ink. In *Rheology of Dispersed Systems*. C.C. Mill (Ed.). Pergamon Press. New York, NY, USA. pp. 82-104.

Cheng D. C-H. 1986. Yield stress: A time-dependent property and how to measure it. *Rheologica Acta,* 25: 542-554.

Creamer L.K. and MacGibbson K.H. 1996. Some recent advances in the basic chemistry of milk proteins and lipids. *International Dairy Journal,* 6: 539-568.

Cross M.M. 1965. Rheology of non-Newtonian fluids: A new equation for pseudoplastic systems. *Journal of Colloid Science,* 20: 417-437.

Cullen P.J., Duffy A.P. and O'Donnell C.P. 2001. On-line rheological characterization of pizza sauce using tube viscometry. *Journal of Food Process Engineering,* 24: 145-159.

De Kee D., Code R.K. and Turcotte G. 1983. Flow properties of time-dependent foodstuffs. *Journal of Rheology,* 27(6): 581-604.

Dickie A.M. and Kokini J.L. 1983. An improved model for food thickness from non-Newtonian fluids mechanics in the mouth. *Journal of Food Science,* 48: 57 – 61, 65.

Doublier J.L. and Launay B. 1981. Rheological galactomannan solutions: Comparative study of guar gum and locust bean gum. *Journal of Texture Studies,* 12: 151-172.

Farrell K. 1985. Spices, Condiments and Seasonings. The AVI Pub. Co., Inc. Westport, Conn., USA.

Genovese D.B. and Lozano J.E. 2000. Effect of cloud particle characteristics on the viscosity of cloudy apple juice. *Journal of Food Science,* 65(4): 641-644.

Geraghty R. and Butler F. 1999. Viscosity characterization of a commercial yogurt at 5C using a cup in bob and a vane geometry over a wide shear rate range. *Journal of Food Process Engineering.* 22: 1 – 10.

Haard N.F. 1985. Characteristics of Edible Plant Tissues. In *Food Chemistry*. (O.R. Fennema, Ed.). Marcel Dekker, Inc. New York, USA. pp. 857-912.

Hayes G.D. 1987. Food Engineering Data Handbook. Longman Group. London, UK.

Hultin H.O. 1985. Characteristics of muscle tissue. In *Food Chemistry*. (O.R. Fennema, Ed.). Marcel Dekker, Inc. New York, USA. pp. 725-789.

Ibarz A. and Barbosa-Cánovas G.V. 2002. Unit Opeations in Food Engineering. CRC Press. Boca Raton, FL, USA.

Kaur S., Kaler R.S.S. and Aamarpali. 2002. Effect of starch on the rheology of molasses. *Journal of Food Engineering,* 55: 319-322.

Kovalenko I.V. and Briggs J.L. 2002. Textural characteristics of soy-based yogurt by the Vane method. *Journal of Texture Studies,* 33: 105-118.

Marcotte M., Taherian A.R., Trigui M. and Ramaswamy H.S. 2001. Evaluation of rheological properties of selected salt enriched food hydrocolloids. *Journal of Food Engineering,* 48: 157-167.

Maruyama K. and Sakashita T. 2006. Relationship between rheological properties and perceived texture of mayonnaise. In *Proceedings of the 4th International Symposium on Food Rheology and Structure.* (P. Fisher, P. Erni, and E.J. Windhab, Eds.). pp. 695-696.

Mohsenin N.N. 1996. Physical Properties of Plant and Animal Materials. Gordon and Breach Publishers. The Netherlands.

Nindo C.I., Tang J., Powers J.R. and Singh P. 2005. Viscosity of blueberry and raspberry juices for processing applications. *Journal of Food Science,* 69: 343-350.

Núñez-Santiago M.C., Santoyo E., Bello-Pérez L.A. and Santoyo-Gutiérrez S. 2003. Rheological evaluation of non-Newtonian Mexican nixtamalized maize and dryprocessed masa flours. *Journal of Food Engineering,* 60: 55-66.

Powrie W.D and Nakai S. 1985. Characteristics of edible fluids of animal origin: eggs. In *Food Chemistry.* (O.R. Fennema, Ed.). Marcel Dekker, Inc. New York, USA. pp. 829-855.

Rao M.A. 1986. Rheological properties of fluid foods. In *Engineering Properties of Foods.* (M.A. Rao, and S.S.H. Rizvi, Eds). Marcel Dekker. New York, NY, USA. pp. 1-47.

Rao M.A. 2007. Flow and functional models for rheological properties of fluid foods. In *Rheology of Fluid and Semisolid Foods. Principles and Applications.* 2nd ed. Springer. New York, NY, USA. pp. 27-58.

Rao M.A., Cooley H.J., and Vitali A.A. 1984. Flow properties of concentrated juices at low temperatures. *Food Technology,* 38(3): 113-119.

Rha C. 1975. Theory, Determination and Control of Physical Properties of Food Materials. D. Reide Publishing Company. USA.

Skriver A., Roemer H., and Qvist K.B. 1993. Rheological characterization of stirred yogurt: viscometry. *Journal of Texture Studies,* 24: 185-198.

Stroshine R. 2000. Physical Properties of Agricultural Materials and Food Products. Dept. of Agricultural and Biological Engineering. Purdue University. West Lafayette, Indiana, USA.

Steffe J.F. 1996. Rheological Methods in Food Processing Engineering. 2nd edition. Freeman Press. East Lansing, MI, USA.

Swaisgood H.E. 1985. Characteristics of edible fluids of animal origin: milk. In *Food Chemistry.* (O.R. Fennema, Ed.). Marcel Dekker, Inc. New York, NY, USA. pp. 791-827.

Tavakolipour H. and Ghandi M. 2006. Rheological characterization of some Iranian honeys. In *Proceedings of the 4th International Symposium on Food Rheology and Structure.* (P. Fisher, P. Erni, and E.J. Windhab, Eds.). pp. 725-726.

Tehrani M.M. and Ghandi A. 2006. Modifications of Bostwick method to determine tomato concentrate consistency. In *Proceedings of the 4th International Symposium on Food Rheology and Structure.* (P. Fisher, P. Erni, and E.J. Windhab, Eds.). pp. 697-698.

Tiban N.N., Kopjar M.K., Babic J., Pilizota V. and Subaric D. 2006. Effect of fructose syrup addition on the rheological behaviour of Croatian honeys. In *Proceedings of the 4th International Symposium on Food Rheology and Structure.* (P. Fisher, P. Erni, and E.J. Windhab, Eds.). pp. 741-742.

Toledo R.T. 1999. Fundamentals of Food Process Engineering. Aspen Publishers, Inc. Gaithersburg, Maryland, USA.

Toğrul H. and Arslan N. 2004. Mathematical model for prediction of apparent viscosity of molasses. *Journal of Food Engineering,* 62: 281-289.

White J.W. 1985. Composition and properties of honey. In *Beekeeping in the United States.* (S.E. McGregor, Ed.) USDA. USA. pp. 57-80.

Whistler R.L. and Daniel J.R. 1985. Carbohydrates. In *Food Chemistry*. (O.R. Fennema, Ed.). Marcel Dekker, Inc. New York, USA. pp. 69-137.

Witczak M. and Juszczak L. 2006. Non-Newtonian behaviour of heather honey. In *Proceedings of the 4th International Symposium on Food Rheology and Structure*. (P. Fisher, P. Erni, and E.J. Windhab, Eds.). pp. 587-488.

Yoo B. and Rao M.A. 1995. Yield stress and relative viscosity of tomato concentrates: effect of total solids and finisher screen size. *Journal of Food Science*, 60(4): 777-785.

Yu C. and Gunasekaran S. 2001. Correlation of dynamic and steady flow viscosities of food materials. *Applied Rheology,* 11(3): 134-140.

In: Food Processing and Engineering Topics
Editors: M. E. Sosa-Morales and J. F. Vélez-Ruiz
ISBN: 978-1-60741-788-0

Chapter 3

PARTICLE SIZE DISTRIBUTION FUNCTIONS FOR FOOD POWDERS

José A. Guerrero-Beltrán[1], María Teresa Jiménez-Munguía[1], Jorge Welti-Chanes[2] and Gustavo V. Barbosa-Cánovas[3]

[1] Departamento de Ingeniería Química y Alimentos
Universidad de las Américas, Puebla, Mexico
[2] Departamento de Biotecnología e Ingeniería en Alimentos
Instituto Tecnológico y de Estudios Superiores de Monterrey, Mexico
[3] Biological Systems Engineering Department
Washington State University, Washington, USA

1. INTRODUCTION

Particle size distribution (*PSD*) is one of the most important characteristics of food powders, the distribution depends on the kind of food powder or mixture of different food powders. There are several factors that may modify the *PSD* in a food powder system such as processing, handling, storing, and transportation. Changes in *PSD* may bring up alteration on their appearance, physical properties, functionality and nutritional characteristics caused by attrition, segregation or agglomeration. Several methods are used to assess *PSD* (microscopy observations, resistive pulse technique, optical methods, sieve analysis, sedimentation-photometry, and centrifuging method). Further, various distribution models can be used to fit *PSD* (normal function, log-normal distribution, Rosin-Rammler or Weibull distribution, Gates-Gaudin-Schuhmann distribution, modified Gaudin-Meloy function, and modified Beta function). However, information in literature, regarding *PSD* models, is limited. Reported data are only presented in a table or in graphical representation, but no fitting of particle size using distribution models are available. Reported particle size data and experimental data of selected foods (freeze-dried coffee, spray-dried coffee, wheat flour, granulated sugar and starch) were plotted and modeled using selected distribution functions.

2. Particle Size Distribution

Importance

Powder technology deals with particles, their behaviour, and particle size. Particles range from submicrometer to large size grains and gravels, and from liquid mist or droplets to bubbles, as well as solid particles and aggregates. Particles are present in the medical, biological, pharmaceutical, chemical, mining, metallurgical, mechanical, civil, food technology, agricultural science, and engineering fields (Iinoya *et al.*, 1991). Powder flowability is influenced by particle size, *PSD*, particle morphology, moisture content, and hygroscopic properties of the powder (Walton, 2000). Powder flowability and flow behavior under pressure, temperature and moisture content may affect handling and processing operations, such as storage in silos, transportation, formulation and mixing, compression and packaging (Teunou *et al.*, 1999). The particle size distribution measurement is important in segregation since fine particles migrate to the bottom of the container while large particles may keep their distribution or migrate to the top of the container. Separation of particles could affect appearance, functionality, and bring up nutritional consequences because some particles could not be well mixed in the whole powder (Barbosa-Cánovas *et al.*, 1985).

During attrition (undesirable breakage of food agglomerates), formation of fines may occur and this may change the bulk properties of the system. As a consequence, changes on appearance, reduction of the dispersability and wettability may occur, as well as changes in particle size distribution (Barletta and Barbosa-Cánovas, 1993a, 1993b). Shattering or fragmentation and erosion or abrasion may occur as a consequence of the breaking of large particles. Friability (percentage of particle reduction) and size stability (percentage of size retention) are measured by *PSD* (Barletta and Barbosa-Cánovas, 1993a).

During spray drying, operational conditions influence the final powder characteristics such as particle size, *PSD*, and moisture content among other characteristics. All of them can affect bulk density and powder flowability. Increasing the air-drying temperature may produce a decrease in bulk and particle density as a consequence of the increasing of particle size and a tendency to be hollow (Walton, 2000). Powders usually have agglomerate, skin-forming and crystalline morphology; from these physical characteristics, crystalline morphologies are rapidly soluble in water, while agglomerates, in general, are insoluble or partially soluble in water. On the other hand, organic materials are predominantly skin-forming such as slim milk, yogurt powder, and co-dried egg. In spray-dried coffee, for instance, smaller particles are formed inside of larger particles (Walton, 2000). Some examples of problems related with *PSD* are explained next.

In the extraction of coffee compounds, the fines are not desirable because they difficult the percolation process. The coffee fines, for instance, are usually pelleted from ground coffee to avoid the retardation of the percolation process. Thus, the coffee soluble solids minimize losses in quality of the extract caused by too long percolation time and high extraction temperatures (Pintauro, 1975). In freeze dried coffee, particles classified as fines may cause quality losses in the final product. Also, fines may confer a light color appearance to the final product and they may affect the sublimation process during freeze-drying (Pintauro, 1975).

In cloudy food products, particularly particle size in cloudy juices, Genovese and Lozano (2000) pointed out that after a large period of storage time only a very small part of the particles should precipitate. Some of these particles might form aggregates and flocculate and eventually precipitate due to gravity. This process is irreversible and could change the appearance of the cloudy juice in which the cloudy stability is of paramount importance.

3. PARTICLE SIZE EQUIVALENT DIAMETER

The classification of the size of a single particle is not easy when the particle is non-spherical. Most of the time, the size of an individual particle is given in terms of its geometric characteristics, its silhouette or its hypothetical equivalent diameter related to some particle properties. The diameter of a rectangular prism, for instance, is calculated using its length, width and height of each side. The diameter may be measured by a microscope to assess the average diameter. The geometric diameter can be measured from the particle silhouette using the Feret´s diameter, the Martin´s diameter or the sieve diameter (table 1). On the other hand, the equivalent diameters of non-spherical particles are the diameters of spheres, which have the same geometric or physical properties (table 1) (Allen, 1981; Iinoya *et al.*, 1991). The average particle diameters can be obtained by mean of the equations reported by Allen (1981) and Iinoya *et al.* (1991).

Table 1. Particle size diameters (Allen, 1981; Iinoya *et al.*, 1991)

Definition of characteristic diameter		Physical meaning and corresponding measuring method
Geometric size		
DF	Feret's diameter	The mean value of the distance between pairs of parallel tangents to the projected outline of the particle.
DM	Martin's diameter	Unidirectional diameter: diameter of particles as the length of a chord dividing the particle into two equal areas.
DA	Sieve diameter	The width of the minimum square aperture through which the particle may pass.
Equivalent diameter		
Da	Equivalent projection area diameter $(4A/\pi)^{1/2}$	Diameter of the circle having the same area as the projection area of a particle (A), corresponding to diameter obtained by light extinction.
Ds	Equivalent surface area diameter (specific diameter)	Diameter of the sphere having the same surface as that of a particle surface (S), corresponding to $(S/\ A/\pi)^{1/2}$ diameter obtained by light extinction.
Dv	Equivalent volume diameter $(6V/\pi)^{1/3}$	Diameter of the sphere having same volume (V) as that of a particle corresponding to a diameter obtained by Coulter counter.

Table 1. (Continued)

Definition of characteristic diameter		Physical meaning and corresponding measuring method
Equivalent diameter		
D_{Stk}	Stokes' diameter	Diameter of the sphere having the same gravitational settling velocity as that of a particle, $D_{Stk} = [18\ \eta\ v/g(\rho_p - \rho)Cc]^{1/2}$ obtained by sedimentation.
Dae	Aerodynamic diameter	Diameter of the sphere having units in specific gravity and having the same gravitational settling velocity as that of a particle, Dac = [18 $\eta v/gCc$], obtained by the same method as above.
Dls	Equivalent light-scattering diameter	Diameter of the sphere having the same intensity of light scattering as that of a standard particle obtained by the light-scattering method.
Dp	Perimeter diameter	Diameter of a circle having the same perimeter as the projected outline of the particle.

C_c = drag coefficient; η = liquid viscosity; v = terminal velocity; g = gravity; ρ_p = particle density; ρ = liquid density.

4. METHODS FOR MEASURING PARTICLE SIZE

4.1. Microscopy

Optical microscopy and electron microscopy can be used to measure particle shape and size. Two electron microscopy equipments can be used: the Transmission Electron Microscopy (TEM) and the Scanning Electron Microscopy (SEM). The resolution of the microscope will be the most important aspect to determine if a microscope is useful to assess particle size. In the case of optical microscopy, the simplest method, particles are wide spread on a glass slide to measure particle size. In the case of transmission microscopy the particles are spread on a very thin film and in the scanning microscopy method the particles have to be mounted in a substrate and coated with carbon, silver or gold as conducting materials. A series of selected microscopic fields are needed during the process. The number of particles and particle size will give information to apply statistical methods (Iinoya *et al.*, 1991). The Scanning Electron Microscopy is used to determine shape, size and *PSD* in most of the particulated foods. The images obtained with this method are analyzed statistically using specialized software. In food suspensions, particles have to be washed and placed in slices to be dried before they are analyzed by microscopy. Particle diameters are calculated as the projected area on a surface (Genovese and Lozano, 2000).

4.2. Resistive Pulse Technique (Coulter Counter)

This method is used to measure particles with several micrometers. Particles are suspended in an electrolyte solution which is passed all the way through an aperture, two electrodes will measure the electrical resistance. The number of recorded pulses will be proportional to the number of particles and the height will correspond to the particle volume. The sensitivity of this device increases as the aperture diameter decreases. Calibration of the device is required using particles with known diameter before the particle analysis (Iinoya *et al.*, 1991).

4.3. Optical Methods

a) Optical Blockage Technique

In this method, the particles dispersion is flown through an illuminated sensitive zone. The reduction of the light is detected by a photodiode. An equivalent diameter of the particle will be obtained when particles are passing through the cross section because of the voltage output of the photodiode (Barbosa-Cánovas *et al.*, 2005). For this technique particles should be diluted in water to avoid duplicity of counts in the sensitive zone.

b) Light Diffraction Method

This method is based on the Fraunhofer diffraction pattern. The intensity of the diffracted light does not depend on the refractive index of the particles. The method uses a beam that comes from a laser source. The diffractive light passes through a system of lenses and is finally recorded by a photodetector. This method is also used to analyze aerosol particles (Barbosa-Cánovas *et al.*, 2005).

c) Photon Correlation Spectroscopy

Particles should be suspended in a liquid to be measured. The scattered light from the particles is a function of the position of the particles, which undergoes Brownian motion. This motion will cause destructive and constructive interferences due to the difference in the optical path length to the photon counter. The fluctuation of the response will depend on particle size (Iinoya *et al.*, 1991). The Photon Correlation Spectroscopy method is used to measure the size of colloidal particles from nanometers to a few micrometers. It measures the fluctuations of the light scattered due to the particles size whereas diffraction measures the angular distribution of the scattered light. A laser beam illuminates particles in suspension and the scattered light is detected by a photomultiplier. The photomultiplier detects a signal, which varies with time. The variation of the light level gives information of the particle size. In this technique, the hydrodynamic diameter (cm) (D_h) is calculated using the Stokes-Einstein equation:

$$D_h = \frac{kT}{3\pi\eta D} \qquad (1)$$

where k is the Boltzmann constant (cm·g_f/K), η is the viscosity of the liquid (g_f s/cm^2), T is the absolute temperature (K) and D is the diffusion coefficient (cm^2/s) for particles moving into the liquid (Genovese and Lozano, 2000).

d) Light Scattering Method

The light scattering property of individual particles is a function of the size, shape and refractive index. At the same time the intensity of the scattered light is a function of the scattering angle, the intensity and the wavelength of the incident light. The optical equivalent diameter is referred to a calibrated particle of known size. The actual diameter cannot be evaluated unless particles have spherical shape (Barbosa-Cánovas *et al.*, 2005). Koc *et al.* (2003) evaluated the *PSD* of spray-dried whole milk using this method. They pointed out that the *PSD* of the milk was unimodal with an average particle size of 138.9 μm. Grabowski *et al.*, (2006) also used this method to measure the mean particle diameter (58.1 μm) of spray-dried amylase hydrolyzed sweet potato puree.

4.4. Sieve Analysis

The sieving method is based on weighting the mass of particles passing through a space of known diameter. The size is defined by the aperture in which a particle will or will not pass through. This method uses several sieves with standard opening sizes based on a root-two progression. The sieves are stacked in a decreasing mesh number. The stack of sieves is shaken mechanically during a time previously selected. The number of sieves needed to assess particle size will depend on the width of the particle distribution. After separation of the bulk of powder, each fraction is weighed and the percentage is calculated. This method is the only method that operates with dry powders (Rietema, 1991; Barbosa-Cánovas *et al.*, 2005).

4.5. Sedimentation-Photometry

The method is based in the particles settling rate under gravitational or centrifugal acceleration of the liquid. In this case, it is assumed that particles have a spherical shape. The larger size limiting the use of this method is given by the Reynolds number of the particle:

$$\mathrm{Re} = \frac{Dv\rho_p}{\eta} \qquad (2)$$

where D is de particle diameter (cm), v is the terminal velocity (cm/s), ρ_p is the particle density (g/cm^3) and η is the viscosity of the liquid (poise, g/cm s). The terminal velocity is given by the Stokes' law:

$$v = \left(\frac{\rho_p - \rho}{18\eta} \right) \left(gD^2 \right) = \frac{h}{t} \tag{3}$$

where ρ is the liquid density for particle (ρ_p) and liquid (ρ, g/cm^3), g is the gravitational acceleration (cm/s^2) and t is the time (s) needed to reach a falling distance h (cm) (Genovese and Lozano, 2000).

4.6. Centrifuging Method

The centrifugal particle size analyzer uses the sedimentation method to determine particle concentration photometrically. The absorbance changes are converted to particle size. The centrifugal sedimentation is given by:

$$\frac{dv}{dt} = \left(\frac{\rho_p - \rho}{\rho_p} \right) r\omega^2 - \frac{18\eta v}{\rho_p D^2} \tag{4}$$

where ω is the angular velocity (1/s), r is the distance from particle to the rotational center (cm) and v (= dr/dt) is the sedimentation velocity of particles (cm/s). If the intensity of the transmitted light increases, the particle concentration decreases (Genovese and Lozano, 2000).

Evaluation of *PSD*, in the case of suspensions, in general, uses the Stokes' law; therefore, the sedimentation time of particles is required. Direct methods such as the scanning and the transmission electron methods for measuring particle size, shape and *PSD* of food dispersions are used. The method to be used will depend on the physical properties of the suspension. However, indirect methods are used as well: sedimentation and centrifugation, light scattering, X-ray diffraction, conductimetric methods, gas and solute adsorption, ultrafiltration, and diffusimetric methods (Genovese and Lozano, 2000).

5. Particle Size Distribution Functions

Particle size distribution functions have been applied for describing behavior of particle size by many researchers. Some of these functions are the normal distribution function, the log-normal function, the Gates-Gaudin-Schuhmann function, the Rosin-Rammbler function, the Modified Gaudin-Meloy function and the Modified-Beta function among other methods. However, not all of them can describe the actual size distribution of food powders (Yu and Standish, 1990; Barbosa-Cánovas *et al.*, 2005). Mean, median and mode of the particle size are the statistical parameters frequently used to represent the *PSD* of a powder. The mean diameter is the sum of all sample sizes divided by the total sample size; the modal diameter is the size most frequently found in the sample, and the median diameter is the particle size in which size diameters are below 50% of this size (Allen, 1981; Genovese and Lozano, 2000). There are several ways to present particle size data in a *PSD*: volume, surface area, and length or number basis among other methods. This presentation depends on the method used to assess particle size (Yu and Standish, 1990). Some distribution functions use the cumulative undersize fraction (f) as a function of the particle size (X) and other methods use a characteristic distribution function which may be specific for the powder.

5.1. The Normal Distribution Function

The normal, or Gaussian, distribution function is defined as:

$$f(D_p) = \frac{1}{\sigma\sqrt{(2\pi)}} \exp\left[-\frac{(D_p - D_{pm})^2}{2\sigma^2}\right] \tag{5}$$

where $f(D_p)$ is the particle size distribution, σ is the standard deviation of the distribution, D_p is the diameter of the particle, and D_{pm} is the mean particle diameter. This distribution function is a bell-shaped curve where D_{pm} and σ are obtained as:

$$D_{pm} = \frac{\Sigma(n_i D_{pi})}{N} \tag{6}$$

$$\sigma = \frac{\left[\Sigma n_i (D_{pi} - D_{pm})\right]^{1/2}}{N} \tag{7}$$

where n_i is the number of particles with diameter D_{pi} and N is the total number of particles. Equation (5) is the representation of a bell-shaped curve centered at $D_p = D_{pm}$. D_{pm} also corresponds to the number median diameter having 50% undersize (Allen, 1981; Iinoya *et al.*, 1991).

5.2. Log-Normal Distribution

This distribution function is very useful in particle size analysis (Iinoya *et al.*, 1991; Barbosa-Cánovas *et al.*, 2005).

$$f(\ln D_p) = \frac{1}{\ln\sigma_g \sqrt{(2\pi)}} \exp\left[-\frac{(\ln D_p - \ln D_g)^2}{2(\ln\sigma_g)^2}\right] \tag{8}$$

where $f(\ln D_p)$ is the particle size distribution, σ_g is the standard deviation of the distribution, D_p is the particle diameter, and D_g is the mean diameter.

The log-normal distribution could be represented in a cumulative distribution way:

$$\ln \sigma_g = \frac{\left[\sum n_i \left(\ln D_{pi} - \ln D_g\right)^2\right]^{1/2}}{N} \tag{9}$$

The log-normal distribution represents a straight line on a log-normal diagram. This distribution is frequently applied to powders and aerosols. In a cumulative fashion, the particle size at the 50% is called number median diameter (NMD) or geometric mean diameter (GMD) (Iinoya *et al.*, 1991). The *PSD* on the number basis differs significantly from that of the mass basis. The conversion from number median diameter (NMD) to mass median diameter (MMD) can be made by the Hatch's equation if the distribution obeys the log-normal form:

$$\ln(MMD) = \ln(NMD) + 3(\ln \sigma_g)^2 \tag{10}$$

Peleg *et al.* (1986) represented the log-normal distribution as:

$$f(x) = \frac{1}{x \ln \sigma_g \sqrt{(2\pi)}} \exp\left[-\frac{\left(\ln x - \ln x_g\right)^2}{2\left(\ln \sigma_g\right)^2}\right] \tag{11}$$

where x_g is the geometric mean of the distribution and σ_g is the standard deviation of ln x. However, the log-normal distribution has an infinite range which depends on the mode (or mean size), this affects the distribution of the particle size (Peleg and Normand, 1986).

5.3. Rosin-Rammler (Weibull) Distribution

This function is widely used in *PSD* (Barbosa-Cánovas *et al.*, 2005). Some products of the milling process often fit the Rosin-Rammler or Weibull distribution. The cumulative distribution is:

$$R = \exp\left(-\frac{D_p}{D_e}\right)^n \tag{12}$$

where D_e is the absolute size constant, defined as the diameter which gives the cumulative oversize of R = 36.8%, and n is the distribution constant expressing the polydispersity of the particle size (Iinoya *et al.*, 1991). Other way of describing this model is as follow:

$$Y = 1 - \exp\left(-\frac{x}{x_R}\right)^n \tag{13}$$

where Y is the weight fraction of material finer than the particle size x, x_R is a constant giving a measure of the present particle size, and n is a constant characteristic of the analyzed material (Barbosa-Cánovas *et al.*, 2005).

5.4. The Gates-Gaudin-Schuhmann Distribution

The Gates-Gaudin-Schuhmann function is presented as:

$$Y = \left(\frac{x}{k}\right)^m \tag{14}$$

where Y is the cumulative weight fraction undersize, x is the particle size, k is a characteristic size of the distribution, and m is a measurement of the distribution spread (Schuhmann slope) (Barbosa-Cánovas *et al.*, 2005).

5.5. The Modified Gaudin-Meloy Function

The modified Gaudin-Meloy function is presented as:

$$Y = \left[1 - \left(1 - \frac{x}{x_o}\right)^r\right]^m \tag{15}$$

where Y is the cumulative weight fraction undersize, x is the particle size, x_o is a parameter related to the maximum particle size, m is the Schuhmann slope, and r is the ratio of x_o to the size modulus (Barbosa-Cánovas *et al.*, 2005).

5.6. The Modified Beta Function

Peleg and Normand (1986) gave a modified version of the Beta distribution to avoid some difficulties that other functions may have, such as the log-normal distribution:

$$f_{am}(x) = \frac{x^{am}(1-x)^m}{\int_0^1 x^{am}(1-x)^m\,dx} \tag{16}$$

where a, and m are constants, and x is the normalized length given by:

$$x = \frac{X - X_{\min}}{X_{\max} - X_{\min}} \tag{17}$$

where X_{min} is the smallest particle size, X_{max} is the largest particle size; therefore $X_{min} < X < X_{max}$, $0 < x < 1$ (Popplewell *et al.*, 1988; Barbosa-Cánovas *et al.*, 2005). This function can be solved using a numerical integration procedure such as those based on the Romberg's technique. The distribution mode x_m can be determined using the constant a alone as:

$$x_m = \left[\frac{a}{(a+1)}\right] \tag{18}$$

and the spread can be determined, independently, by the constant m. The variance is given by:

$$\sigma^2 = \frac{(am+1)(m+1)}{[(a+1)m+2]^2[(a+1)m+3]} \tag{19}$$

6. Particle Size and Some Physical Properties of Selected Food Powders

Several size distribution functions have been mentioned above to describe the behavior of powders. However, in the literature, the researchers only give information about particle size presenting their data in a table or in a graph fashion. Few of them use distribution function models to describe their data. Barbosa-Cánovas *et al.* (2005) applied several mathematical models to obtain the distribution functions that may fit experimental data obtained from a sieve analysis of sugar, corn meal and instant milk. They found that the Rosin-Rammler function (figure 1) and the modified Gaudin-Meloy function fit very well the particle size data of the studied foods. The log-normal function and modified Beta function fit the sugar particle size data very well (figure 2). Narayanan and Vu (2000) determined some physical characteristics (table 2) and the particle size of selected food powders using U.S. standard sieves. The cumulative undersize particle distributions are shown in (figure 3). These researchers did not apply any *PSD* function to their experimental data.

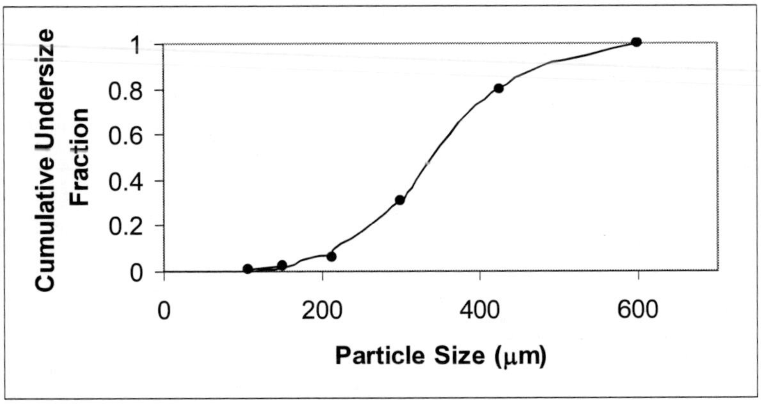

(a)

Figure 1. Continued on next page

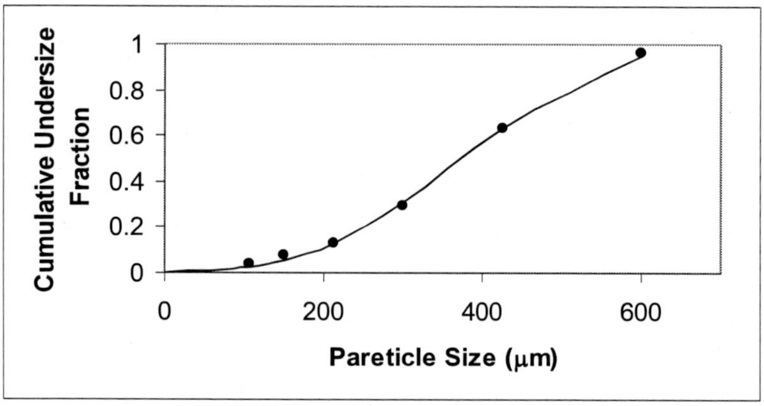

(b)

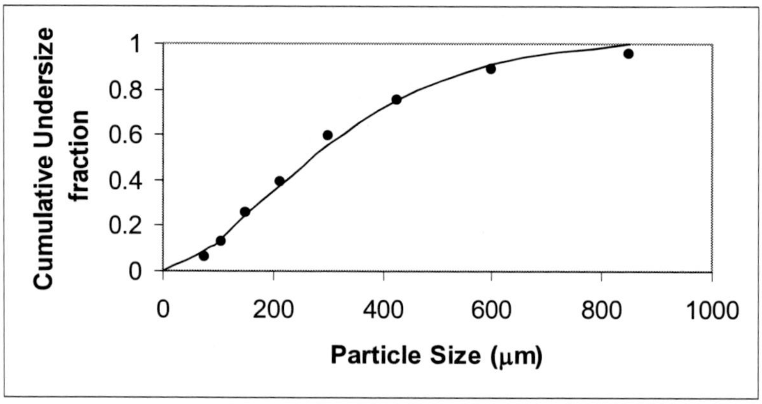

(c)

Figure 1. Fitting of the Rosin-Rammler function to the particle size distribution of (a) sugar, (b) corn meal, and (c) instant milk (Barbosa-Cánovas *et al.*, 2005).

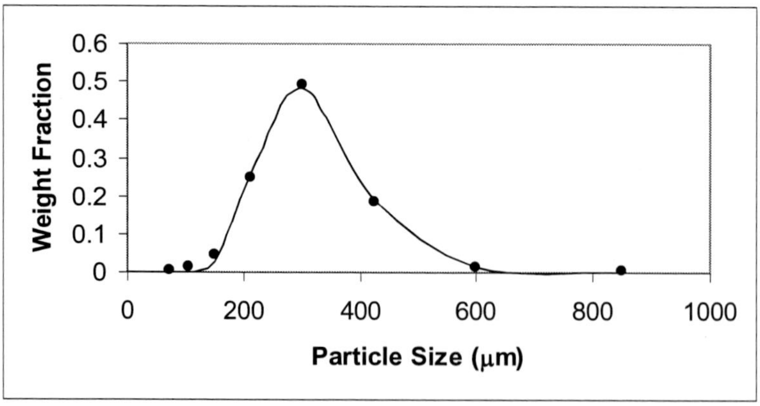

(a)

Figure 2. Continued on next page

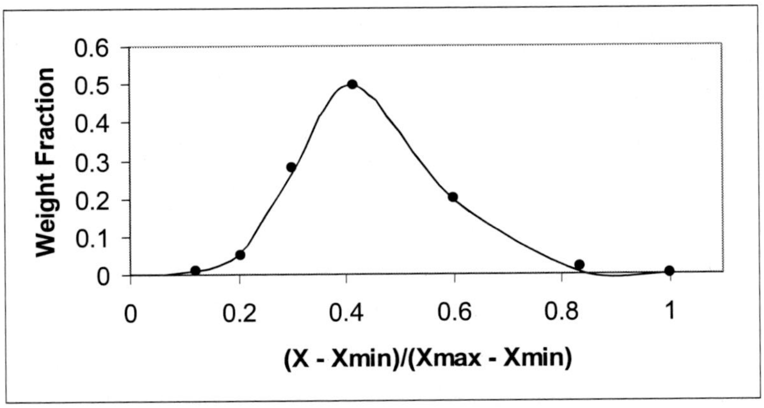

(b)

Figure 2. Fitting of the particle size distribution of sugar. (a) log-normal function and (b) Beta function (Barbosa-Cánovas *et al.*, 2005).

Table 2. Physical characteristics of some food powders (Narayanan and Vu, 2000)

Powder Sample	Type	Size distribution (μm)	Mean particle Size (μm)	Moisture content (%)[a]	Bulk density (g/cm^3)
Coffee	Filter brand	100 - 1000	800	6.5 - 18.2	0.324
Coffee creamer	Non-dairy	100 - 400	200	5.1 - 12.4	0.465
Wheat flour	Pre-sifted	60 - 150	90	12.2 - 40.3	0.491
Dry milk	Non-fat	100 - 2000	900	8.2 - 30.4	0.592
Sugar	Granulated	170 - 600	350	0 - 10.1	0.817
Salt	Iodized	170 - 420	300	0 - 5.1	1.095

[a]: Wet basis.

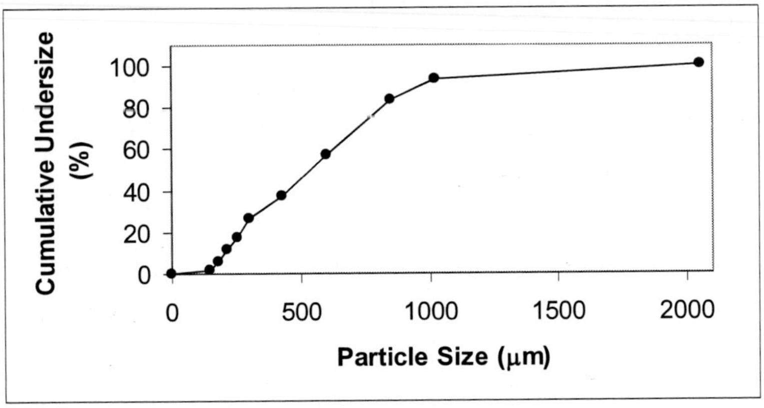

(a)

Figure 3. Continued on next page

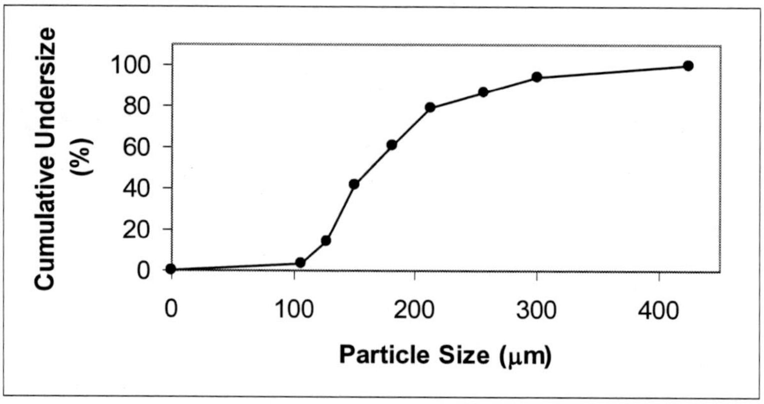

(b)

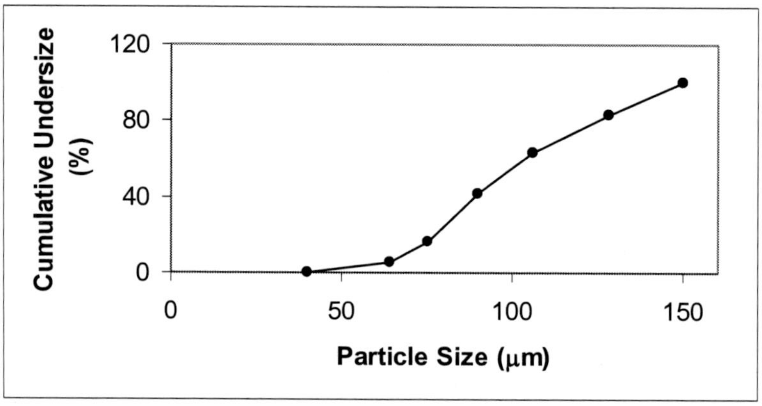

(c)

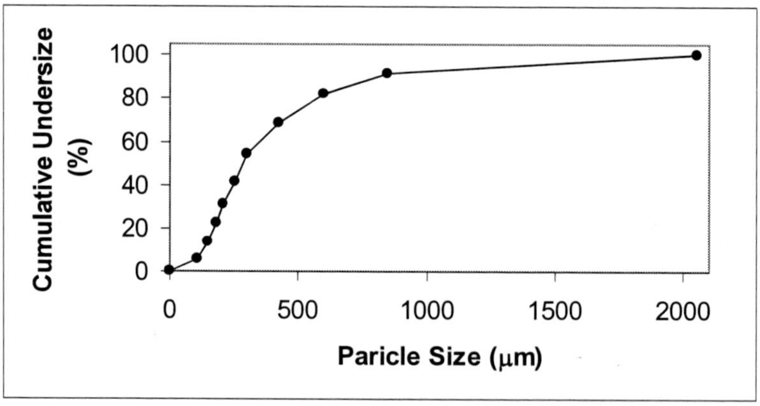

(d)

Figure 3. Continued on next page

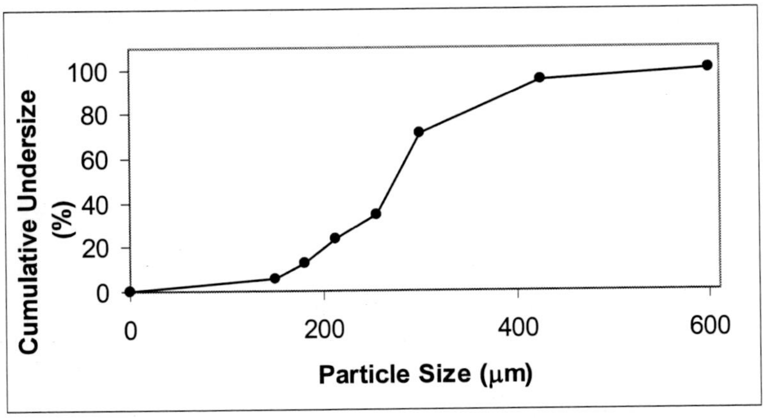

(e)

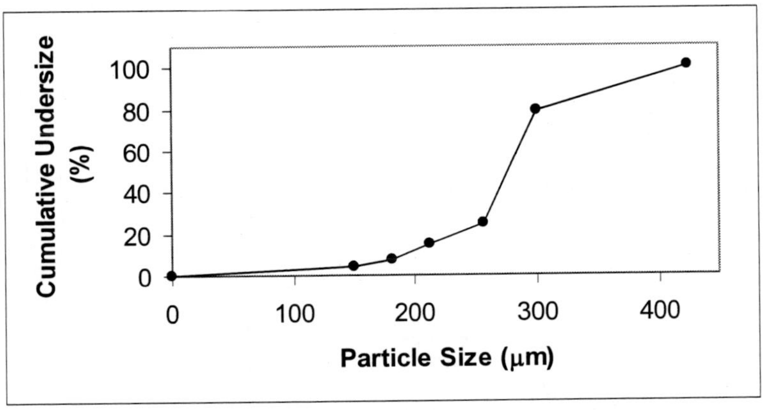

(f)

Figure 3. Particle size distribution of (a) coffee, (b) coffee creamer, (c) flour, (d) dry milk, (e) sugar, and (f) salt (Narayanan and Vu, 2000).

Table 3. Characteristics of cocoa powder and three commercial products (Omobuwajo *et al.*, 2000)

Sample	Bulk density (kg/m^3)	Tap density (kg/m^3)	Average size (μm)	Frequent size (μm)
Cocoa powder	317.46 ± 5.22	408.16 ± 4.90	570 ± 20	730
Brand A	434.78 ± 4.98	625.07 ± 5.17	570 ± 60	730
Brand B	644.19 ± 7.09	800.00 ± 6.83	340 ± 50	180
Brand C	645.16 ± 6.87	833.33 ± 7.38	2,380 ± 80	730

Omobuwajo *et al.* (2000) assessed the particle size distribution of cocoa powder and three instant chocolate beverage commercial products by sieve analysis. Authors used a set of Endecotts laboratory sieves with mesh sizes of 2,360, 1,700, 1,180, 850, 600, 425, 300, 212, 150, and 75 μm in an Endecotts sieve shaker. Some physical characteristics of these powders are showed in table 3. The particles size distributions are depicted in figure 4. $Y_i = (\Delta m_i/m)$

*(1/Δdp_i); where Y_i is the relative distribution of a particular powder, Δm_i is amount of powder retained in each sieve, *i*, *m* is the total mass of powder, Δdp_i is the mesh size increment.

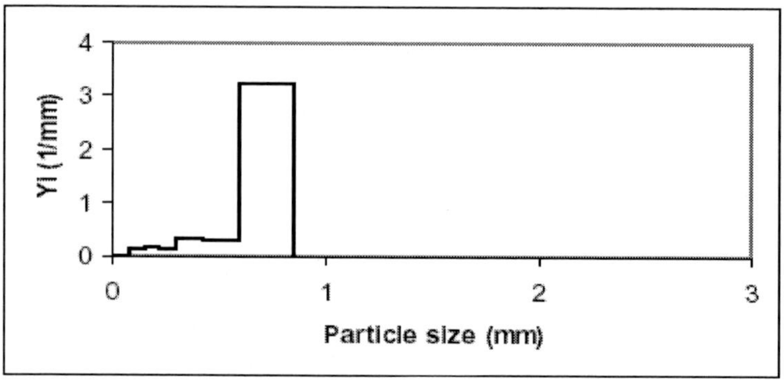

(a)

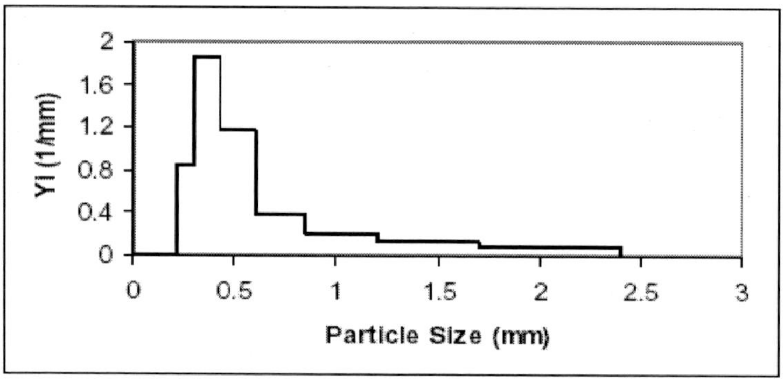

(b)

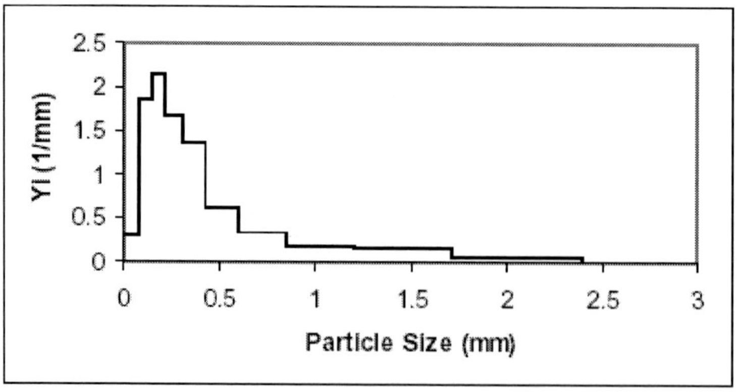

(c)

Figure 4. Continued on next page

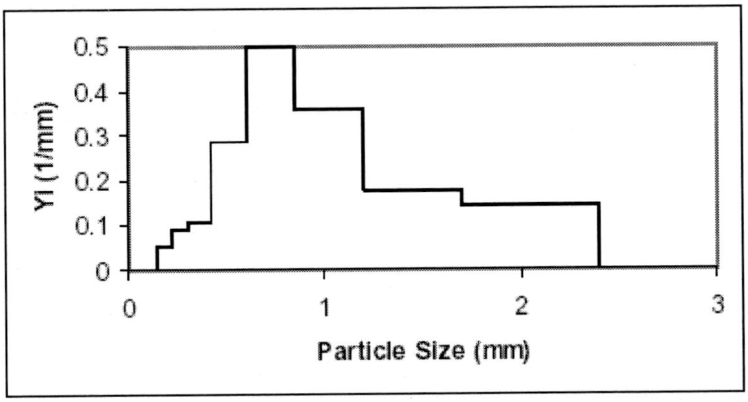

(d)

Figure 4. Particle size distribution of (a) cocoa powder, (b) Brand A, (c) Brand B, and (d) Brand C chocolate beverages (Omobuwajo *et al.*, 2000).

Teunou and Fitzpatrick (2000) and Teunou *et al.* (1999) analyzed some physical properties on wheat flour, tea, whey, and milk to assess powder flowability. The particle size distribution was measured using the laser diffraction method (table 4). However, they only reported the mean particle size of the food products.

Table 4. Physical properties of flour, tea, whey and milk (Teunou and Fitzpatrick, 2000; Teunou *et al.*, 1999)

Product	Mean size (μm)	Bulk density (kg/m^3)	Tap density (kg/m^3)	Particle density (kg/m^3)
Wheat lour	73	601	695	1,467
Tea	25	617	913	1,568
Whey	98	516	622	1,216
Milk	197	-	593	1,133

Lima *et al.* (2000) measured the particle size of three different brands of peanut flour using a Coulter LS Instrument. The instrument uses a light scattering array. The particle size was found in the range 0.04 – 200 μm. Sample A was obtained mixing peanut flour (containing 12% of fat) with deionized water a 65°C to obtain slurry with 30 % of solids. Afterward, the slurry was dried in a drum dryer. The peanut flour was processed through a pin mill for better particle dispersability. Samples B and C were commercial peanut flours. These researchers only present the mean particle size diameter and the particle size diameters that they found below a given percentage. Commercial peanut flours showed multimodal particle size distribution whereas sample A showed a monomodal (bell shape) distribution. Lima *et al.* (2000) assure that a narrow particle size band distribution around the mean particle size is necessary to reduce both stickiness and grittiness when peanut butter is manufactured.

Trezza and Krochta (2000) analyzed the particle size of whey protein isolate-carnauba wax (WPI-CW), whey protein isolate-anhydrous milk fat (WPI-AMF), Shellac-carnauba wax (Shellac-CW) and whey protein concentrate (WPC) dispersion coatings using a Malvern MS 20 particle size analyzer. They used the Sauter mean diameter (SMD) as the characteristic dimension of the particle size (table 5). The SMD is called the volume-surface mean, which is the average size of dispersed particles based on the specific volume per unit surface. Trezza and Krochta (2000) assured that the shape of the particle size distribution of dispersed particles may affect the dispersion coatings gloss. This means that the more uniform the size distribution the higher the gloss values of coatings. On the other hand, when the SMD is smaller than 0.39 μm more stable dispersions are obtained and a higher gloss is observed as well.

Table 5. Particle size of WPI-AMF, WPI-CW, Shellac-CW and WPC dispersion coatings (Trezza and Krochta, 2000)

	Particle size (μm)		
Coating	Sauter mean diameter	50%	90%
WPI-AMF	0.39	0.44	2.40
WPI-CW	1.40	2.50	23.00
Shellac-CW	1.60	2.00	5.20
WPC	0.87	1.20	6.90

7. Experimental Sieve Analysis and Particle Size Modeling of Selected Food Powders

7.1. Moisture Content

Moisture content and sieve analysis were performed in commercial spray-dried instant coffee, freeze-dried instant coffee, wheat flour, potato starch, and cane sugar. The moisture content, evaluated using a vacuum oven (70°C), is showed in table 6.

Table 6. Measured food powders moisture content

Product	Moisture content[a] (%)	
	Experimental	Reported[b]
Spray dried Coffee	5.4	2.5
Freeze dried Coffee	6.7	-
Wheat flour	9.7	12.0
Potato starch	10.4	12.0
Cane sugar	1.7	0.5

[a]: Wet basis.; [b]: Hayes (1987).

7.2. Sieve Analysis and Histograms

The sieving analysis was performed using an ATM sonic sifter (ATM Corp., Milwaukee, WI). The stack was placed in the sieving machine and it was turned on using pulse features for 3 min. Oscillation amplitude was 6 for coffee and 8 for sugar, flour and starch. After 3 minutes, the stack of sieves was removed from the machine and each sieve (including fines collector) was weighted. Experimental results were fitted using the Gates-Gaudin-Schuhmann (GGS), the Rosin-Rammler (RR), and the Modified Gaudin-Meloy (MGM) distribution functions. Microsoft excel "Solver" procedure was used to adjust the parameters in the models by minimizing the sum of squares of the error (sse). Frequency histograms showing particle size distribution of the selected food powders are shown in figures 5 and 6.

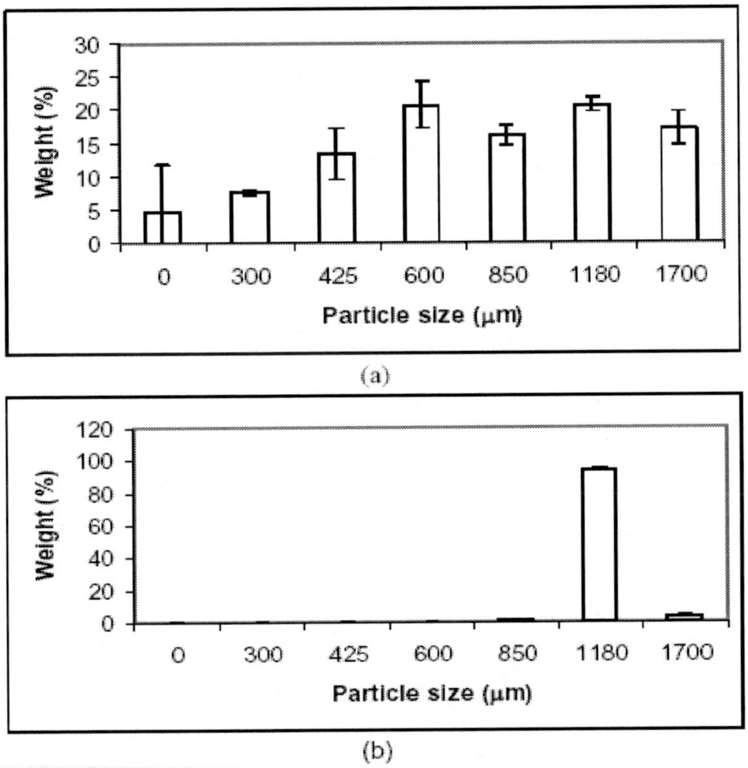

Figure 5. Particle size distribution histogram for spray dried (a) and freeze dried (b) coffee.

It can be observed, in figures 5 and 6, that spray dried coffee shows a behavior close to a normal or bimodal distribution with modes at 600 and 1180 μm. Freeze dried coffee, on the other hand, clearly shows an unimodal distribution with mode at 1180 microns (figure 5). Freeze dried coffee is highly homogeneous in particle size, since more than 90% of the particles have a size ranging from 1180 to 1700 microns. The particle analyses conducted for wheat flour and wheat starch cannot be directly interpreted since the *PSD* for those products lies outside the range of the set of the used sieves (figure 6). In the case of flour, a size of 75 microns appears to be the median of the size distribution, whereas in the case of starch, the median is located at an even lower particle size. No sieves with opening smaller than 75 microns where available and no further information regarding particle size distribution of these products was obtained. The *PSD* of cane sugar closely approaches a normal distribution, having a mode of 300 microns (figure 6). Powders with particle size smaller than 45 microns have to be studied by alternative methods since this size is very small.

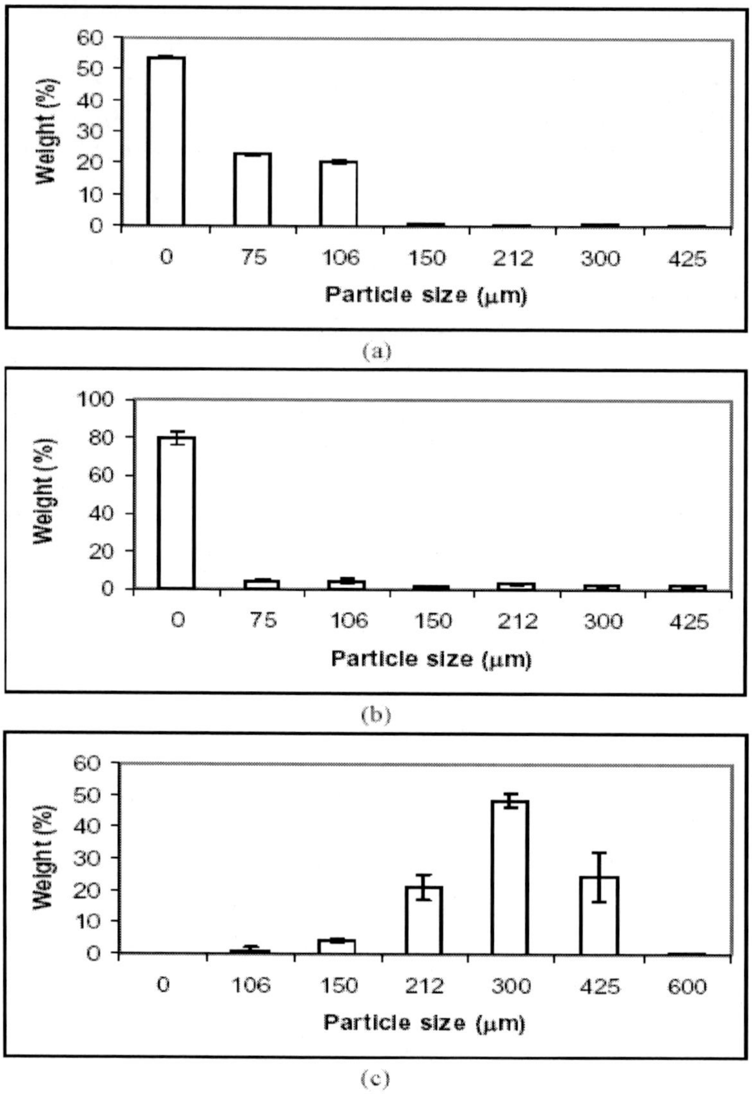

Figure 6. Particle size distribution histogram for (a) wheat flour, (b) potato starch, and (c) cane sugar.

7.3. Practical Size Modeling for Spray-Dried Coffee

A. Gates-Gaudin-Schuhmann model

Spray-dried coffee particle size distribution is shown in table 7. The cumulative undersize (CU) data was computed summing the mesh powder fractions ranging from 0 to 1180 μm (table 7). Afterward, the GGS model (Equation 16) was applied looking for the k and m values to fit the model. A "Solver" system from the Excel program was used to determine the k and m constants. At the same time, the sum of the squares of the error (SSE) was computed until minimizing the error. This means that the CU experimental data (Y) along with the particle size (μm) was listed in a column in an Excel file. Then, selected cells were used to each constant (in this case k and m). The constants were "modified" by giving them values continuously until getting the best values that adjust the experimental data to the model. This was obtained when the SSE value was minimized. Therefore, the k and m values were "choose" as the ones that best adjust the prediction of the experimental data to the model. Finally, a trend line was obtained using the experimental and predicted CU values to obtain the correlation coefficient.

Table 7. Particle size modeling with the Gates-Gaudin-Schuhmann distibution for spray-dried coffee

Mesh size (μm)	Particle fraction (%)	CUF Exp (%)	CUF Pred (%)	SSE
0	4.82 ± 6.82	0.00	0.00	0.00
300	7.60 ± 0.38	4.82	11.94	0.507
425	13.31 ± 3.79	12.42	17.81	0.289
600	20.61 ± 3.54	25.74	26.44	0.005
850	16.04 ± 1.52	46.35	39.42	0.450
1180	20.61 ± 1.03	62.38	57.42	0.246
1700	17.00 ± 2.68	82.99	87.27	0.183

CUF = cumulative undersize data (%), SSE = sum of square error.

The Gates-Gaudin-Schuhmann distribution function with its constant values computed with the "Solver" system is presented next:

$$Y = \left(\frac{x}{34.47} \right)^{1.1464} \tag{20}$$

where Y is the cumulative undersize fraction (CU) (%) data, and x is the particle size (%). The SSE, on the other hand, was computed using the next equation:

$$SSE = \sum \frac{\left(CUF_{Exp} - CUF_{Pred} \right)^2}{100} = 1.71 \tag{21}$$

Data obtained with the GGS model are listed in table 9.

B. Modify Gaudin-Meloy model

The cumulative undersize fraction (CUF) for spray-dried coffee particle size, presented in table 7, was computed summing the mesh powder fractions ranging from 0 to 1180 μm (table 8). The MGM model (Equation 17) was applied looking for the x_o, k and m values which may best adjust the model. Again, a "Solver" system from the Excel program was used to determine the x_o, k and m constants as described before.

Table 8. Particle size modeling with the Modify Gaudin-Meloy (MGM) function and the Rosin-Rammler (RR) distribution for spray-dried coffee

	Model					
	MGM			RR		
Mesh size (μm)	CUFExp	CUFPred	SSE	CUFExp	CUFPred	SSE
0	0.000	0.000	0.0000	0.000	0.000	0.0000
300	0.048	0.067	0.0004	0.048	0.082	0.0011
425	0.124	0.136	0.0001	0.124	0.146	0.0005
600	0.257	0.255	$7x10^{-6}$	0.257	0.255	$4x10^{-6}$
850	0.463	0.433	0.0009	0.463	0.423	0.0016
1180	0.623	0.636	0.0002	0.623	0.628	$2x10^{-5}$
1700	0.829	0.842	0.0002	0.829	0.851	0.0004

MGM = Modify Gaudin-Meloy (MGM) function, RR = the Rosin-Rammler model, CUF = cumulative undersize fraction, SSE = Sum square error.

The Modified Gaudin-Meloy equation with its constant values obtained with the “Solver” system is presented next:

$$Y = \left[1 - \left(1 - \frac{x}{8000}\right)^{11.55}\right]^{2.62} \tag{22}$$

where Y is the cumulative undersize fraction (CFU), and x is the particle size fraction. The SSE for this model was 0.0017. The spary-dried data from the MGM modeling are listed in table 9.

Table 9. Parameter values for the functions used to describe particle size distributions and their correlation coefficient (R^2)

		Products				
Function	Parameter	Spray dried coffee	Freeze dried coffee	Wheat flour	Potato starch	Cane sugar
GGS	k	34.470	1107.440	$5x10^{-6}$	$5x10^{-5}$	46.870
	m	1.140	10.650	0.255	0.290	1.820
	R^2	0.970	0.999	0.930	0.935	0.947
RR	xR	1186.700	1187.700	86.120	25.770	394.020
	n	1.790	614.720	1.990	0.440	4.410
	R^2	0.994	0.999	0.998	0.999	0.999
MGM	xo	8000.100	7997.970	320.000	380.000	7999.800
	r	11.550	58.370	5.000	5.000	91.530
	m	2.620	44136.200	1.800	0.900	4283.000
	R^2	0.997	0.999	0.995	0.974	0.998

GGS = Gates-Gaudin-Schuhmann, RR = Rosin-Rammler, MGM = Modified Gaudin-Meloy.

C. Rosi-Rammler (Weibull) Distribution

The cumulative undersize fraction (CUF) data are listed in table 8. The RR model (Equation 15) was applied to experimental data for calculating the x_R and n values which may best adjust to the model. As described before, the "Solver" system from the Excel program was used to determine the x_R and n constants. The Rosin-Rammler model containing its constantans to describe the particle size distribution of pray day coffee is presented next:

$$Y = 1 - \exp\left(-\frac{x}{1187}\right)^{1.79} \qquad (23)$$

where Y is the cumulative undersize fraction (CFU), and x is the particle size fraction. The SSE for this model was 0.0037. In table 9 are listed the RR model parameters.

7.4. Data from Particle Size Distribution Modeling

The Rosin-Rammler (RR), the Modified Gaudin-Meloy (MGM), and the Modified Gaudin-Meloy (MGM) functions were selected for modeling the experimental particle size data of the selected food powders. These models were selected because of their simplicity. In table 9 are listed the parameter values and the correlation coefficients (R^2) for the three selected distribution functions applied to the selcted food powders. It can be observed that the Rosin-Rammler (RR) and the Modified Gaudin-Meloy (MGM) models fitted the experimental data of the selected food powders adequately since the correlation coefficients ($R^2 > 0.974$) are closely to 1.0; however, the Gates-Gaudin-Schuhmann (GGS) model also provided high correlation coefficients ($R^2 > 0.930$). On the other hand, the SSE values for the RR (0.0037) and the MGM (0.0017) models were lower that that for the GGS model (1.71). Therefore, the RR and the MGM models fit the experimental particle size data in a better way than the GGS model.

Figures 7 to 8 illustrate the cumulative undersize percentage experimental values along with the fitted data using the Rosin-Rammbler model. The information that can be obtained from the cumulative undersize fraction plot is the number median diameter. The particle size at the 50% point of the cumulative axis is called the geometric mean diameter or number median diameter (Iinoya *et al.* 1990). The median is the particle size for which 50% of the material is smaller and 50% is larger than the mean diameter. Since the freeze-dried coffee had almost only one particle size, it is not possible to take advantage of the information coming from these plots. The same reason applies to the cases in which a large fraction of the powder passed all the way to the fine's collector. Regarding the spray dried coffee and the sugar, the medians showed a value of 850 and 300 microns respectively.

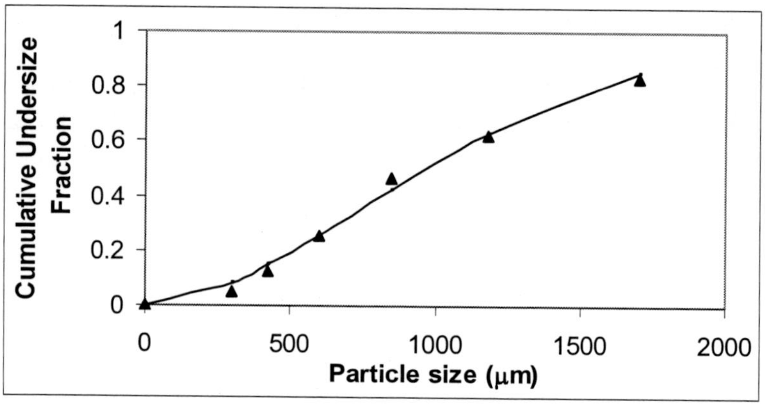

(a)

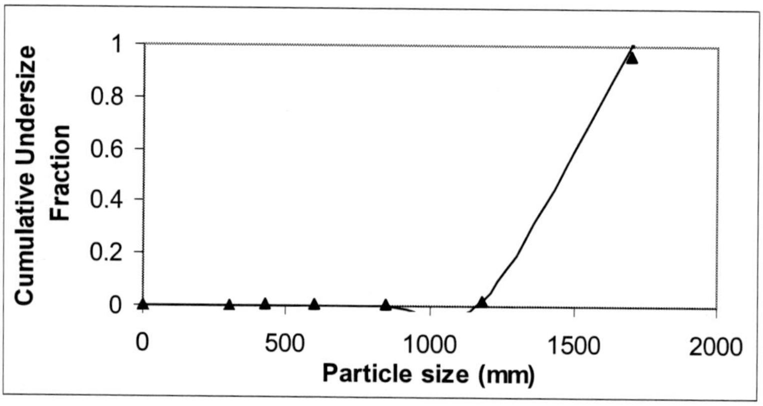

(b)

Figure 7. Modeling (-) of the cumulative undersize distribution data (▲) for (a) spray dried (a) and freeze dried (b) coffee using the Rosin-Rammler model.

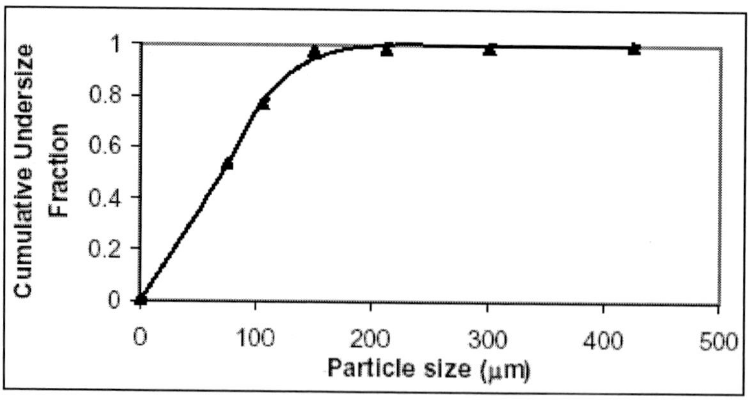

(a)

Figure 8. Continued on next page

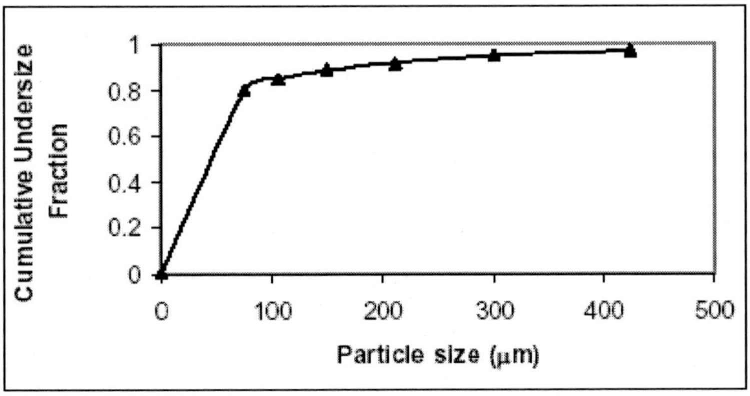

(b)

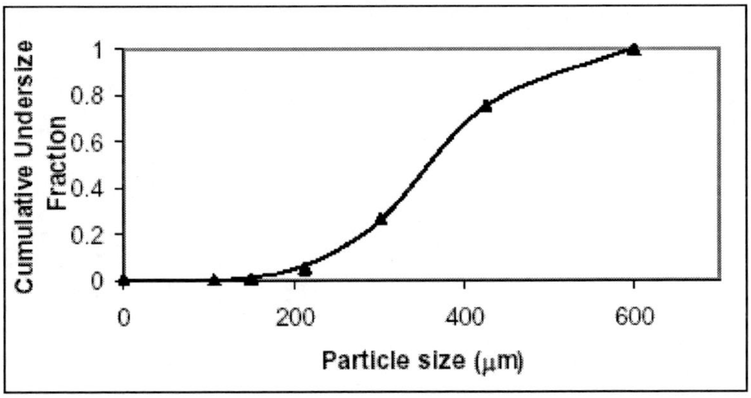

(c)

Figure 8. Modeling (-) of the cumulative undersize distribution data (▲) for (a) wheat flour, (b) potato starch, and (c) cane sugar using the Rosin-Rammler model.

All selected food products showed different *PSD*; however, it can be chosen any of the three *PSD* functions for modeling purposes. The distribution data and particle size can be used as a control quality parameter to deliver always the same particle size with a characteristic media o median particle size. Storage conditions are also important since temperature and moisture content of the surrounding environments can modify the particle size and produce caking and consequently microbial growth or other deteriorative phenomena.

8. Final Remarks

It is a matter of fact that the particle size of powdered foods and its control is very important in the food powder industry. Information regarding particles size and particle size distribution models, from data reported elsewhere and experimental data, was analyzed in this chapter. There exist a number of techniques to assess the particle size in the food powder area; however, the sieve analysis is a fast and reliable method to determine particle size distribution in food powders. Also, the particle size analysis of food powders and its quality control, when obtained by its specific process, is of paramount importance since the particle size can influence some physical properties during storage, handling and transportation.

References

Allen T. 1981. Particle Size Measurement. 3rd ed. Chapman and Hall. USA. pp. 103-147.

Barbosa-Cánovas G.V., Malave-Lopez J. and Peleg M. 1985. Segregation in food powders. *Biotechnology Progress,* 1(2): 140-146.

Barbosa-Cánovas G.V., Ortega E., Juliano P. and Yan H. 2005. Food Powders: Physical Properties, Processing, and Functionality. Springer. USA.

Barletta B.J. and Barbosa-Cánovas G.V. 1993a. An attrition index to assess fines formation and particle size reduction in tapped agglomerated food powders. *Powder Technology,* 77: 89-93.

Barletta B.J. and Barbosa-Cánovas G.V. 1993b. Fractal analysis to characterize ruggedness changes in tapped agglomerated food powders. *Journal of Food Science,* 58(5): 1030-1046.

Genovese D.B. and Lozano J.E. 2000. Particle size determination of food suspensions: Application to cloudy apple juice. *Journal of Food Process Engineering,* 23: 437-452.

Grabowski J.A., Truong V.D. and Daubert C.R. 2006. Spray-drying of amylase hydrolyzed sweet potato puree and physicochemical properties of powder. *Journal of Food Science,* 71(5): E209-E217.

Iinoya K., Gotoh K. and Higashitani K. 1991. Powder Technology Handbook. Marcel Dekker. USA. pp. 3-39.

Hayes G.D. 1987. Food Engineering Data Handbook. Longman Group. UK.

Koc A.B., Heinemann P.H. and Ziegler G.R. 2003. A process for increasing the free fat content of spray-dried whole milk powder. *Journal of Food Science,* 68(1): 210-216.

Lima I.M., Guraya H.S. and Champagne E.T. 2000. Improved peanut flour for a reduced-fat peanut butter product. *Journal of Food Science,* 65(5): 854-861.

Narayanan R.M. and Vu K.T. 2000. Free-space microwave measurement of low moisture content in powdered foods. *Journal of Food Processing Preservation,* 24: 39-56.

Omobuwajo T.O., Busari O.T. and Osemwegie A.A. 2000. Thermal agglomeration of chocolate drink powder. *Journal of Food Engineering,* 46: 73-81.

Peleg M. and Normand M.D. 1986. Simulation of size reduction and enlargement processes by a modified version of the Beta distribution function. *The American Institute of Chemical Engineers Journal,* 32(11): 1928-1930.

Peleg M., Normand M.D. and Rosenau J.R. 1986. A distribution for particle populations having a finite size range and a model independent of the spread. *Powder Technology,* 46: 209-214.

Pintauro N.D. 1975. Coffee Solubilization: Commercial Processes and Techniques. Noyes Data Corporation. New Jersey. USA.

Popplewell L.M., Campanella O.H., Normand M.D. and Peleg M. 1988. Description of Normal, Log-Normal and Rosin-Rammler particle populations by a modified version of the Beta distribution function. *Powder Technology,* 54: 119-125.

Rietema K. 1991. The Dynamics of Fine Powders. Elsevier Applied Sci. Great Britain. pp. 19-38.

Teunou E., Fitzpatrick J.J. and Synnott E.C. 1999. Characterization of food powder flow ability. *Journal of Food Engineering,* 39: 31-37.

Teunou E. and Fitzpatrick J.J. 2000. Effects of storage time and consolidation on food powder flowability. *Journal of Food Engineering,* 43: 97-101.

Trezza T.A. and Krochta J.M. 2000. The gloss of edible coating as affected by surfactants, lipids, relative humidity, and time. *Journal of Food Science,* 65(4): 658-662.

Walton D.E. 2000. The morphology of spray-dried particles: a qualitative review. *Drying Technology,* 18(9): 1943-1986.

Yu A.B. and Standish N. 1990. A study of particle size distributions. *Powder Technology,* 62: 101-118.

In: Food Processing and Engineering Topics
Editors: M. E. Sosa-Morales and J. F. Vélez-Ruiz
ISBN: 978-1-60741-788-0

Chapter 4

RHEOLOGY AND TEXTURE OF CHEESE

Jorge F. Vélez-Ruiz
Chemical and Food Engineering Department
Universidad de las Americas, Puebla, Mexico

1. INTRODUCTION

Cheese as dairy product, is one of the first, most popular and universal elaborated food item. Cheese represents a product in which the milk components are preserved. Cheesemaking practice has changed from an incipient and experimental art to an industrial technology within strong scientific support. Since Rheology studies the deformation of solids and the flow of fluids, when the instrumental characterization is done on cheese, is essentially a quantitative expression of its physical behavior, particularly it is a measurement of its mechanical properties. Cheese as many solid foods has a viscoelastic nature.

In 1995, more than 2000 types of cheese already existed (Gunasekaran and Ak, 2003), although only 500 have been recognized by the International Dairy Federation (Fox, 2003), and the variety still is augmenting every time, due to the presence of milk from different mammalian species, the diversity of manufacturing methods, and the consumption amplitude by human population.

1.1. Cheese Definition

This food item, known as cheese (in English), "fromaggio" (in Italian), "fromage" (in French), "kase" (in German), and "queso" (in Spanish), due to its huge variety of types, makes a single definition of cheese very difficult. Many definitions have been cited by world organizations, book authors and official institutions of each country, trying to facilitate the understanding of the concept. Only some of them are cited next:

- FAO/WHO definition: "cheese is the fresh or matured product obtained by the drainage (of liquid) after the coagulation of milk, cream, skimmed or partly skimmed milk, butter milk or a combination thereof" (Scott, 1986; Walstra *et al.*, 2006).

- National Dairy Council definition: "cheese is a concentrated dairy food made from milk, it may be a fresh or matured product obtained by draining the whey (the moisture or serum of the original milk) after coagulation of casein, milk's major protein" (NDC, 2005).
- "Cheese is the result of milk coagulation by the action of coagulating enzymes known as rennet, followed by the partial elimination of whey (draining), resulting in a curd from which cheese originates" (Eck, 2000)
- "Cheese is the fresh or ripened product composed of varying amounts of coagulated milk proteins, fat, water and salts" (Rosenthal, 1991).
- Larousse definition: "cheese is a dairy product made from coagulated milk, cream, skimmed milk, or a mixture of any of these, drained in a mould" (Lang, 1989).
- Cheese is a dairy product made to preserve most of the milk components, including fat, protein and minor constituents from the milk, eliminating water and adding salt, with a special flavor and a solid or semisolid consistency (Vélez-Ruiz, 2002).

Independently of these and other definitions, some basic aspects are characteristic of cheese:

- milk as the raw material comes from any mammalian specie,
- milk is subjected to a biochemical transformation, forming a coagulum,
- whey is eliminated, in minor or major intensity, contributing to form a curd,
- characteristic consistency/texture that is developed through the last manufacturing steps

Thus, there are an enormous variety of cheese types, from different world regions, multiple manufacturing processes, diversity of consumption traditions, and many formulations, generating a huge variation of flavors, tastes, textures and shelf lives. Therefore, from the physical viewpoint, the cheese represents a rheological system in which many studies have been developed and many more are being completed.

1.2. Cheesemaking

Though there are a lot of cheese types, the elaboration process of most of them follow the same common stages in which the variations in some of the steps contribute to generate a diversity of cheese products. Those common stages or unit operations are briefly commented next (Scott, 1986; Rosenthal, 1991; Eck and Gilis, 2000; Vélez-Ruiz, 2002; Fox, 2003; Gunasekaran and Ak, 2003; Walstra *et al.*, 2006), in which some specific equipments and process conditions may vary.

1. Milk recollection. Milk is recollected, clarificated and cooled down, previous to the cheese manufacturing to ensure a hygienic raw material.
2. Milk preparation. Basic processes such as, standardization, homogenization, heating and/or microorganisms addition may be carried out. The fat-protein ratio is frequently standardized, $CaCl_2$ is normally added, and pH is sometimes adjusted. On the other hand, pasteurization destroys pathogenic microorganisms and many enzymes.

3. Milk coagulation. Addition of rennet, coagulant or acid in order to transform milk to coagulum. The enzyme acts on a specific amino acid of the casein, whereas the acid generates protein precipitation.
4. Whey elimination. The formed coagulum contracts and expel part of the entrapped liquid, constituting the syneresis phenomenon. Whey elimination from the cheese is favored by cutting, scalding, and/or stirring.
5. Curd brining/salting. Salt is added to the curd, as solid material or in solution, to favor whey elimination, flavor developing, and preserving effects.
6. Final Treatments. Agitation, milling, thermal treatment, pressing, moulding, turning, packing, waxing, wrapping, ripening, are some of the final operations than may be utilized as part of the cheese making, in order to reach the specific characteristics of each type. Of all these treatments, ripening is the most important due to the biochemical and microbial modifications happened during the storage.

Each basic operation contributes to the milk/curd/cheese transformation in which the biochemical (enzymatic, microbial acidification, hydrolysis, lipolysis, proteolysis, etc) and physical (homogenization, shearing, mixing, gelation, syneresis, curd fusion, solids diffusion, etc) changes are shown out by the rheological behavior of this raw/intermediate/final material. In summary, through the manufacturing process, the three most important modifications affecting cheese characteristics are i) the milk formulation for the starting, with proper ingredients and others such as calcium chloride, cream, lactoperoxidase, ropy microorganisms, milk powder, etc; ii) the applied operational variables, including rennet type, salt, forces or stresses (centrifuge, pressure, shear), temperatures of cooling and heating, treatment times, and shear rates, among others; and iii) the biochemical and physicochemical transformations developed during the elaboration and maturing stages.

1.3. Cheese Classification

Grouping of cheese types is extremely complicated due to the enormous variety of them. In addition to aforementioned intrinsic and extrinsic factors, there are variants due to size, shape, as well as region of manufacture, among others. A number of efforts have been realized to classify cheese, taking different points of view in order to meaningfully group them. Some classifications are based on the cheese origin (animal, country), involved process, empirical rheological parameters, moisture content, and others.

A very general and useful, but limited classification because many factors are ignored, is that cited by Scott (1986) based on composition and shown in table 1. This classification ignore if cheese is ripe or not, it does not indicate if microorganisms are present or not, if they are bacteria or molds, neither it includes size and external appearance.

Fox (2003) cites a cheese classification based on the coagulation method into four groups:

A. Rennet coagulated cheeses, including most of the cheese types. This big group may be subclassified in many others.
B. Acid coagulated cheeses, such as cottage, cream and quark
C. Heat-acid coagulated cheeses, like the Ricotta cheese.

D. Concentrated/crystallizated cheese in which very few types are included, Mysost cheese by example.

Table 1. Grouping of cheese on a compositional base

Cheese group	Water (% of fat free matter)	Fat (% of dry matter)	Descriptive type
Extra hard	< 51	> 60	High fat cheese
Hard	49 – 55	< 60 and > 45	Whole milk cheese
Half fat	53 - 63	< 45 and > 25	Half fat cheese
Semi soft	61 – 68	< 25 and > 10	Low fat cheese
Soft	> 61	> 10	Skim milk cheese

From Scott, 1986.

A more complete classification, as a combination of the previous two, is that showed out by Gunasekaran and Ak (2003), it is summarized in figure 1. These groups take account of important aspects pointed out by the Codex Alimentarius, such as moisture content, fat content and ripening agent.

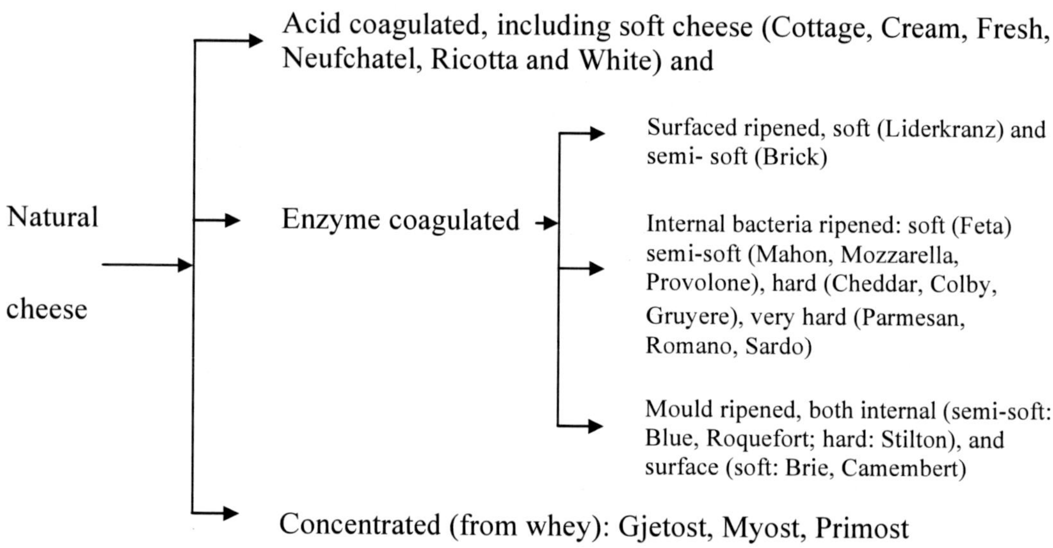

Figure 1. Classification of cheese based on the maturation agent (From Gunasekaran and Ak, 2003)

In addition, several other groupings may be included, but no one is universal and totally accepted. Thus a more physical or reological base is needed in order to group them and to get a very good one.

Table 2. Characteristics of the rheological/textural measurement instruments

Empirical	Imitative	Fundamental
Measure of a physical parameter under well defined conditions. Practical experience is related to cheese texture.	Reproduce those conditions to which the food material are exposed in practice, simulating mouth features and processing conditions.	Measuring of well defined physical properties with fundamental support.
Often simple, inexpensive and portable. Generally poor precision and reproducibility.	Great flexibility of design for different cheese types	Instrumental measurements are very valuable in research and new product development.
Mainly used for quality control purposes	Simulates the jaw action during eating.	Fundamental parameters such as viscosity, flow properties, viscoelastic parameters, moduli of compression, elastcity, relaxation and shear, among others, may be obtained.
They may lack of a scientific basis, but this does not disqualify their utlity	All the instruments contain a load cell vertically driven at different speeds.	
Some examples:		Data are expressed in known scientific units. Predictions of some properties are possible.
	It can cycle over a fixed distance or load range and probes can be attached.	
Compression devices (apply force to achieve a compression effect), such as the ball indenter.		Rheometers are commercial instruments, with very fine and precise measuring features, but they are expensive. Texturemeters are also very commercial.
	Some examples:	
Cutting fittings (by using wires or blades and measuring the applied force or time to cut a standard sample).	General Foods Texturometer	In this group, there are included a significant number of tests and features, such as: bending, compression, creep, dynamic, extensional, fracture, melting, shear, squeezing, stress relaxation, tension, torsion, vane and nonlinear viscoelasticity, among others.
Flow and mixing devices (by measuring the viscosity or consistency of the liquid or semisolid sample), such as the Bostwick Consistometer.	Instron Universal Testing Machine	
	Lloyd and Stevens Instruments	

Table 2. (Continued)

Empirical	Imitative	Fundamental
Penetration devices (needed force to push a probe into the food at a specific depth), such the cone penetrometer. Shearing fittings (by measuring the force needed to shear through the food), such the Kramer Shear Cell.	Stable Micro Systems A very useful and popular test, is the Texture Profile Analysis. In it, a two-bite cycle is employed to measure textural attributes (cohesiviness, elasticity, hardness, for instance) These equipments can be adapted for some fundamental texture measures.	Anton Paar (Physica, Rheolab, Rheoplus), Bohlin, Control Stress, Texas Instrument Rheometers. Texture Analyzer. Tests and calculations of material properties are based on well defined rheological concepts.

2. Rheological and Textural Determinations

The increasing importance for understanding the relationship among food rheology, food microstructure and consumer perception of texture is out of doubts. Food companies work very hard to produce texturally attractive food products, and scientists research extensively to find rheological and textural properties that contribute to the understanding of cheese changes and their effects on the final product. Sensory specialists try to correlate textural attributes with human perception. Therefore, too much research has been published on cheese texture and rheology.

Properly, the instrumental methods for food texture and rheology determinations started in the XIX century (Segards and Kapsalis, 1987). Since that time, numerous instruments and methods have been developed to evaluate rheological properties of foods, and particularly for cheese texture (Guinee, 2003). The diversity of instruments may be grouped in three recognized groups: empirical, imitative and fundamental, in which their main charactetistic are presented in table 2. In cheese characterization, from the practical point of view the three approaches are utilized, but only two of them, the empirical and well defined tests are broadly applied (Segards and Kapsalis, 1987; Rosenthal, 1999; Bourne, 2002; Gunasekaran and Ak, 2003; Kilcast, 2004b; Vélez-Ruiz, 2008). The imitative determinations appear in a good number of works, but they are less valuable in cheese rheological characterization.

Two of the more commercial instruments, texture analyzers and rheometers belong to the fundamental group. Rheometers allow continuous determinations, commonly at a selected shear rate and less frequently at a given shear stress, the time of testing may be short or long, allowing flow characterization of time independent and time dependent cheese materials. For cheese studies, they can be used in dynamic mode, in which a sinusoidal varying stress or strain is applied for viscoelastic characterization. Rheometers have become widespread at the present, and commercial types have incorporated constant improvements and technological advances, such as electric motors, transducers, higher speeds and memories, digital processors and microprocessors (TA Instruments, 2008).

Texturemeters are instruments that allow the characterization of texture and textural identification of solid foods of any type through mechanical loading. These instruments allow the viscoelastic characterization through creep and stress relaxation tests. Of course other measures on cheese samples may be carried out with these instruments, such as adhesivity, compression, penetration, and torsion; being uniaxial compression one of the most applied tests for cheeses. They normally are unable or very limited for flow characterization.

Instron was an integrated test system that performed all common cheese tests by using popular fixtures. It incorporated a microprocessor data system that presented values in engineering units, and the drive system was synchronously controlled, making it possible to characterize the viscoelastic response of cheese products (Instron Corporation, 1980). It was able to handle calculations of stress and strain automatically, to integrate the load elongation curve and evaluate energy (Instron Corporation, 1981), among other aspects.

A Texture Analyzer measures force or distance in compression and tension modes, providing eight different forms of test. It quantifies the cheese texture, characterizing adherence, coherence, compressibility, firmness, fracture, hardness, softness, spreadability, true stress among others (Texture Technologies Corporation, 1990). It may include over 30 standard probes that allow a complete textural characterization of cheese items.

2.1. Physical Cheese Characteristics

Cheese is a food material with a wide rheological behavior, in which the applied force/stress and obtained deformation are function of three main characteristics, including various aspects: i) composition of the cheese, where the percentages of fat, moisture and protein are determinant; ii) physicochemical state, such as gas sites, liquid whey, semisolid fat, and/or solid fat, minerals and NaCl trapped or occluded inside of a proteinic gel; and, iii) microstructure arrangement, which represents the spatial distribution and attractions among components and particles.

Cheese consists of a three dimensional matrix of paracasein salt (with calcium and phosphate) trapping fat globules, moisture, and other materials (enzymes, peptides, microorganisms, minerals and sodium chloride). The matrix network is constituted by fused aggregates of casein/para-casein, fat globules and water solution as the three main components (Vélez-Ruiz and Barbosa-Cánovas, 1997).

As a consequence of the many mentioned factors of variation, cheese consistency varies from fluid to solid behaviors, being almost pourable (Cammembert, Cottage), passing by spreadable (Cream, Goat cheese), soft and curd like (Fresh, White), meltable (Cheddar, Chihuahua, pasta filata, Manchego, Mozzarella, Oaxaca), elastic or rubber-like (Cheddar, Emmental, Gruyere, Manchego, Oaxaca, Processed), up to reach a rigid, almost stony consistency (Edam, Parmesano) (Velez-Ruiz, 2002; Guinee, 2003; Walstra *et al.*, 2006).

Thus, a considerable quantity of rheological and/or textural measurements have been employed for cheese characterization. Empirical tests have been explored since cheesemaking appeared, and still are used, whereas than imitative determinations are limited in cheese, texture profile analysis is frequently applied as an imitative test, and the fundamental measures provide the best physical characterization of cheese at this time.

Empirical and imitative tests share several characteristics, a device for mechanical deformation of cheese; a couple of devices for recording of the applied forces and deformation, respectively; a time measurer and a cell to hold the cheese sample. Additionally and due to the influence of local variations, it has been necessary to complete multiple determinations to reach a representative average magnitude (Konstance and Holsinger, 1992).

2.2. Influence of Biochemical Changes on Physical Nature of Cheese

For the specific case of cheese, the physical nature is strongly related to biochemical changes of cheese components. In according with Lucey *et al.* (2003), some cheese properties such as color, melt, stretch and texture are determined by casein interactions, as well as extent and type of proteolysis.

Those forces and interactions leading and contributing to the formation and stability of casein micelles importantly influence the functional cheese properties. The curd as a first stage involves the micelles desestabilization to generate the characteristic network. Cheeses manufactured by the rennet enzyme are susceptible of aggregation as a consequence of the reduction in steric and charge repulsion by the loss of macropeptide from the κ-casein (Lucey *et al.*, 2003).

Lately cheese curd suffer additional transformations due to the final steps of the cheese manufacture; stirring helps to maintain a granular appearance, cutting, pressing and salting favor syneresis, whereas heating and cooking confer plastic consistency. In all these changes, pH is an important factor, and cheeses elaborated by rennet coagulation have low moisture content. On the other side, cheeses prepared by acid coagulation contain higher water content attributed to a less degree of modifications in the protein network.

Natural, processed and matured cheeses are not entirely homogeneous or isotropic, from the engineering point of view, therefore all the rheological results should be interpreted with caution.

2.3. Main Objective Tests

Even though many rheological and textural properties may be measured, it is very difficult to simulate the complexity of physiological activities involved in food consumption. Sensory evaluation of some properties, such as chewiness, crispness, and toughness are complex processes occurring within the mouth, involving interactions among teeth, tongue, cheeks, auditory signals, and saliva; in addition to the food material subjected to forces/stresses and under complex deformations. None mechanical test is able to reproduce the combination of sensations that make up our texture perception.

The rheological characterization and objective textural determinations of cheese are of broad importance. They are necessary to know the flow properties of liquid milk, cream or dispersion system, the physical behavior of the curd/paste through the process, the mechanical properties of the curd/cheese previous to the packaging, ripening or storing, the consistency or texture of the cheese when is melted and placed on shelf, or when it is touched by the consumer, the rheology and sensory relationships for consumer satisfaction, and the eating or texture nature when is masticated or studied as well. Thus, every body is involved in the cheese rheology and texture: consumers, distributors, engineers, industrial processors, retailers, scientists, and technologists, each one in his specific area of incidence.

Rheological and textural cheese characteristics may be determined by various instrumental ways generating results as a function of different variables in the selected test. A representative set of characteristics for the most common rheological/textural determinations is summarized next, in order of importance.

Dynamic characterization. Due to its viscoelastic nature, dynamic measurements to evaluate cheese characteristics are well suited. Small amplitude oscillation tests are applied within the linear viscoelasticity. Many rheological parameters may be obtained, such as complex, storage, viscous moduli, and phase angle, among others. It is useful for structural understanding. It has been utilized in coagulating milk, curd, Camembert, Cheddar, Cheshire, Cream, Feta, Fresh, Goat, Gouda, Manchego, Montereym Mozzarela, Oaxaca, Processed, Quarg and White cheeses, practically in all type of cheese.

Uniaxial Compression is the most popular measure for cheese rheological and textural characterization. Cylindrical samples are preferred and a wide range of crosshead speeds may be used, in which oil may be used for friction decreasing. Force-time data are converted into stress and strain, and different stress-strain data are expressed as a function of crosshead speeds. Brie, Camembert, Cheddar, Danbo, Feta, Goat, Gouda, Gruyere, Manchego, Mozzarella, Processed and White cheeses have been characterized with this test.

Other tests such penetration, fracture, torsion and extensional determinations have been applied for specific types of cheeses. For penetration or puncture determinations, a selected geometry penetrates the sample cheese to a given depth; cones, simple and multiple probes may be utilized. Fracture, bending and tension are common methods of characterization and fracture or breaking stress is normally estimated from a stress-strain curve obtained from these tests. The energy required for fracture may be obtained from the area under the curve of stress vs. deformation. In torsion test the sample piece is subjected to a twisting effect on the axis, this test has several limitations. Extensional test is relatively new in comparison with shear reometry, in it a failure strain or stress is measured as indicator of stretchability. Compression, flow, melting, tension, and torsion, among other measures, can provide useful values and patterns for cheeses.

From the immense number of options for cheese characterization, the selection of the most convenient cheese test will depend of three main considerations, 1) the available instrument, 2) the purpose of the test, and 3) the particular characteristics of the cheese. Specific recommendations and more details for each test maybe found in specialized literature (Rosenthal, 1999; Bourne, 2002; Gunasekaran and Ak 2003; Lucey *et al.*, 2003), and will be commented in next sections.

3. Empirical Tests on Cheese

Empirical devices in cheese plants have been used for many years. Ball compressors, curd tension meters, hardness tester, and borer and needle penetrometers, among others, have provided rheological and textural information from cheese (Bourne, 2002). These instruments are not considered as precision tools and frequently the obtained results are of difficult comparison with other values coming from fundamental equipments. Some cheese characterizations, in which empirical tests were applied in technical and research studies, are included in table 3.

Among the analyzed factors, in which empirical tests were applied as a physical indicator, we can find many works,such as the effect of formulation and storage time (Ak *et al.*, 1993), milk homogenization and fat content (Tunick *et al.*, 1995) on Mozzarella cheeses; milk pasteurization temperature (Rynne *et al.*, 2004), fat content (Ustunol *et al.*, 1995), and acid development (Lee *et al.*, 2005) on Cheddar cheeses; the effect of whey protein on cream cheese (Sanchez *et al.*, 1996b),and as part of the processed cheese (Lee *et al.*, 2006).

The influence of milk formulation: composition (Leach *et al.*, 2003; Fagan *et al.*, 2007), rate of cooling and effect of maturity of the raw material (Piska and Stetina, 2004). Fat content on melting of processed cheese (Drake *et al.*, 1999), on stretchability of Manchego cheese (Fife *et al.*, 2002), and of White cheese (Madadlou *et al.*, 2005). Influence of fat replacers, preacidification and starters (Zisu and Shah, 2005a), microbial, acidification and whey protein concentrate (Zisu and Shah, 2005b), pH and calcium (Guinee *et al.*, 2002), Ca, pH and moisture (McMahon *et al.*, 2005) on Mozzarella cheeses. Effect of emulsifying salts on pasta filata cheeses (Mizuno and Lucey, 2005).

Table 3. Empirical tests used in cheese characterization

Test	Characteristics and conditions	Type and objective	Measures and observations	Reference
Flow	Rheology with a Rheometrics Mechanical Spectrometer	Whipped cream cheese	K=776 Pa s^n, n= 0.061	Bistany & Kokini, 1983
Flow	Flow properties by shear flow with aWells-Brookfield	Soft cheeses (Neufchatel)	K: 1416-5352 dyne s^n/cm^2, n: 0.22-0.37, τ_y: 715-1144 dyne/cm^2	Massaguer-Roig *et al.*, 1984
Elongation	Elongational viscosity by compression	Process American	Viscosity: 0.42-5.5x10^5 Pa s, as a function of temperature	Campanella *et al.*, 1987
Extension	Cheese specimens stretched among clamps	Mozzarella, storage and formulation	Fracture Stress: 30-48 kPa, deformability: 60-111 kPa, strain: 0.4-0.5	Ak *et al.*, 1993
Extrusion	Capillary extrusion divided in three phases: compression, unsteady and constant	Cream, effect of homogenization, and cooling rate	Shear stress (400-1200 Pa) was related to homogenenization pressure and cooling rate	Sanchez *et al.*, 1994a, 1994b
Meltability	Arnott test in a Petri dish	Cheddar, fat	Melt: 2.1-6.0 cm, related with G*	Ustunol *et al.*, 1994
Elongation	Stretching under constant weight	Mozzarella, storage	Viscosity: 10^4 -750 x 10^3 Pa s, was not affected by storage	Ak & Gunasekaran, 1995
Flow	Flor characterization with a Haake	Commercial brands	K: 12-41 Pa s^n, n: 0.27-0.64	Omar *et al.*, 1995
Meltability	Shreiber test by expansion of a disc of 18 mm, after 5 min at 232°C	Mozzarella, effect of fat and homogenization	Meltability: 1.0-3.2 (dimensionless), increased with fat content and storage time	Tunick *et al.*, 1995
Viscosity & firmness	Viscosity by on-line and with a coaxial Rheometers. Firmness by penetrometry	Effect of processing and WPC (whey protein conc)	Viscosity: 1-2.2 Pa s, firmness: 3-27 N Samples exhibited time dependency	Sanchez *et al.*, 1996a, 1996b
Strechability & viscosity	A fixed quantity of grated cheese on a rectangular pizza base	Cheddar and Mozzarella, effect of test and ripening	Mozzarella: 52-97 cm, viscosity: 200-1200 Pa s, Cheddar: 10-75 cm.	Guinee & O´ Callaghan, 1997
Melting	Shreiber melt test with discs of 35 mm and 15 mm thick, 10 min-100°C	Processed, effect of fat and lecithin	Melt diameter: 40-63 cm, was higher in full fat cheeses	Drake *et al.*, 1999

Table 3. (Continued).

Test	Characteristics and conditions	Type and objective	Measures and observations	Reference
Melt, flow, stretch	Melt time at 280°C, flow of a disc, stretch by extension and viscosity in a Brookfield	Cheddar, effect of fat and ripening	Melt time: 80-280 s, stretch: 10-74 cm, viscosity: 240-1000 Pa s	Guinee *et al.*, 2000
Hardness	Penetration 15 mm, cubic samples of 4 cm	Gaziantep, effect of brining	Hardness: 3-45 N, varied with salt	Kaya, 2002
Stress, strain and rigidity	Fracture by torsion in a Torsion Gelometer (four shaped plugs at 20-22°C)	16 different samples of cheeses	Stress: 8.3-151 kPa, rigidity: 8-297 kPa, strain: 0.04-1.86, varied with cheese type	Tunick & Van Hekken, 2002
Strechability & viscosity	Strecht with a tensile tester, viscosity and melt with a tube test method, 60-90°C	Manchego affected by fat	Melt strength: 40-279 g_f, stretch length: 4-30 cm, melt distance: 6-20 cm	Fife *et al.*, 2002
Viscosity & flow props.	Shear tests with an Universal Testing Machine, during heating & cooling	Processed affected by formulation and age	Viscosity: 60-937 Pa-s, n: 0.12-0.67, K: 252-7448 Pa s^n	Leach *et al.*, 2003
Apparent viscosity	Shear method at 15 s^{-1} with a Bohlin CVO, samples warmed at 80°C	Processed typical & fat-free	Typical: 8-18.6, fat-free: 0.6-6.6 Pa s, as a function of shearing time	Lee *et al.*, 2003
True shear stress and strain	Torsional methods, cylinders (19 mm D, 28.7 mm length), twisted with a Haake 550	Mozzarella and Monterey Jack	Ranged of strain: 0.001-2.0, range of stress: 0.018-40 kPa	Brown *et al.*, 2003
Apparent viscosity	Pressure drop and volumetric flow as flow curves in which slippage was not evident	Processed, effect of temperature (55-95°C)	Apparent viscosity: 1-23 Pa s, as a function of temperature, moisture and protein contents	Dimitreli & Thomareis, 2004
Flowability Stretchability Viscosity	Increase in diameter of cheese discs (45 mm D, 6.5 mm thick) on melting (280°C-4 min) Length of strings at failure (0.066 m/s) Molten cheese at 80°C, with a Brookfield	Cheddar, effect of milk pasteurization temperature	Flowability: 3-63%, and apparent viscosity increased with maturation Stretchability increased in the first 30 days and then decreased	Rynne *et al.*, 2004

Table 3. (Continued).

Test	Characteristics and conditions	Type and objective	Measures and observations	Reference
Stretch Yield point	Vertical distance before breaking. Sudden change of slope from a force-distance curve	Mozzarella, effect of formulation	Distance of 200-450 mm as a function of fat replacers and starter	Zisu & Shah, 2005a
Flowability Stretchability	Increase in diameter of cheese discs Uniaxial extension at a 0.066 m/s	Mozzarella, effect of pressure	Flowability: 2-12%, stretch: 21-45 cm, augmented with storage time	Sheehan *et al.*, 2005
Meltability	Distance inside of a tube of melted sample, 3 g of sample at 100°C-90 min	White, effect of fat and rennet	Melt distance: 45-68 mm, full-fat cheese melted to a greater extent	Madadlou *et al.*, 2005
Meltability Stretchability	Difference between final and initial length. Distance by pulled sample before breaking	Mozzarella, effect of formulation	Melt distance: 42-48 mm for control, and 30-63 mm for modified. Stretch: 280-420 mm depended also of time	Zisu & Shah, 2005b
Stretchability (Fork test)	Melted cheese is lifted up using the fork. Average of the strand length, 1 minute after the pizza was taken out of the oven	Pasta filata, effect of emulsifying salts	Length of stretch: 0.2-0.6 m, as a function of fat and salt, in comparison to control cheese (0.4 m)	Mizuno and Lucey, 2005
Meltability and flow	Distance by molten cheese after heating (16 min at 90°C)	Mozzarella, effect of formulation	Meltability: 60-220 mm, was affect ed by Ca. Higher flow for lower Ca	McMahon *et al.*, 2005
Meltability and flow	Measures at 72°C of cheese discs (30 mm D, 7 mm thick)	Cheddar, effect of acid	Flowability: 36-81%) changed during the first week	Lee *et al.*, 2005
Firmness	Force required to penetrate (depth 10 mm, 1 mm/s) with a cone, cubes of 25 mm. TA.XT2 Texture Analyzer	Processed, effect of treatment on whey protein	Firmness of treated whey protein cheese was equivalent to commercial low fat process cheese	Lee *et al.*, 2006
Meltability	Metl CV (degree of melting) was a ratio of areas: $(A_1/A_0)x100$, A_1: Area after melting, A_0: original area	Processed, effect of composition	CV: 102-600, allowed to measure low and high melt values	Fagan *et al.*, 2007

Process variables have been analyzed, such as the effect of curd homogenization (Sanchez *et al.*, 1994a), cooling rate (Sanchez *et al.*, 1994b), combined homogenization and cooling (Sanchez *et al.*, 1996a) on rheology of double cream cheeses. Effect of temperature (Leach *et al.*, 2003; Dimitreli and Thomareis, 2004), influence of cooking time (Lee *et al.*, 2003) on viscosity of processed cheeses. Campanella *et al.* (1987) tested a compression measure to quantify the elongational viscosity of processed American cheese as a function of temperature The effect of salting on hardness of a semi-hard cheese (Gaziantep) was studied by Kaya (2002). And the effect of ripening on stretchability of Mozarella and Cheddar cheeses was studied by Guinee and O´Callagahn (1997).

Some determinations are not empirical tests, such as viscosity and rheological properties carried out on cream, and most commonly on melted cheeses at high temperatures. Similarly, textural characterizations such as torsion gelometry have been applied on some cheese types (Tunick and Van Hekken, 2002). The founded experimental results and additional information about these specific tests are also included in table 3.

4. Imitative Tests on Cheese

Instruments including force-time curves based on compression bite-sized samples were introduced in 1960-1963. Based on them and combining with sensory data, industrial people were able to develop the TPA (Texture Profile Analysis) as the most recognized texture test of that moment and many years more; and even though the test has been strongly criticized, still is widely used (table 4). Those instruments were substituted by heavy universal testing machines in the end of 70´s and 80's; and lately, in the 90´s, by light texturemeters, which are now utilized for TPA test.

Seven textural parameters may be extracted from the force-time curve obtained by TPA, fracturability (originally brittleness), hardness, adhesiveness, cohesiveness, springiness (originally elasticity), guminess and chewiness (Bourne, 2002). Some of them are more useful depending of the cheese type. Topcu and Saldamli (2006) studied the TPA characteristics for White cheese affected by ripening, a semi-soft brined in which the samples were compressed 33% at a speed of 0.5 mm/s. The data obtained from curves were interpreted and then converted to textural parameters in the next way: the compression force recorded as the maximum of the first cycle was the hardness; cohesiveness as the ratio of the second and first areas represented the strength of the internal bonds making up the body of the cheese; and the distance recovered by the sample between end of the primary cycle and start of the second, was a measure of the springiness. Furthermore, gumminess constituted the energy needed to disintegrate the cheese sample until it is ready for swallowing, its value corresponded to hardness times cohesiveness; whereas the energy utilized to chew the cheese sample (gumminess x springiness) was expressed as chewiness; and finally, that work used to overcome attractive forces between the cheese surface and other material surface in contact with the sample, was a measure of adhesiveness (a negative area, commonly identified as A3).

Table 4. Texture Profile Analysis applied to cheese items

Objective and cheese type	Hardness (N)	Cohesiveness (A2/A1)	Springiness (mm)	Gumminess (N)	Chewiness (N mm)	Adhesiviness (A3, N mm)	Instrument/Reference
Ripening on Argentino	35-95	0.08-0.18	---------	---------	---------	0.05-20	Universal Testing Machine Bertola *et al.*, 1991
	Hardness decreased, cohesiveness and adhesiveness increased with ripening						
Lactobacilli and maturation on Cheddar	0.11-0.22. Firmness: 200-850 kPa, fracturability: 55-200 kPa. Three parameters decreased with maturation time						Universal Testing Machine Trépanier *et al.*, 1991
Fat and homoge-nization on Mozzarella	26-198	----------	5.2-8.9	----------	----------	----------	Universal Testing Machine Tunick *et al.*, 1993
	Hardness increased with homogenization and decreased with fat content						
Fat and ripening on Kefalograviera	1.4-3.8	0.17-0.25	3.9-5.8	0.26-0.67	1.01-3.70	----------	Universal Testing Machine Katsiari & Voutsinas, 1994b
	Textural parameters decreased with ripening						
Fat on Mozzarella	49-105	------------	5.5-8.9	------------	------------	------------	Tunick *et al.*, 1995
Fat and ripening on Cheddar	194-762	0.13-0.22	58-79 %	----------	----------	0.49-1.15	Universal Testing Machine Bryant *et al.*, 1995
	As fat decreased, hardness and springiness increased, the other two parameters decreased						
Fat and substitute on Manchego	0.29-0.78	0.21-0.28	----------	0.09-0.22	0.20-0.92	----------	Universal Testing Machine Robles-Martínez *et al.*, 1997
Fat and lecithin on processed	27-32	0.23-0.41	0.31-0.42	6.5-12.6	---------	6.5-11.2	XT2 Texture Analyzer Drake *et al.*, 1999
	Reduced fat cheeses with lecithin were more similar to full fat cheeses that reduced						
High pressure on Gouda	49-114	0.15-0.40	5.7-11.6	14.7-30.4	---------	---------	TA500 Texture Analyzer Messens *et al.*, 2000
	Samples treated at 225 and 400 MPa did not show macroscopical breakage						
Formulation on Processed	1.1-2.6	0.61-0.76	9.5-9.8	0.79-1.58	6.7-14.5	116-323	XT2 Texture Analyzer Awad *et al.*, 2002
	Chewiness, gumminess, adhesiviness and hardness showed significant diference						
Formulation on Manchego	0-24.3	0.20-0.30	3.20-5.80	---------	10.6-32.2	0.10-0.50	XT2i Texture Analyzer Lobato-Calleros *et al.*, 2002
Maturation on natural Cheddar	-------	0.65-0.71	0.77-0.82	---------	---------	---------	XT2 Texture Analyzer Pollard *et al.*, 2003

Table 4. (Continued).

Objective and cheese type	Hardness (N)	Cohesiveness (A2/A1)	Springiness (mm)	Gumminess (N)	Chewiness (N mm)	Adhesiviness (A3, N mm)	Instrument/Reference
Maturity and cooling on Processed	4.9-8.2	----------	----------	0.36-0.53	----------	0.30-0.52	XT2i Texture Analyzer Piska & Stetina, 2004
	Cooling was more significant than influence of maturity degree						
Fat replacers on Mozzarella	3-156	0.60-0.75	16.0-18.0	----------	----------	----------	Universal Testing Machine Zisu & Shah, 2005a
Freezing on soft caprine	10.8-13.8 7.50-11.7 10.5-13.0	0.08-0.11 0.08-0.15 0.10-0.24	control cheese (refrigerated) frozen control (stored at -20°C 3 and 6 months frozen		9.00-13.8 5.00-13.8 11.0-24.0	(4°C for 28 d) (for 2 d) ---------	Universal Testing Machine SM25-155 Van Hekken *et al.*, 2005
Ripening on Cheddar	102-233	0.22-0.56	0.44-0.82	---------	8.48-74.1	---------	XT2i Plus Texture Analyzer O'Mahony *et al.*, 2005
	These four textural attributes decreased with the ripening time (1-180 d)						
Formulation on Mozzarella	115-145 75-123	0.71-0.73 0.74-0.75	18.5-18.7 18.3-18.6	control cheese modified cheeses	1520-1950 1020-1720	---------- ----------	Instron Testing Machine Zisu & Shah, 2005b
Emulsifying on pasta filata	8.6-10.1 8.5-10.1	0.77-0.81 0.73-0.81	nonfat cheese with trisodium citrate nonfat tetrasodium pyrophosphate		28-33 24-33	0.02-0.31 0.03-0.31	XT2 Plus Texture Analyzer Mizuno & Lucey, 2005
Formulation on Mozzarella	1.08-7.95	Hardness depended on moisture and calcium contents, low adhesiveness was related to high calcium content				0.05-0.95	Stevens Texture Analyzer McMahon *et al.*, 2005
Ripening on White	3.37-5.74	0.23-0.41	4.17-5.63	0.97-1.62	4.89-9.05	0.06-0.43	TA Plus Texture Analyzer Topcu & Saldamli, 2006
	Texture attributes were significantly affected by the ripening period						
Formulation on Processed	5.54-263	0.12-1.08	0.17-0.88	---------	1.24-13.2	5.20-37.8	TA HDi Texture Analyzer Fagan *et al.*, 2007
	The five attributes were successfully modeled with varying degrees of accuracy						
Formulation on Processed	15.7-100	0.19-0.21	1.51-3.86	6.70-32.9	10.2-127	6.00-16.0	Instron Testing Machine Dimitreli & Thomareis, 2007
	Fat acted as a lubricant reducing textural and viscoelastic properties						

A1: area for the first byte, A2: area for the second byte, A3. negative area.

Among the most studied factors, through TPA test on cheese, various may be mentioned: effect of homogenization and fat content (Tunick *et al.*, 1993), milk homogenization (Tunick *et al.*, 1995) on Mozarella cheeses; influence of formulation on Manchego cheeses (Robles-Martínez *et al.*, 1997; Lobato-Calleros *et al.*, 2002), on pasta filata cheese (Mizuno and Lucey 2005), on Mozzarella cheeses (Tunick *et al.*, 1995; Zisu and Shah, 2005a, 2005b; McMahon *et al.*, 2005), also on processed cheeses (Drake *et al.*, 1999; Awad *et al.*, 2002; Dimitreli and Thomareis, 2007; Fagan *et al.*, 2007); effect of lactobacilli addition (Trépanier *et al.*, 1991), fat reduction and ripening (Bryant *et al.*, 1995) on Cheddar; influence of maturity and cooling rate on processed cheese (Piska and Stetina, 2004); effect of ripening on White cheese through 60 days (Topcu and Saldamli, 2006), on processed cheese during 90 days (Awad *et al.*, 2002), during 180 days on Cheddar cheese (O'Mahony *et al.*, 2005), through 11 months of maturation on natural Cheddar cheese (Pollard *et al.*, 2003); influence of post cheesemaking treatments, such as refrigeration and freezing on soft caprine milk cheese (Van Hekken *et al.*, 2005).

Also, the effect of cheese characteristics, of several types, such Brick, Colby, Cheddar, Gouda, Mozzarela and Romano, have been analyzed by Tunick and Van Hekken (2002), finding that TPA and dynamic parameters were not correlated to torsion results, with exception of torsion shear stress to TPA hardness ($R^2 > 0.8$) and torsion shear strain to TPA cohesiveness. A study of Moiny *et al.* (2002) on commercial Cheddar cheese was focused to study the loading rate and its correlation to sensory texture profiles, finding very high correlations between sensory and instrumental measurements, except for hardness.

5. Fundamental Tests on Cheese

Fundamentals studies on cheese include dynamic and transient tests, as well as compression. Samples used in these determinations are of specific shape and are deformed in a controlled manner, allowing systematical analyzes of the rheological response.

5.1. Compression Test

In the simplest test, an uniaxial compression stress is applied downward on the cheese sample, the resulting deformation is expressed as Cauchy or Hencky strain (Tunick, 2000). This test is useful for providing mechanical properties like fracture, true and yield stresses, fracture and yield strains, elastic modulus and yield work per volume, among others. Cauchy and Hencky strains may be evaluated from equations 1 and 2 (Konstance and Holsinger, 1992).

$$\varepsilon_C = \frac{\Delta L}{L_0} = \frac{L - L_0}{L_0} = \frac{L}{L_0} - 1 \tag{1}$$

$$\varepsilon_H = \ln\left(\frac{L}{L_0}\right) \tag{2}$$

where: ε_C es the Cauchy strain (dimensionless), L is the final length (m), L_0 is the original length (m), and ε_H is the Hencky strain (dimensionless).

During the force-compression tests on cheese, a comparison between bonded and lubricated determinations is carried out, the bonded and lubricated stresses, as well as, the corrected bonded stress can be computed from the next relationships (Tunick and Nolan, 1992). Casiraghi *et al.* (1985) characterized three type of cheeses (Cheddar, Processed and Mozzarella) by bonded and lubricated uniaxial compression, finding clear differences.

$$\sigma_b = \frac{F}{\pi r_0^2} \tag{3}$$

$$\sigma_l = \frac{F}{\pi r_0^2}\left(\frac{h}{h_0}\right) \tag{4}$$

$$\sigma_{bc} = \sigma_b\left(1 + \frac{r_0^2}{2h^2}\right)^{-1} \tag{5}$$

where: σ_b is the bonded stress (Pa), F is the total force applied to a cheese sample (N), r_0 is the original radius (m), σ_l is the stress in lubricated condition (Pa), h is the final height (m), h_0 is the original height (m), and σ_{bc} is the corrected bonded stress (Pa).

According to Luginbühl (1996), the determination of stress and strain from the compression curves, and particularly the evaluation of the apparent fracture point, presents some controversial aspects with respect to the appropriate stress representation.

Investigations on cheese, where compressions tests have been applied and monitored as a function of process variables of the cheesemaking, are included in table 5. Commercial Cheddar cheeses were studied to correlate instrumental and sensory results by Jack *et al.* (1993); in it, the elasticity and cohesiveness were correlated significantly. The effect of milk relation (sheep´s/goat´s) and ripening on Feta cheese were analyzed by Mallataou *et al.* (1994). The influence of fat content and ripening on Feta cheese (Katsiari and Voutsinas, 1994a), on Cheddar cheese (Fenelon and Guinee, 2000; Guinee *et al.*, 2000), fat content on Danbo cheese (Madsen & Ardo, 2001), confirmed that low-fat samples had a firmer and more elastic texture, and the proteolysis was altered. The effect of pH and calcium concentration on Mozzarella cheese were analyzed by Guinee *et al.* (2002) in which firmness decreased with the ripening time; the influence of pasteurization temperature of milk on the Cheddar cheese rheology has been studied by Rynne *et al.* (2004), finding that three textural parameters decreased with increasing of temperature. Lee *et al.* (2005) corroborated the effect of acid development on compression force of Cheddar cheese, manufactured with and without milk concentrated by reverse osmosis. Buffa *et al.* (2001) studied the effect of milk treatment and ripening on goat cheese, founded differences among treatments became less notable through ripening. Schlesser *et al.* (1992) followed the chemical and physical changes of Camembert cheese through ripening. Bachmann *et al.* (1999) monitored the ripening of Emmental over 12 months; Hort and Le Grys (2002) followed the texture of Cheddar cheese during 64 weeks of maturation that was related with a decrease in strain at fracture.

Table 5. Compression data related to rheese items characterization

Cheese type	Objectives and method	Observations	Instrument/Reference
Cheddar, Mozzarella, Processed	Effect of cheese type and test conditions (bonded and lubricated)	-True stress: 0-4.2 for Cheddar, 0-2 for Mozzarella, 0-1.8 x10^4 Pa for Processed), total force: 0-470 for Cheddar, 0-420 for Mozzarella, 0-250 N for Processed	Instron Universal Testing Machine Casiraghi *et al.*, 1985
Mozzarella	Effect of formulation and manufacture	Compressive modulus of elasticity: 0.13-0.48 kg/cm^2, dcreased with fat content and augmented with pH	Instron Machine Masi & Addeo, 1986
Camembert	Effect of ripening and compression degree	Deformation force: 1.5-27 N, elasticicty: 250-1100 kPa, bioyield: 1.8-8.1 N, depended of both variables	FTC Shear Press Schlesser *et al.*, 1992
Feta	Effect of fat content	Fracture force: 1.0-3.0 kg$_f$, hardness: 2-5 kg$_f$, increased compression: 15-19%, decreased with fat decreasing	Universal Machine Katsiari & Voutsinas, 1994a.
Feta	Effect of milk and ripening	Fracture force: 0.4-1.6 kg, compression: 16-25%, hardness: 1.1-2.5 kg, varied with milk relation	Univ Instron Machine Mallatou *et al.*, 1994
Emmental	Effect of ripening	Fracture stress: 313-516 kPa, strain: 46-69 %,, decreased with ripening, stress at 33%: 130-241 kPa, augmented	Universa l Machine Bachmann *et al.*, 1999
Cheddar	Effect of fat and ripening	Fracture stress: 210-840 kPa, strain: 31-48%, firmness: 270-705 N, decreased significantly with ripening time	Univ Instron Machine Fenelon & Guinee, 2000
Cheddar	Effect of fat and ripening	Fracture stress: 300-800 kPa, firmness: 280-600 N, decreased significantly with ripening time, and decreased with fat	Univ Instron Machine Guinee *et al.*, 2000
Danbo	Effect of fat	Fracture stress: 92-194 kPa, strain: 1.1-1.3, work: 55-127 kJ/m^3, and Young's modulus: 143-208 kPa, varied with fat	Univ Testing Machine Madsen & Ardo, 2001
UK Cheddar	Effect of ripening	Fracture stress: 30-90 kPa, strain: 0.27-0.80, energy: 7-61 kJ/m^3, and Young's modulus: 0.29-0.73 MPa, varied with age	Univ Testing Machine Hort & Le Grys, 2001
Goat	Treatment and ripening	Fracture stress: 25-191 kPa,and strain: 0.24-0.76, varied Raw and pressured cheeses were firmer and less fracturable	XT2 Texture Analyzer Buffa *et al.*, 2001
Mozzarella	Effect of pH, Ca and T	Firmness: 150-500 N, varied with the three parameters Firmness was not significantly affected by storage time	Univ Testing Machine Guinee *et al.*, 2002
Cheddar	Effect of formulation and maturation. Cylinders of 2 cm	Fracture stress: 7.4-104 Pa, decreased with maturation Fracture strain: 0.42-1.20, witht 20-80% of compression	Univ Testing Machine Everett & Olson, 2003

Table 5. (Continued)

Cheese type	Objectives and method	Observations	Instrument/Reference
Natural Cheddar	Effect of maturation Cylinders 25 mm diameter, 25 mm height	Force of fracture (F): 45-261 N, F = 197.06-14.45t Degree of compression at point of fracture (C): 42-98 %, C = 97.93 -5.28t (t: time in months)	XT Texture Analyzer Pollard *et al.*, 2003
Cheddar half fat	Effect of milk pasteurization temperature (PT) Cubes 25 mm, 30% of deformation	Fracture stress: 150-600 kPa, as the minimum slope Fracture strain; 0.25-0.60, as fraction displacement at fracture Firmness as force: 220-450 N, for 30% of compression The three parameters decreased as PT increased	HDi Texture Analyzer Rynne *et al.*, 2004
Mozzarella reduced-fat	Effect of high pressure Cubes 25 mm 30% of deformation	Firmness as force: 540-630 N, for 30% of compression Fracture stress: 750-990 kPa, without significant difference	HDi Texture Analyzer Sheehan *et al.*, 2005
White with low and full fat	Influence of rennet concentration Cubes 15 mm 57% of deformation	Fracture stress: 80-161 kPa, fracture strain: 0.19-0.32 Work: 8-32 kJ/m^3, Young´s modulus: 730-1173 kPa The four parameters increased with fat reduction	Univ Testing Machine Madadlou *et al.*, 2005
Cheddar	Acid development. Cylinders 17.5 mm D, 19.2 mm H 80% of deformation	A lower and a more distinct fracture: 105 N, for low pH and reverse osmosed milk cheeses compared with others (60-90 N)	XT2 Texture Analyzer Lee *et al.*, 2005
White	Effect of brine concentration Cylinders 24 mm D, 16 mm H 62.5% of deformation	Fracture stress: 7.98-33.0 kPa, increased with brine content Fracture strain: 0.36-0.40, augmented significantly, because the treatment favored a higher texture of the gel	HTE Testing Machine Madadlou *et al.*, 2007
Manchego	Maturation effect (36-336 d) Cylinders 20 cm D, 10 cm H, 3 kg of weight 40% of deformation	Maximum force: 300-576 N, for 72-200 d of maturation Dowstroke compression work: 100-660 N mm, for the same maturation period	TA.XT2i Testing Machine Conde *et al.*, 2007

5.2. Dynamic Tests

Dynamic experiments must be carried out in the linear viscoelasticity range to keep the cheese structure intact. These tests are conducted under small amplitude oscillatory shear conditions, placing the cheese sample between two parallel plates and subjected to sinusoidal oscillations. Three forms of dynamic determinations can be used to get cheese properties: frequency sweep test (G′ and G" are plotted as a function of frequency (ω) at the selected temperature; temperature sweep in contrary, G′ and G" are plotted as a function of temperature at a controlled frequency; and, time sweep in which G′ and G" are determined as a function of time at constant frequency and temperature. Dynamic testing offers very rapid results with minimal changes in the sample, and it is supported on equations 6 to 14 (Vélez-Ruiz and Barbosa-Cánovas, 1997).

$$\gamma = \gamma_0 \sin(\omega t) \quad (6)$$

$$\tau = \tau_0 \sin(\omega t - \delta) \quad (7)$$

$$\tau = G'\gamma + \eta'\gamma \quad (8)$$

$$G' = (\tau_0/\gamma_0)\cos\delta \quad (9)$$

$$\eta' = G''/\omega \quad (10)$$

$$G'' = (\tau_0/\gamma_0)\sin\delta \quad (11)$$

$$G^* = \tau_0/\gamma_0 = \left[(G')^2 + (G'')^2\right]^{1/2} \quad (12)$$

$$\eta^* = G^*/\omega = \left[(\eta')^2 + (\eta'')^2\right]^{1/2} \quad (13)$$

$$\eta'' = G'/\omega \quad (14)$$

where: γ is the strain as a function of time (t), γ_0 is the amplitude of the strain (degree); ω is the frequency of oscillation, τ is the shear stress, τ_0 is the amplitude of the shear stress; δ is the phase angle; $G'\gamma$ represents the elastic nature of the cheese, $\eta'\gamma$ represents the viscous nature of the cheese, G' is the storage (or elastic) modulus, η' is the dynamic viscosity, τ_0/γ_0 is the amplitude ratio, G" is the loss (or viscous) modulus, G* is the complex modulus, η^* is the complex viscosity; and η'' is the out-of-phase component of complex viscosity.

Some reported data and experimental conditions are summarized in table 6, where different aspects of cheesemaking have been studied. The viscoelastic properties of coagulating milk were pioneered by Bohlin *et al.* (1984), whereas the development of dynamic moduli of curd as a function of oscillation, temperature and time were measured by Bistany and Kokini (1983) concluding with a nonlinear relationship between steady shear rheological properties and small-amplitude dynamic viscoelastic properties of several foods, including a whipped cream cheese. Dejmek (1987) characterized the dynamic response of cheese curd. The influence of pressure homogenization and cooling rate on storage and loss moduli of double cream cheese was studied by Sánchez *et al.* (1994a, 1994b). The effect of

Table 6. Fundamental Analyses Applied on Cheese Affected by Different Factors

Cheese type	Specific test	Measurement characteristics and observations	Instrument/Reference
Coagulating milk	Dynamic in a coaxial cylinder geometry	G':(56-73 Pa, G'' (15-21 Pa) as a function of added CaCl, coagulation time and rennet concentration	Dynamic Instrument Bohlin *et al.*, 1984
Curd	Dynamic, Couette fixture	G' (15-145), G" (10-42 Pa) as a function of temperature	Dejmek, 1987
Mozzarella, formulation	Dynamic (D = 25 mm, 4 or 8 mm thick, 20°C)	$G' = 2.27 \times 10^5\ \omega^{0.17}$, $G'' = 1.03 \times 10^5\ \omega^{0.19}$ for natural, $G' = 1.59 \times 10^5\ \omega^{0.21}$, $G'' = 1.98 \times 10^5\ \omega^{0.16}$ for modified	Dynamic Analyzer Nolan *et al.*, 1989a
Cheddar and Processed	Dynamic (D = 25 mm, 22.5°C)	G' (100-200), G" (30-100 kPa) for Cheddar, G' (200-300), G" (90-110 kPa) for processed, at 0.69% strain	Dynamic Analyzer Nolan *et al.*, 1989b
Cheddar and Cheshire	Dynamic (D = 25 mm, 4 mm thick, 20°C)	G' ($1x10^4$-3 $x10^7$), G" (5 $x10^3$-2 $x10^6$ dyne/cm2) for Chesire, G' (0.2-2 x 10^5), G" (0.2-1.2 $x10^5$dyne/cm^2) for Cheddar	Rheometrics Dynamic Tunick *et al.*, 1990
Mozzarella, cooking	Dynamic (D = 12.5 mm, 22.5 mm thick, 22°C)	G' (3-10 x 10^5 dyne/cm2) and G" (1.5-3 dyne/cm^2), Cooking temperature (38-44° C) had little effect	Rheometrics Yun *et al.*, 1994
Cream, homogenization, cooling rate	Dynamic with a concentric cylinder (gap, 3.5 mm)	G' (3.1-31 kPa) increased and G" (0.56-8.9 kPa) decreased with homogenization pressure. Increase in curd cooling rate decreased G' and G''	Rheomat 120 Sanchez *et al.*, 1994a Sanchez *et al.*, 1994b
Mozzarella,	Dynamic	G' (4.5-7.7 Pa), G" (1.5-2.4 Pa) no affected by homogenization	Tunick *et al.*, 1995
Cheddar, fat and ripening	Dynamic (D = 60 mm, 1-2 mm thick, 25-75°C)	G' (0.10-30 kPa), G" (0.1-6.0 kPa) varied with the fat content, ripening time and temperature	Rheometrics Spectrom Ustunol *et al.*, 1995
Cream, processing	Dynamic in triplicate, 20°C	G' (2.5-25 kPa), G" (0.4-6.3 kPa) varied with the manufacture stage, exhibiting viscoelastic-plastic behavior	Rheomat 120, Mettler Sanchez *et al.*, 1996a

Table 6. (Continued).

Cheese type	Specific test	Measurement characteristics and observations	Instrument/Reference
Mozzarella, storage	Dynamic (D = 30 mm, 3.7 mm thick, 10 and 20°C)	G' (30-630 kPa), G" (14-53 kPa) varied with test temperature and storage time	Bohlin VOR Ak & Gunasekaran, 1996
Curd	Dynamic (10 mL)	G' (0-30 Pa), G" (0-8Pa) varied with rennin level	O ´Callaghan *et al.*, 1999
Gouda, milk treatment	Dynamic (D = 25 mm, 5 mm thick, 14°C)	Decreased of G' (80-400 Pa) and G" (40-120 Pa) depended of pressure. Pressure treatment was of limited value	Bohlin CVO Messens *et al.*, 2000
Camembert, ripening	Dynamic (D = 40 mm, 5mm thick, 23°C)	Decreased of G' (158-1585 kPa) and G" (90-316 kPa)depended on aging, softening attributed to proteolysis	Rheometer ARES Guizani *et al.*, 2002
Mozzarella, pH, Ca and T	Dynamic (D = 41 mm, 2.0 mm thick, 20-80°C) at 1 Hz	The magnitude of G' (0-190 Pa) decreased, and phase angle □ (14-60°) augmented as a function of temperature	Carri-Med CSL500 Guinee *et al.*, 2002
Cheddar, formulation and aging	Dynamic (D = 20.3 mm, 3.6 mm thick, 25-60-25°C) at 1 Hz	The magnitude of G' (200-20000 kPa) varied with T-Reduction in: fat increased, and moisture decreased the viscoelasticity, G' decreased significantly with aging	Venugopal & Muthukumarappan, 2003
Mozzarella and Monterey	Dynamic (D = 30. mm, 2 mm thick, 25°C)	The magnitude of G' for Mozzarella (10-60 kPa) and Monterey Jack (1.2-6.9 kPa) cheeses changed slightly with age	Bohlin VOR Brown *et al.*, 2003
Manchego, formulation and T	Dynamic (D = 50 mm, 2.0 mm thick, 25-70°C) at 1 Hz and 0.2 kPa	The magnitude of G' (0.01-70 kPa), and G'' (0.2-20 kPa) varied with T. Control exhibited higher G' but lower tan □ The emulsifiers affected contrarily the viscoelasticity	Physica DSR 400 . Physica. Lobato- Calleros *et al.*, 2003
Processed, maturity and cooling	Yield stress in control stress Dynamic (D = 35. mm, 1 mm thick, 20° C)	Range of linear viscoelastic region: 0-200 Pa, G' for slow (10-25 kPa) was higher than G' for rapid cooling (4-11 kPa) Rapid cooling of the molten mix decreased the complex modulus	Rheometer Reostress RS 80 (Haake) Piska & Stetina, 2004

Table 6. (Continued).

Cheese type	Specific test	Measurement characteristics and observations	Instrument/Reference
Processed, effect of moisture	Dynamic (D = 40 mm, 20°C), 1 week of storage	The frequency sweeps of samples showed the difference in viscoelasticity parameters, G' (10-600 Pa) decreased, whereas G'' (400-600 Pa) augmented as a function of moisture content	Carri-Med CSL100 (TA Instruments) Lee *et al.*, 2004
Soft goat, effect of freezing	Dynamic (D = 25.4 mm, 4-5 mm thick, 22°C) (measures by triplicate)	G′ was lower for frozen control (11-15 kPa) than the refrigerated control (16-19 kPa); G' for 3 and 6 months ranged: 14-18 kPa Freezing and long-term frozen storage had minimal impact	Rheometer RDA-700 (Rheometrics) Van Hekken *et al.*, 2005
White with levels of fat and rennet	Dynamic (D = 25 mm, 1 mm thick, 20°C) (six measurements)	G' increased with decreasing of fat concentration G': 400- 1100 for normal rennet, 800-1700 kPa for double rennet in full fat cheese, G' augmented with fat decreasing	Rheometer Physica UDS 200 (Physica) Madadlou *et al.*, 2005
Cheddar affected by acid	Dynamic (D = 25 mm, 2.2-2.4 mm thick, 3 T) (three samples)	G' at 5 and 40°C did not change significantly during ripening The high pH of non osmosed exhibited a large change in loss tangent (0.3-1), in other cheeses did not change (0.3-0.5)	Rheometer UDS 200 (Paar Physica) Lee *et al.*, 2005
Processed, formulation and T	Dynamic (D = 20 mm, 5 mm thick, 10-50-10°C)	The linear viscoelastic region decreased with increasing T G′ was greater for 80% reduced-fat compared to regular, Activation energy for regular: 44.1, fat reduced: 32.1 kJ/mol	Rheometer Bohlin Subramanian *et al.*, 2006
Goat, affected by ripening	Dynamic (D = 40 mm, 5 mm thick, 25°C) (two samples)	Solids and proteolysis showed and opposite effect on texture G' (50-450 kPa) and G" (15-75 kPa) depended on aging Structure were weaker at the beginning of ripening Rigidity augmented almost ten fold at the end of ripening	Rheometer ARES (Rheometrics Scientific) Guizani *et al.*, 2006

Table 6. (Continued).

Cheese type	Specific test	Measurement characteristics and observations	Instrument/Reference
Fresh with low and full fat content	Dynamic (D = 50 mm, 3 mm thick, 5 °C) (nine measurements)	G' (18-26 kPa) and G" (3.6-6.0 kPa) depended on strain Non-lineal (tan δ_B) and lineal (tan δ_A) regions were explored Reduced fat cheeses exhibited non-significant difference of the tan δ_B/ tan δ_A ratio of the full-fat white cheeses	Rheometer Rheoplus/32 V2.62 (Anton Paar) Lobato- Calleros *et al.*, 2006
Cheddar made of milk HP and pasteurized	Dynamic (D = 25 mm, 1 mm thick, 20° C) (six samples)	G' (80-324 kPa) and G" (32-142 kPa) depended on frequency At 2.75, 33 and 133 rad/s, G' of chees HP > pasteurized > raw milk Cheese made with pasteurized milk were more deformable Cheese from milk treated with 483 MPa-10°C behaved more solid-like that others	Rheometer UM/MC 120 (Paar Physica) San Martín-González *et al.*, 2007
White made from different brines	Dynamic (D=25 mm, 5 mm thick, 23° C) (six measurements)	G' increased with brine concentration; G': 12- 21 kPa for 9% brine, G': 24-36 kPa for 13% brine, and G': 60-114 kPa for 17% brine	Rheometer Physica UDS 200 (Physica) Madadlou *et al.*, 2007

T: temperature, ω: angular velocity.

milk homogenization was not significant on rheology of Mozzarella cheese, in contrary to fat content (Tunick *et al.*, 1995). Homogenization and cooling were detected as important on the rheology and structure of double cream cheese by Sanchez *et al.* (1996a). No significant effect of cooking temperature on the rheology of Mozzarella cheese was observed by Yun *et al.* (1994). The evolution of milk coagulation was compared, between the dynamic approach against online techniques, by O´Callaghan *et al.* (1999). Chemical composition and its effect was studied by Nolan *et al.* (1989a) in Mozzarella cheese, Dimitreli and Thomareis (2007) in Processed cheese, to describe the dynamic response. Guinee *et al.* (2002) determined the effect of pH and calcium concentration on Mozzarella cheese, in which the elastic modulus decreased as a function of ripening time.

Subramanian *et al.* (2006), investigated the effect of heating and cooling on rheological response of processed cheese manufactured with different formulations, fat content changed more the cheese rheology than moisture content. Piska and Stetina (2004) studied the influence the degree of maturity of raw materials and cooling velocity on processed cheese, finding that the rapid of cooling on texture and rheology of processed cheese was more significant than the influence of maturity. Lee et al (2004) examined the influence of moisture on processed cheese spreads, with increasing moisture content, the rheological behavior of cheese samples changed, from solid-like (G' > G'') to liquid liked (G'' > G'). The effect of emulsifying salts on pasta filata cheese was examined by Mizuno and Lucey (2005), finding that cheese made with tetrasodium pyrophosphate exhibited a G' higher than the control. The effect of acid development on dynamic parameters of Cheddar cheese was researched by Lee *et al.* (2005). The influence of fat and moisture content, as well as heating and cooling on Cheddar cheese were analyzed by Venugopal and Muthukumarappan (2003), finding opposite effects of these components on the viscosity.

Lobato-Calleros *et al.*, (2003, 2006) completed two similar researches for comparison.In the first case, they studied the effect of two emulsifiers on the rheology of Mexican Manchego cheese; and in the other one, fresh white cheese made with full and reduced fat contents.They concluded that the presence of emulsifiers decreased the storage modulus of Manchego type; whereas carboxymethylcellulose cheese showed a similar rheology and texture than full-fat cheese. A study on the influence of fat and rennet concentrations, and brine concentration on white cheese properties were carried out by Madadlou *et al.* (2005, 2007). The effect of fat reduction and ripening on Cheddar resulted in an increase in viscoelasticity due to fat decreasing in contrary to ripening time (Ustunol *et al.*, 1995).

A comparative study of dynamical characteristics of Cheddar and processed cheeses, in which a superior meltability was reported for the second type, was completed by Nolan *et al.* (1989b). A similar comparison of Cheddar and Cheshire cheeses was completed by Tunick *et al.* (1990), finding clear differences between both British dairy products.

New technologies have been incorporated to cheese manufacture. A work of Sheehan *et al.* (2005) about the effect of high pressure (HP) treatment on low-fat Mozzarella cheese, did not show significance of the treatment; San Martín-González *et al.* (2007) compared Cheddar cheese made by pasteurized and pressurized milks, in which high pressure treatment of milk causes casein disruption and also, lactoglobulins denaturation; dynamic and creep tests confirmed a higher association of protein and moisture in cheese elaborated with milk HP.

The effect of ripening through 30 days in Camembert cheese was investigated by Guizani *et al.*, (2002) in which cheeses suffered a softening due to proteolysis; and in goat cheese by Guizani *et al.* (2006), in which a transformation of soft and elastic nature to a hard and brittle consistency was detected. A comparative analysis between Mozzarella and Monterey Jack was carried out by Brown *et al.* (2003) including the effect of age (4-38 d) on those young cheeses, finding that storage modulus and fracture decreased during aging. Everett and Olson (2003) compared the rheology of Cheddar cheese at 35 and 70 days of maturation, finding a reduction in fracture parameters; similar analysis were carried out by Venugopal and Muthukumarappan (2003) through 168 days of storage on the same cheese. Conde *et al.* (2007) followed the effect of maturation on Manchego cheese during 336 days and compared two measuring techniques. Textural changes developed during maturation of natural Cheddar cheese through 11 months were followed by Pollard *et al.* (2003). They concluded that fracture properties confirmed a reduction in firmness and increasing in crumbliness during cheese maturation.

Other studies have analyzed the influence of post cheesemaking treatments, the effect of frozen storage (1 month) on viscoelasticity of Mozzarella cheese was studied by Ak and Gunasekaran (1996) , in which G' and G'' were lower at 20 than at 10°C, and both moduli decreased with storage time. Van Hekken *et al.* (2005) researched the effect of refrigerated storage on soft caprine cheese and they concluded that rheological techniques were capable of determine very low values of oscillatory strain properties exhibited by caprine cheese and also, the freezing had a minimal impact on cheese rheology.

5.3. Creep and Relaxation Tests

In transient measures the stress or strain is followed during the time of testing. For creep test, a constant stress is applied to cheese samples, and the strain is measured as a function of time. In many cases, creep experiments are coupled with a recovery determination, in which the load is removed and the recuperation of the sample is recorded. It may be performed in bending, compression, shear, and tension modes. In the case of a stress relaxation test, the stress is measured after a sudden step strain. It also, may be performed in different configuration modes. Transient tests in cheese are conducted at higher strains than in dynamic determinations (Vélez-Ruiz, 2008).

If cheese item behaves as a linear viscoelastic solid during the creep test, the creep strain increase with time and may be related to the stress by equation 15 (Tunick and Nolan, 1992).

$$\varepsilon_{xy} = J(t)\sigma_{xy} \qquad (15)$$

where: ε_{xy} is the creep strain, J(t) is the creep compliance function, and σ_{xy} is the creep stress, being x the perpendicular direction to the plane of the stress, whereas y, refers to the direction of the stress component.

The creep compliance function may be rearranged into three parts, corresponding to particular phases of creep: instantaneous deformation, retarded elastic deformation, and flow response (equation 16; Tunick and Nolan, 1992).

$$J(t) = J_0 + J\phi(t) + \frac{t}{\eta_3} \qquad (16)$$

where: J_0 is the compliance at time zero, J is the decay compliance, $\phi(t)$ is a retardation function, with an initial time (t) equal to 0 ($\phi(0) = 0$), and η_3 is the viscosity in steady flow.

Studies about transient measurements on cheese are scarce, relaxation determinations appear more frequently than creep analysis. Related with relaxation test, there are some reports: in Italian cheeses (Mozzarella, Silano, Provolone and Caciocavallo) by Masi and Addeo (1986); in Cheddar curd by Nolan (1987); in Gouda cheese by Goh and Sherman (1987); in Galvanino, an unripened pasta filata cheese, by Masi (1989); in acid fresh cheese by Korolczuk and Mahaut (1990); in Argentino by Bertola *et al.* (1991); in Gouda by Messens *et al.* (2000); in Cheddar by Venugopal and Muthukumarappan (2001), and by Hort and Le Grys (2001); in goat cheese by Buffa *et al.* (2001). While for creep analysis have been reported less works: in Cheddar cheese by Purkayastha *et al.* (1984, 1985), in curd by Halim and Shoemaker (1990), in processed by Drake *et al.* (1999), in Gouda by Messens *et al.* (2000), in Mozzarella and Monterey Jack by Brown *et al.* (2003), in Cheddar by San Martín-Gonzalez *et al.* (2007). Both tests are not usually used in rheological characterizations of cheese due to the complexity of the mathematical modeling and lack of utility of the model parameters.

6. Final Remarks

At this time, there are numerous valuable contributions furthering the understanding of cheese physical properties; specifically, rheological and textural properties. A great number of references and some chapters have been consulted, covering 25 years of investigations; they may be analyzed in order to get support to specific needs and facilities.

From the historical and practical point of view, all type of rheological and textural tests has been completed since the 60's, transient determinations were majorly applied during three decades (1960-1990). Unfortunately the utility of the results has been limited due to the complexity of the mechanical model and the dependency of the creep and relaxation parameters to compression degree and surface friction, among others factors. Surprisingly, the empirical and imitative tests have resulted very useful, even in a limited manner, and still they are used with relative frequency; certainly the TPA has lost credibility. The compressive and dynamic characterization have been very well , as may be observed by the great quantity of researches reporting these tests, and they will continue as the most utilized and fundamental tests, although the applied strain for dynamic test is too low.

The complexity of cheese items makes particularly important to carry out the experiments with too much care to be able and sure of assessing the data reliability, as well as the conclusions and limitations obtained from them.

The food industry is demanding by rapid on-line methods to predict the rheology and texture of foods, and particularly of cheeses, to get a better control of the transformation process, and to produce a final item with optimal and uniform quality. Therefore the rheological or textural characterization for cheeses could be subjected to two trends: in one of them, a great effort will be done to homogenize those instrumental tests that allow the knowledge of rheological nature and texture attributes, that's to say, the instrumental tests for cheeses should be standardized. On the other hand, the so called new instrumental techniques, such as NIR (Near Infrared), NMR (Nuclear Magnetic Resonance), and acoustic devices, will be incorporated to the cheese industry.

REFERENCES

Ak M. M. and Gunasekaran S. 1995. Measuring elongational properties of Mozzarella cheese. *Journal of Texture Studies,* 26: 147-160.

Ak M. M. and Gunasekaran S. 1996. Dynamic rheological properties of Mozzarella cheese during refrigerated storage. *Journal of Food Science,* 61 (3): 566-568, 584.

Ak M. M., Bogenrief D., Gunasekaran S. and Olson N.F. 1993. Rheological evaluation of Mozzarella cheese by uniaxial horizontal extension. *Journal of Texture Studies,* 24: 437-453.

Awad R.A., Abdel-Hamid L.B. and El-Shabrawy S.A. 2002. Texture and microstructure of block type processed cheese with formulated emulsifying salt mixtures. *Lebensmittel-Wissenschaft und Technologie,* 35: 54-61.

Bachmann H.P., Bütikofer U. and Meyer J. 1999. Prediction of flavour and texture development in Swiss-type cheeses. *Lebensmittel-Wissenschaft und Technologie*, 32: 284-289.

Bertola N.C., Bevilacqua A.E. and Zaritzky N.E. 1991. Changes in rheological and viscoelastic properties and protein breakdown during the ripening of 'Port Salut Argentino' cheese *International Journal of Food Science and Technology*, 26: 467-478.

Bistany K.L. and Kokini J.L. 1983. Dynamic viscoelastic properties of foods in texture control. *Journal of Rheology,* 27 (6): 605-620.

Bohlin L., Hegg P.O. and Ljusberg-Wahren H. 1984. Viscoelastic properties of coagulating milk. *Journal Dairy Science,* 67: 729-734.

Bourne M.C. 2002. Food Texture and Viscosity, Concept and Measurement. Academic Press. New York, USA.

Brown J.A., Foegeding E.A., Daubert C.R., Drake M.A. and Gumpertz M. 2003. Relationships among rheological and sensorial properties of young cheeses. *Journal Dairy Science,* 86: 3054-3067.

Bryant A., Ustunol Z. and Steffe J. 1995. Texture of Cheddar cheese as influenced by fat reduction. *Journal of Food Science,* 60 (6): 1216-1219, 1236.

Buffa M.N., Trujillo A.J., Pavia M. and Guamis B. 2001. Changes in textural, microstructural, and colour characteristics during ripening of cheese made from raw, pasteriuzed or high-pressure treated goats' milk. *International Dairy Journal,* 11: 927-934.

Campanella O.H., Poppewell L.M., Rosenau J.R. and Peleg M. 1987. Elongational viscosity measurements of melting American process cheese. *Journal of Food Science,* 52 (5): 1229-1251.

Casiraghi E.M., Bagley E.B. and Christianson D.D. 1985. Behavior of Mozzarella, Cheddar and processed cheese spread in lubricated and bonded uniaxial compression. *Journal of Texture Studies,* 16: 281-301.

Conde T., Cárcel J.A. García-Pérez J.V. and Benedito J. 2007. Non-destructive analysis of Manchego cheese texture using impact force-deformation and acoustic impulse-response techniques. Jour*nal of Food Engineering,* 82: 238-245.

Dejmek P. 1987. Dynamic rheology of curd. *Journal Dairy Science,* 70: 1325-1330.

Dimitreli G. and Thomareis A.S. 2004. Effect of temperature and chemical composition on processed cheese apparent viscosity. *Journal of Food Engineering,* 79: 265-271.

Dimitreli G. and Thomareis A.S. 2007. Texture evaluation of block-type processed cheese as a function of chemical composition and in relation to its apparent viscosity. *Journal of Food Engineering,* 79: 1364-1373.

Drake M.A., Truong V.D. and Daubert C.R. 1999. Rheological and sensory properties of reduced-fat processed cheeses containing lecithin. *Journal of Food Science,* 64 (4): 744-747.

Eck A. 2000. What is a cheese ?. In *Cheesemaking. From Science to Quality Assurance*. (A. Eck and J.C. Gilis; Eds). Lavoisier Publising. 661-662.

Eck A. and Gilis J.C. 2000. Cheesemaking. In *Science to Quality Assurance*. Lavoisier Publising, Andover, Hampshire, UK.

Everett D.W. and Olson N.F. 2003. Free oil and rheology of Cheddar cheese containing fat globules with different proteins. *Journal Dairy Science,* 86: 755-763.

Fagan C.C., Everard C., O'Donell C.P., Downey G., Sheehan E.M., Delahunty C.M., O'Callahan D.J. and Howard V. 2007. Prediction of processed cheese instrumental texture and meltability by mid-infrared spectroscopy coupled with chemometric tools. *Journal of Food Engineering,* 80: 1068-1077.

Fenelon M.A. and Guinee T.P. 2000. Primary proteolysis and textural changes during ripening in Cheddar cheeses manufactured to different fat contents. *International Dairy Journal,* 10: 151-158.

Fife R.L., McMahon D.J. and Oberg C.J. 2002. Test for measuring the stretchability of melted cheese. *Journal Dairy Science,* 85: 3539-3545.

Fox P.F. 2003. Cheese. In *Encyclopedia of Dairy Sciences* (H. Roginski, J.W. Fuquay and P.F. Fox; Eds). Academic Press. 252-261.

Goh H.C. and Sherman P. 1987. Influence of surface friction on the stress relaxation of Gouda cheese. *Journal of Texture Studies,* 18: 389-404.

Guinee T.P. 2003. Cheese rheology. In Encyclopedia of Dairy Sciences (H. Roginski, J.W. Fuquay and P.F. Fox, eds). Academic Press. New York, USA. pp.341-349.

Guinee T.P. and O´Callagahn D.J. 1997. The use of a simple empirical method for objective quantification of the stretchability of cheese on cooked pizza pies. *Journal of Food Engineering,* 31: 147-161.

Guinee T.P., Mark A.E., Auty M.A.E. and Fenelon M.A. 2000. The effect of fat content on rheology, microstructure and heat-induced functional characteristics of Cheddar cheese. *International Dairy Journal,* 10: 277-288.

Guinee T.P., Feeney E.P., Auty M.A.E. and Fox P.F. 2002. Effect of pH and calcium concentration on some textural and functional properties of Mozzarella cheese. *Journal Dairy Science,* 85: 1655-1669.

Guizani N., Kasapis S., Al-Attabi Z. and Al-Ruzceiki M.H. 2002. Microbiological, physicochemical, and biochemical changes during ripening of Camembert cheese made of pasteurized cow´s milk. *International Journal of Food Properties,* 5 (3): 483-494.

Guizani N., Al-Attabi Z., Kasapis S. and Gaafar O.M. 2006. Ripening profile of semi-hard standard goat cheese made from pasteurized milk. *International Journal of Food Properties,* 9: 523-532.

Gunasekaran S. and Ak M.M. 2003. Cheese Rheology and Texture. CRC Press. New York, USA.

Halim H.K. and Shoemaker C.F. 1990. Effect of addition of α–, β-, and κ-casein, and Na-caseinate on viscoelastic properties of skim milk curd. *Journal of Texture Studies,* 21: 323-337.

Hort J. and Le Grys G. 2001. Developments in the textural and rheological properties of UK Cheddar cheese during ripening. *International Dairy Journal,* 11: 475-481.

Instron Corporation. 1980. A new system for objective food textural measurements. Commercial brochure. Canton, MA. USA.

Instron Corporation. 1981. Materials test processors. Commercial brochure. Canton, MA. USA.

Jack F.R., Pateson A. and Piggott J.R. 1993. Relationships between rheology and composition of Cheddar cheeses and texture as perceived by consumers. *International Journal of Food Science and Technology,* 28: 293-302.

Katsiari M.C. and Voutsinas L.P. 1994a. Manufacture of low-fat Feta cheese. *Food Chemistry,* 49 (1): 53-60

Katsiari M.C. and Voutsinas L.P. 1994b. Manufacture of low-fat Kefalograviera cheese. *International Dairy Journal,* 4 (6): 533-553.

Kaya S. 2002. Effect of salt on hardness and whiteness of Gaziantep cheese during short-term brining. *Journal of Food Engineering,* 52: 155-159.

Kilcast D. 2004a. Measuring consumer perceptions of texture: an overview. In Texture in Foods. V2 *Solid Foods* (D. Kilcast ed.). CRC Press. New York, USA.

Kilcast D. 2004b. Texture in Foods. Vol. 2 Solid Food. CRC Press. New York, USA.

Konstance R.P. and Holsinger V.H. 1992. Development of rheological test methods for cheese. *Food Technology,* Jan.: 105-109.

Korolczuk J. and Mahaut M. 1990. Relaxation studies of acid type cheese texture by a constant speed cone penetrometric method. *Journal of Texture Studies,* 21: 107-122.

Lang J.H. 1989. Larousse Gastronomique. Crow Pub. Inc. New York, USA. pp. 231.

Leach M.R., Farkas B.E. and Daubert C.R. 2003. Rheological characterization of process cheese using tube viscometry. *International Journal of Food Properties,* 6 (2): 259-267.

Lee S.K., Buwalda R.J., Euston S.R., Foegeding E.A. and McKenna A.B. 2003. Changes in rheology and microstructure of processed cheese during cooking. *Lebensmittel-Wissenschaft und Technologie*, 36: 339-345

Lee S.K., Anema S. and Klostermeyer H. 2004. The influence of moisture content on the rheological properties of processed cheese spreads. *International Journal of Food Science and Technology,* 39: 763-771.

Lee W., Clark S. and Swanson B. G. 2006. Low fat process cheese food containing ultrahigh pressure-treated whey protein. *Journal of Food Processing and Preservation,* 30: 164-179.

Lee M.R., Johnson M.E. and Lucey J.A. 2005. Impact of modifications in acid development on the insoluble calcium content and rheological properties of Cheddar cheese. *Journal Dairy Science,* 88: 3798-3809.

Lobato-Calleros C., Ramirez-Santiaho J., Osorio-Santiago V.J., Vernon-Carter E.J. and Hornelas-Uribe Y. 2002. Microstructure and texture of Manchego cheese-like products made with canola oil, lipophilic and hydrophilic emulsifiers. *Journal of Texture Studies,* 33: 165-182.

Lobato-Calleros C., Velásquez-Varela J., Sanchez-García J. and Vernon-Carter E.J. 2003. Dynamic rheology of Mexican Manchego cheese-like products containing canola oil and emulsifier blends. *Food Research International,* 36: 81-90.

Lobato-Calleros C., Rodríguez E, Sandoval-Castilla O., Vernon-Carter E.J. and Alvarez-Ramirez J. 2006. Reduced-fat white fresh cheese-like products obtained from $W_1/O/W_2$ multiple emulsions: Viscoleastic and high-resolution image analyses. *Food Research International,* 39: 678-685.

Lucey J.A., Johnson M.E. and Horne D.S. 2003. Perspectives on the basis of the rheology and texture properties of cheese. *Journal Dairy Science,* 86: 2725-2743.

Luginbühl W. 1996. The effect of stress correction on fracture point coordinates in uniaxial compression tests of cheese. *Lebensmittel-Wissenschaft und Technologie*, 29: 433-437.

Madadlou A., Khosrowshahi A. and Mousavi M.E. 2005. Rheology, microstructure, and functionality of low-fat Iranian white cheese: Made with different concentration of rennet. *Journal of Food Engineering,* 81: 330-335.

Madadlou A., Khosrowshahi A., Ebrahimzadeh M. and Farmani J. 2007. The influence of brine concentration on chemical composition and texture of Iranian white cheese. *Journal of Food Engineering,* 81: 330-335.

Madsen J.S. and Ardo Y. 2001. Exploratory study of proteolysis, rheology and sensory properties of Danbo cheese with different fat contents. *International Dairy Journal,* 11: 423-431.

Mallataou H., Pappas C.P. and Voutsinas L.P. 1994. Manufacture of Feta cheese from sheep´s milk, goat´s milk or mixtures of these milks. *International Dairy Journal,* 4: 641-664.

Masi P. 1989. Characterization of history-dependent stress relaxation behaviour of cheeses. *Journal of Texture Studies,* 19: 373-388.

Masi P. and Addeo F. 1986. An examination of some mechanical properties of a group of Italian cheeses and their relation to structure and conditions of manufacture. *Journal of Food Engineering,* 5: 217-229.

Massaguer-Roig S., Rizvi S.S.H. and Kosikowski F.V. 1984. Characterization of thixotropic behavior of soft cheeses. *Journal of Food Science,* 49: 668-670, 684.

McMahon D.J., Paulson B. and Oberg C.J. 2005. Influence of calcium, pH, and moisture on protein matrix structure and functionality in direct-acidified nonfat Mozzaella cheese. *Journal Dairy Science,* 86: 3754-3763.

Messens W., Van de Walle D., Arevalo J., Dewettinck K. and Huyghebaert A. 2000. Rheological properties of high-pressure-treated Gouda cheese. *International Dairy Journal,* 10: 359-367.

Mizuno R. and Lucey J.A. 2005. Effects of two types of emulsifying salts on the functionality of nonfat pasta filata cheese. *Journal Dairy Science*, 86: 3411-3425.

Moiny V., Meullenet J.F. and Xiong R. 2002. Uniaxial compression of Cheddar cheese at various loading rates and its correlation to sensory texture profiles. *Journal of Texture Studies,* 33: 237-254.

NDC. National Dairy Council. 2005. Available at: www.nationaldairycouncil.org (May 26, 2006)

Nolan E.J. 1987. Stress relaxation of stored stirred Cheddar curd. *Journal of Texture Studies*, 18: 273-280.

Nolan E.J., Holsinger V.H. and Shieh J.J. 1989a. Dynamic rheological properties of natural and imitation Mozzarella cheese. *Journal of Texture Studies,* 20: 179-189.

Nolan E.J., Shieh J.J. and Holsinger V.H. 1989b. A comparison of some properties of Cheddar and pasteurized process American cheese. *Proceeding of the Fifth International Congress of Engineering and Food,* Cologne, Germany. Pp. 370-379.

O´Callaghan D.J., O´Donell C.P. and Payne F.A. 1999. A comparison of on-line techniques for determination of curd setting time using cheesemilks under different rates of coagulation. *Journal of Food Engineering,* 41: 43-54.

O'Mahony J.A., Lucey J.A. and McSweeney P.L.H. 2005. Chymosin-mediate proteolysis, calcium solubilization, and texture development during the ripening of Cheddar cheese. *Journal Dairy Science*, 88: 3101-3114.

Omar Z.B., Raphaelides S. and Kesteloot R. 1995. Texture evaluation of French acid-type fresh cheese. *Journal of Texture Studies,* 26: 325-338.

Piska I. and Stetina J. 2004. Influence of cheese ripening and rate of cooling of the processed cheese mixture on rheological properties of processed cheese. *Journal of Food Engineering,* 61: 551-555.

Pollard A., Sherkat F., Seuret M.G. and Halmos A.L. 2003. Textural changes of natural Cheddar cheese during the maturation process. *Journal of Food Science*, 68 (6): 2011-2016.

Purkayastha S., Peleg M. and Normand M.D. 1984. Presentation of the creep curves of solid biological materials by a simplified mathematical version of the generalized Kelvin-Voigt model. *Rheologica Acta,* 23: 556-563

Purkayastha S., Peleg M., Jonson E.A. and Normand M.D. 1985. A computer aided characterization of the compressive creep behavior of potato and Cheddar cheese. *Journal of Food Science,* 50 (1): 45-50.

Robles-Martínez J.C., Lobato-Calleros C., Aguirre-Mandujano E., Santos-Moreno A. and Vernon-Carter E.J. 1997. Textural characteristics of low fat Manchego cheese: effect of fat substitutes. *Revista Mexicana de Ingeniería Química,* 7 (1): 49-54. (In Spanish).

Rosenthal I. 1991. Milk and Dairy Products. Properties and Processing.VCH Pub. Inc. New York, USA.

Rosenthal A.J. 1999. Relation between instrumental and sensory measures of food texture. In *Food Texture. Measurement and Perception* (A.J. Rosenthal, Ed.) Aspen Pub. Inc. New York, USA.

Rynne N.M., Beresford T.P., Kelly A.L. and Guinee T.P. 2004. Effect of milk pasteurization temperature and in situ whey protein denaturation on the composition, texture, and heat-induced functionality of half-fat Cheddar cheese. *International Dairy Journal*, 14: 989-1001.

Sanchez C., Beauregard J.L., Chassagne M.H., Bimbenet J.J. and Hardy J. 1994a. Rheological and textural behaviour of double cream cheese. Part I: Effect of curd homogenization. *Journal of Food Engineering,* 23: 579-594.

Sanchez C., Beauregard J.L., Chassagne M.H., Duquenoy A. and Hardy J. 1994b. Rheological and textural behaviour of double cream cheese. Part II: Effect of curd cooling rate. *Journal of Food Engineering,* 23: 595-608.

Sanchez C., Beauregard J.L., Chassagne M.H., Bimbenet J.J. and Hardy J. 1996a. Effects of processing on rheology and structure of double cream cheese. *Food Research International,* 28 (6): 547-552.

Sanchez C., Beauregard J.L., Bimbenet J.J. and Hardy J. 1996b. Flow properties, firmness and stability of double cream cheese containing whey protein concentrate. *Journal of Food Science,* 61 (3): 840-843, 846.

San Martín-González M.F., Rodríguez J.J., Gurram S., Clark S. Swanson B., and Barbosa-Cánovas G.V. 2007. Yield, composition and rheological characteristics of Cheddar cheese made with high pressure processed milk. *Lebensmittel-Wissenschaft und Technologie,* 40: 697-705

Scott R. 1986. Cheesemaking Practice. Elsevier Applied Science Pub., Publ. London, UK

Segards R.A., and Kapsalis J.G. 1987. Texture, rheology, psychophysics. In *Objective Methods in Food Quality Assesment* (J.G. Kapsalis, Ed.). CRC Press. New York, USA. pp. 155-184.

Schlesser J.E., Schmidt S.J. and Speckam R. 1992. Characterization of chemical and physical changes in Camembert cheese during ripening. *Journal Dairy Science,* 75: 1753-1760.

Sheehan J.J., Huppertz T., Hayes M.G., Kelly A.L., Beresford T.P. and Guinee T.P. 2005. High pressure treatment of reduced-fat Mozzarella cheese: Effects on functional and rheological properties. *Innovative Food Science and Emerging Technologies* 6: 73-81

Subramanian R., Muthukumarappan K. and Gunasekaran S. 2006. Linear viscoelastic properties of regular- and reduced-fat pasteurized process cheese during heating and cooling. *International Journal of Food Properties*, 9: 377-393.

TA Instruments. 2008. Commercial brochure. New Castle, DE. USA

Texture Technologies Corporation. 1990. Commercial brochure. New York, NY. USA.

Topcu A. and Saldamli I. 2006. Proteolytical, chemical, textural and sensorial changes during the ripening of Turkish white cheese made of pasteurized cow´s milk. *International Journal of Food Properties,* 9 (4): 665-678.

Trépanier G., Simmard R.E. and Lee B.H. 1991. Effect of added lactobacilli on composition and texture of Cheddar cheese during accelerated maturation. *Journal of Food Science,* 56 (3): 696-700.

Tunick M.H. 2000. Symposium: Dairy Products Rheology. Rheology of dairy foods that gel, stretch, and fracture. *Journal Dairy Science,* 86: 1892-1898.

Tunick M.H. and Nolan E.J. 1992. Rheology of Cheese. In *Physical Chemistry of Food Processes.* Vol. 1. Fundamental Aspects (I.C. Baianu, Ed.) Van Nostrand Reinhold Pub., New York, USA.

Tunick M.H. and Van Hekken D.L. 2002. Torsion gelometry of cheese. *Journal Dairy Science,* 85: 2743-2749.

Tunick M.H., Nolan E.J., Shieh J.J., Basch J.J., Thompson M.P., Maleef B.E. and Holsinger V.H. 1990. Cheddar and Chesire cheese rheology. *Journal Dairy Science,* 73: 1671-1675

Tunick M.H., Malin E.L., Smith P.W., Shieh J.J., Sullivan B.C., Mackey K.L. and Holsinger V.H. 1993. Proteolysis and rheology of low fat and full fat Mozzarella cheeses prepared from homogenized milk. *Journal Dairy Science,* 76: 3621-3628.

Tunick M.H., Malin E.L., Smith P.W. and Holsinger V.H. 1995. Effect of skim milk homogenization on proteolysis and rheology of Mozzarella cheese. *International Dairy Journal,* 5: 483-491.

Ustunol Z., Kawachi K. and Steffe J. 1994. Arnott test correlates with dynamic rheological properties for determining Cheddar cheese meltability. *Journal of Food Science,* 59 (5): 970-971.

Ustunol Z., Kawachi K. and Steffe J. 1995. Rheological properties of Cheddar cheese as influenced by fat reduction and ripening time. *Journal of Food Science,* 60 (6): 1208-1210.

Van Hekken D.L., Tunick M.H. and Park Y.W. 2005 Effect of frozen storage on the proteolytic and rheological properties of soft caprine milk cheese. *Journal Dairy Science,* 88: 1966-1972.

Velez-Ruiz J.F. 2002. Notes of the Dairy Technology Course. Universidad de las Americas, Puebla. Mexico. Unpublished Notes.

Velez-Ruiz J.F. 2008. Notes of the Physical Properties of Foods Course. Universidad de las Americas, Puebla. Mexico. Unpublished Notes.

Vélez-Ruiz, J.F. and Barbosa-Cánovas, G.V. 1997. Rheological properties of selected dairy products. *Critical Reviews in Food Science and Nutrition,* 37 (4): 311-359.

Venugopal V. and Muthukumarappan K. 2001. Stress relaxation characteristics of Cheddar cheese. *International Journal of Food Properties*, 4 (3): 469-484.

Venugopal V. and Muthukumarappan K. 2003. Rheological properties of Cheddar cheese during heating and cooling. *International Journal of Food Properties,* 6 (1): 99-114.

Walstra P., Wouters J.T.M. and Geurts T.J. 2006. Dairy Science and Technology. CRC Press. Nueva York, USA

Yun J.J., Hsieh Y.L. Barbano D.M. and Rohn C. 1994. Rheological and chemical properties of Mozzarella cheese. *Journal of Texture Studies,* 25: 411-420.

Zisu B. and Shah N.P. 2005a. Textural and functional changes in low-fat Mozzarella cheeses in relation to proteolysis and microstructure as influenced by the use of fat replacers, preacidification and EPS starter. *International Dairy Journal,* 15: 957-972.

Zisu B. and Shah N.P. 2005b. Low-fat Mozzarella as influenced by microbial exopoly saccharides, preacidification, and whey protein concentrate. *Journal of Dairy Science*, 88: 1973-1985.

PART II: TOPICS RELATED TO HEAT TRANSFER IN FOOD PROCESSING

In: Food Processing and Engineering Topics
Editors: M. E. Sosa-Morales and J. F. Vélez-Ruiz
ISBN: 978-1-60741-788-0

Chapter 5

EVALUATION OF CONVECTIVE HEAT TRANSFER COEFFICIENT FOR FOOD PROCESS OPERATIONS

Jorge F. Vélez-Ruiz
Chemical and Food Engineering Department
Universidad de las Americas, Puebla, Mexico

1. INTRODUCTION

Food process operations frequently involve heating or cooling stages between fluids and solid phases, being the heat transfer by conduction and convection the dominant mechanisms, that in many of the processes both happen simultaneously.

The increasing interest of the Food Industry for manufacturing high quality, nutritious and safe food products, as well as modeling of thermal processes and design of food equipment, have resulted in a need for a precise evaluation of the convective heat transfer coefficient. It is necessary to understand the contribution of this convective heat transfer coefficient to the overall heat transfer processes involving liquid foods and solid surfaces, as well as solid foods within Newtonian and non-Newtonian fluids, for an accurate evaluation and supply of the heat transfer process and equipment.

Over the last thirty years, research efforts around the world have been conducted to study, understand, characterize, evaluate and model a diversity of heating/cooling treatments in which the convective heat transfer coefficient for fluids (h) is the key parameter. They include heating and cooling of liquid foods in general, heat exchange for non-Newtonian fluids and regular items or irregular particles, heating/cooling of food particles in Newtonian fluids, blanching of vegetables, frying processes of foods, pasteurization of liquid foods, canning of a diversity of food items, aseptic processing in continuous flow, and thermal treatments during end-over-end agitation of cans and packed containers; among others. Related to thermal processes, several of the studies on h have been related to pasteurization and sterilization treatments of a diversity of food products.

To quantify the convective or surface heat transfer coefficient there are different approaches, in which the temperature of the food sample, liquid or solid, as a function of time is indispensable for most of them. Some methods are very sophisticated, others are of an intermediate degree of difficulty, and some approaches are practical and simple. Whatever is the approach, all the applied methods require accurate thermophysical properties of the fluid and solid foods in order to obtain a satisfactory value of the convective heat transfer coefficient. They will be commented and included in this chapter, together with reasonable data of reported and experimental h values, their measurements and techniques employed for its characterization will be mentioned.

The correct evaluation and precise utilization of h in food process operations will result in shorter processing times, higher production rates, better quality of the products, reduced power needs, environment conservation, improved process control, and lower production costs.

2. Heat Transfer in Food Processing

Heat transfer is the principal transport phenomenon or unit operation in many food processes. Heat transfer may flow by three different mechanisms: conduction, convection, and radiation.

2.1. Heat Conduction

It is the transfer of heat between adjacent molecules without appreciable displacement of the body particles and characteristic of solid materials including solid foods. Heat conduction in foods depends on basic properties of the solid item and rather significantly on the particular geometry. The final expression utilized to describe food conduction depends on whether Cartesian, cylindrical, or spherical coordinates are involved in characterizing the solid system (Welty *et al.*, 1976; Vélez-Ruiz, 2008). This phenomenon is mathematically described by the first Fourier´s law for steady state situations.

$$Q = \rho C_P \frac{dT}{dt} = kA\,\nabla T \tag{1}$$

that it is simplified for one direction:

$$Q = kA \frac{dT}{dL} \tag{2}$$

where: Q is the flow of heat transfer (W), ρ is the density (kg/m^3), Cp is the specific heat (kJ/kg °C), dT/dt is the temperature difference or change (°C or K) of the solid food with respect to time (s, min, h), k is the thermal conductivity (W/m K), A is the solid surface in which the heat flow is happening (m^2), ∇ is the gradient operator expressing the variation of temperature with respect to the rectangular coordinates (d/dx + d/dy + d/dz), T is the solid temperature (°C or K), and dT/dL is the temperature gradient or difference of temperature wit respect to the direction (°C/m).

Therefore, the rate of heat penetration by conduction within the solid phase is a function of several factors including density, specific heat, and thermal conductivity of the food material, and the driving force or temperatures difference, as the most important.

2.2. Heat Convection

Is the transfer of thermal energy from one part to another within a liquid phase, vapor or gas by mixing of one portion of the fluid with another portion. Heat transfer in any fluid material will normally occur by convection and since this transfer process is due to mixing action within the fluid, the rate of heat flow from or to a solid surface will depend on the fluid flow characteristics as well as the thermal properties of both phases (Welty *et al.*, 1976; Vélez-Ruiz, 2008). The basic equation which describes the heat transfer by convection is known as the Newton´s law for cooling.

$$Q = h(T_s - T_\infty) \tag{3}$$

where: Q is the heat flux (Q/A, W/m^2), h is the convective heat transfer coefficient (W/m^2 K), T_s is the solid temperature (°C), and T_∞ is the bulk fluid temperature (°C).

Most of the information leading to the evaluation of h is in the form of empirical relationships, that may be derived from a theoretical basis of the boundary layer, by analogies to the momentum and mass transfer phenomena or may be based on dimensional analysis (Welty *et al.*, 1976; Saunders, 1988; Holdsworth, 1997; Lewis and Heppel, 2000; Krokida *et al.*, 2002; Vélez-Ruiz, 2008). Whatever is the origin of the equation for h evaluation, it requires the determination of dimensionless numbers or groups of variables or properties, some of the most important are presented in table 1.

The heat transfer coefficient (h) is an essential parameter for calculations and modeling of many heat transfer operations encountered in the food industry. In many practical food processing situations, a single, average or global value is assumed to be representative of the complex convection phenomenon occurring over the entire surface on the food/equipment of interest.

Table 1. Dimensionless numbers for evaluation of h

Name	Group of variables	Comments/Observations
Nusselt (Nu)	$h L_c / k$	It includes h as the most important
Reynolds (Re)	$v L_c \rho / \eta$	Used for flow characterization
Generalized (GRe)	$\dfrac{\rho v^{2-n} L_c^n}{\left[2^{n-3} K \left(\dfrac{3n+1}{n}\right)^n\right]}$	Reynolds for non-Newtonian fluids
Prandtl (Pr)	$Cp \eta / k$	Influence of physical properties in thermal processes

Table 1. (Continued). Homogenize size of groups

Name	Group of variables	Comments/Observations
Generalized (GPr)	$\dfrac{CpK\left(\dfrac{3n+1}{n}\right)^{n} 2^{n-1}}{4kv^{1-n}D^{n-1}}$	Used for non-Newtonian fluids
Graetz number	$w\ Cp / k\ L_c$	Influence of moving fluid and heating/cooling
Grashoff (Gr)	$L_c^3\rho^2\beta g\Delta T/\eta^2$	Influence of buoyancy forces in natural convection
Froude (Fr)	$g\ L_c / k$	Influence of inertia and gravity forces
Biot (Bi)	$h\ L_c / k$	For unsteady state heat transfer
Fourier (Fo)	$\alpha\ t / L_c^2$	For unsteady state heat transfer

α: thermal diffusivity (m^2/s), β: thermal expansion coefficient (1/°C), Cp: specific heat (kJ/kg K), ρ: fluid density (kg/m^3), η: fluid viscosity (Pa s), g: gravitational acceleration (m/s^2), h: convective heat transfer coefficient (W/m^2K), k: thermal conductivity (W/mK), K: consistency coefficient (Pa s^n), L_c: characteristic dimension (m), n: flow behavior index (dimensionless), t: time (s), ΔT: temperature difference (°C or K), v: average fluid velocity (m/s), and w: mass flow rate (kg/s).

3. Heating and Cooling Processes in Food Industry

Many food process operations involve heat transfer between a solid food material and a fluid medium or a heated/cooled surface and a liquid food or a particulate food in liquid phase. While heat is transferred to the surface on the food product by convection, heat transfer within the solid food item is governed by conduction. Baking, blanching, canning, chilling, cooking, cooling, drying, evaporation, freezing, frying, pasteurization, ultra-pasteurization and sterilization are the most used food process operations, several of them are grouped as thermal processes. Some characteristics of the main processes that are used in food industry are briefly commented below and some examples of applications are included in table 2.

Table 2. Examples of food processes applied to different food items

Food process	Characteristics of the process	Food Item	Comments	Reference
Pasteurization	72°C - 12 min	"Sous vide" products	Stored and retailed at low T	Holdsworth, 1997
Sterilization	12 D	Non acidified foods	D: decimal reduction time	Holdsworth, 1997
HTST	140°C more than 10 s	Liquid foods	Continuous flow	Barigou *et al.*, 1998
Baking	186-225°C - 6.5 and 8.5 min	Cakes	Air at 0.02-0.44 m/s	Baik *et al.*, 1999
Frying	140 and 180°C, more than 40 s	French fries	Water loss was important	Costa *et al.*, 1999

Table 2. (Continued)

Food process	Characteristics of the process	Food Item	Comments	Reference
Pasteurization	63°C-30 min or 71°C-10 min	Milk/Ice cream	Different dairy foods	Lewis & Heppell, 2000
Pasteriuzation	60-3.5; 62.5-3.0 64.4°C-2.5 min 75°C - 30 s	Whole eggs Beer	US/Australia/GB Beer can be also pasteurized in cans	Lewis & Heppell, 2000
Sterilization	110-116°C – 20 and 30 min	Milk	Static or continuous retorts	Lewis & Heppell, 2000
UHT	145°C - 3 s	Milk	Other UHT were tested	Simpson *et al.*, 2001
Frying	185°C - 60 and 90 s	Chicken & French fries	h was lower for chicken	Pannu & Chinnan, 2003
Cooking	68 and 71°C for 15 s	Ground beef products	Cylindrical geometry	Zorrilla & Singh, 2003
Frying	190°C - 5 to 120 s	Potato dough	Temperatures of 160-190°C	Budzaki & Seruga, 2005
Chilling	Chill slaughter below 10° in 2 h	Chicken meat	Tunnel had 4 segments	Landfield & Houska, 2006
Drying	Natural convection, air at 49°C	Celery slices	Air humidity 12.5%	Bialobrzewski, 2006

T: temperature, HTST: high temperature-short time, UHT: Ultra high temperature.

The main objective of any thermal process is the complete destruction of contaminating microorganisms or enzymes that are present initially, including combinations of temperature and time to reach the purpose of the specific treatment. Under commercial conditions, the initial microbial load is reduced as much as possible to avoid over heating which could cause quality loss of the food, damage in acceptability and nutritional value. For other treatments in which the heat is applied or eliminated, the purpose may be different, some heat treatments such cooking, baking or frying produce major changes in sensory attributes.

Some selected food processes, in which the heat transfer plays an important role, are shortly commented next.

Pasteurization is a thermal process applied to a food item with the purpose of minimizing possible hazards arising from pathogenic microorganisms associated with milk, it is also used to inactive heat-labile microorganisms (moulds, vegetative bacteria, and yeasts) which may cause food spoilage or food poisoning. Whereas, sterilization in a practical sense refers to the achievement of commercial sterility by elimination of those microorganisms which cause illness or are capable of growing in food causing damages. Canning is one of the most effective treatments for preserving foods and it is an important part of the food supply. This industry has been developed widely since the origin of the can, passing by plastic containers and with the development of multilayer packages. The combination of process temperature and time will depend of the specific food item to be preserved and the type of retort to be utilized (Lewis and Heppel, 2000). Cooling and freezing as low temperature treatments are oriented to food preservation, in the cooling process food decrease its temperature over the freezing point and without phase change of the water. In contrary, the freezing treatment involves temperatures below the freezing point of the food and the conversion of liquid water to ice is present, it is used for long periods of storage.

Evaporation is utilized in the food industry to concentrate or increase the solids content of a liquid food by using steam as the heating medium. Commonly the concentration process of the food item is performed under vacuum conditions.During convection drying with hot air, that is also used as a preservation method, heat penetrates from the air stream into the surface of food material resulting in its gradual heating, and consequently, generating the evaporation of water that allows the food dehydration.

Frying is an unique unit operation in food processing. In it, simultaneous heat and mass transfer are involved; heat is transferred from the oil to the food and moisture is evaporated from the foodstuff. During the frying process physical, chemical and sensory properties are modified due to application of high temperature (Sosa-Morales *et al.*, 2006). And cooking as a general treatment refers to the application of heating for preparing of foods until the item is modified in appearance, chemical composition, flavor, and tenderness or texture, generating palatable foods. Baking is a form of cooking, normally carried out inside of an oven, for instance baked bread; in which by mean of heat transfer an unpalatable dough mass is converted into a most appetizing, easily digestible, light, and porous food (Desrosier, 1970).

In all the aforementioned heat transfer treatments, the knowledge of h is indispensable to evaluate the overall heat transfer coefficient,and lately the needed heat rates. These calculations are fundamental for equipment design. In addition, the increased demand for high quality, nutritional and safe food products at affordable prices has resulted in a need to minimize the severity of thermal processing without compromising safety and environment. Some experimental data of h, are included as part of the table 3.

A better understanding of convective heat transfer, followed by the formulation of adequate mathematical models, contribute to the right and precise evaluation of h and consequently the reliable prediction and optimization of food processes and food quality improvements.

4. Evaluation of Convective Heat Transfer Coefficient

Although a good number of studies have been conducted in the last four decades to determine the conductive heat transfer coefficient (h) for heating/cooling processes in food industry, specialized literature on h evaluation provides scant information. Furthermore, uncertainty in the order of 10-30% has been commonly related to h estimations.

Analytical modeling is very difficult due to the discontinuity inherent to phase changes and unsteady process conditions. Numerical modeling is accurate, but is strongly dependent of the food thermal properties undergoing the specific process, and frequently results in an average h during the entire process operation.

Among the cited techniques that are used for h evaluation, are the following: energy balance, calorimetric approach, data reduction technique, heating power; genetic algorithm, holographic interferometry, image analysis, finite difference, dimensional analysis, and direct measurement of heat flux, just to mention a few. These and another techniques are also mentioned in table 3, including some of their characteristics.

Table 3. Experimental determinations and reported data of h in Food Process Operations

h (W/m^2 K)	Food process	Method/Model	Comments/Observations	Reference
400-1200	Blanching process for vegetable tissue	Laplace transform Finite difference	h varied with time, it decayed from high value to an approximately constant value after 20 s.	Rumsey *et al.*, 1980
26-33 183-188	Canning of sucrose syrup in rotating autoclave	Energy balance	Low h for natural, high h for forced convection h was higher in end-over-end than in axial rotation	Naveh and Kopelman, 1980
280-960	Canning of nectars	Numerical integration	h was independent of the radius of rotation and the size of the can	Anantheswaran & Rao, 1985
400-1000	Blanching of green beans	Finite difference simulation	Calculated h was 12 fold lower than predicted by an existing equation	Zhang & Cavalieri, 1991
2.8-6.1	Heating in an oven	Heat balance	h increased with T and processing time	Unklesbay *et al.*, 1991
25– 95	Drying of vegetals	Mass and heat balances	Slices and cylinders of apple, carrot, & potato Air at 40-65°C, 11-50% RH, 1-5 m/s	Ratti & Crapiste, 1995
15-420	Heating of foods	Rate and ratio methods	In general, the rare method gave more consistent and conservative h values	Awuah *et al.*, 1995
36 and 45 260 and 290	Heating in an oven	Heat balance	Oven temperature: 150 and 200°C, without and with ventilation	Duquenoy & Nouafo, 1997
1000-8000 2000-15000	Boiling of corn syrup Boiling water	---	h was related to the rotational velocity and it was not influenced by flow rate	Yanniotis & Kolokotsa, 1997
54-75	Heating with air	Computational	h varied with L/D relationship and Re number	Verboven *et al.*, 1997
10.9-26.6 78.7-358.2	Cooling in air Cooling in water	Inverse problem	Pears and tomatoes as spherical shape, whereas cucumbers as cylinders	Dincer, 1997
443-750	Frying of potato	Direct and indirect	h was related to water loss	Costa *et al.*, 1999
90-200	Frying of mashed potato (5 x 5 x 0.3 cm)	Explicit finite difference	Sunflower oil, 150-190°C, variation in thermal properties (k and Cp) were considered	Sahin *et al.*, 1999
20 - 48	Baking of cakes	Commercial h-monitor	Hot air in a direct-fired tunnel oven	Baik *et al.*, 1999
10-280 10-125	Cooling of apples Cooling of five fruits	Rgular thermal regime	Air cooled flowing through apples, apricots, grapes, peaches, raspberries and strawberries	Fikiin *et al.*, 1999
7-160	Heating with air	Psicrometric and non-stationary	h was related to food shape and dimensions, air velocity, and turbulence	Ghisaberti & Kondjoyan, 1999
150-200 90-140	Heating of cans Cooling of cans	Analytical and Finite elements	The average h was good enough, instead of instantaneous h	Varga & Oliveira, 2000

Table 3. (Continued).

h (W/m^2 K)	Food process	Method/Model	Comments/Observations	Reference
300-1100	Immersion frying of potato	Analytical solution	Low h related with sample heating	Farkas & Hubbard, 2000
250 – 650	Double-sided cooking of hamburguer patties	Heat flux sensor	Three thickness (10.0, 10.5, 11,0 mm) and three temperatures (177, 191, 204°C)	Wichchukit *et al.*, 2001; Zorrilla *et al.*, 2001
1369-2749	Melting of ice spheres	Ablation heat transfer	For Reynolds number of 2766 to 17081	Tessneer *et al.*, 2000
1.5-5.3	Ripening of cheese	Psycrometric	Free convection on Emmental cheese	Pajonk *et al.*, 2001
900	Cooking of meat patties	Assumed value	Development of a 2-D mathematical model	Zorrilla & Singh, 2003
150-225	Jet impingement heating	Lumped mass	h was related to the jet number (3-11)	Nitin & Karwe, 2003
345-550 450-1070	Frying of chicken fillets Frying of French fries	Metal transducer Metal transducer	Degree of oil abuse did not appear to influence the h value	Pannu & Chinnan, 2003
250-650	Cooking of frozen meat patties	Simulation	h as a function of time, different gap thickness and set grill temperatures (177, 191 and 204°C)	Zorrilla *et al.*, 2003
155-297	Frying of donuts	Lumped parameters	155 for 180, 270 for 190, and 297 W/m^2 K for 200°C	Vélez-Ruiz and Sosa-Morales, 2003
197-775 94-195	Frying of potato dough	Heat balance	h at the beginning and as a function of T h at the evaporation and as a function of C	Budzaki & Seruga, 2005
9 – 39 6 - 90	Tunnel freezing Fluidized bed freezing	Heat flux sensor	Experimental h was validated by numerical simulation	Amarante & Lanoisellé, 2005
18-60	Chilling of chicken	Simulation	Tunnel with 4 segments and weight loss	Landfield & Houska, 2006
15-40	Cooking in an oven	Heat flux sensor	Experimental with two plaster cubes of gypsum	Carson *et al.*, 2006
50, 1000	Heating of potato	Assumed values	Potato cubes (1.2 cm) in still and boiling water	Palazoglu, 2006
8.1 ± 0.4	Drying of celery root	Inverse method	Practically constant, it did not depend of moisture	Bialobrzewski, 2006
4.2-39	Beef carcass cooling	Heat flux sensor and h sensor	h varied with the position and air velocity $h = av^n$, a and n, determined for carcass placements	Willix *et al.*, 2006
500 & 1000	Large food particles in cylindrical cans	3D Simulation	Mass, momentum and energy conservation were solved by using the finite volume method	Rabiey *et al.*, 2007

h: convective heat transfer coefficient, T: temperature, C: wheat flour concentration, v: air velocity (m/s).

Correlations are very useful as a first approach, but its application in real engineering situations finds limitations due to the particular system configuration and partial available information. The dimensionless analysis equations gathering the important and significative parameters of the physical phenomena, has the next general relation between dimensionless numbers:

$$Nu = a\,Re^{b}\,Pr^{c}\,Gr^{d} OG^{e} \tag{4}$$

where: Nu is the Nusselt number, Re is the Reynolds number, and Pr is the Prandtl nuber, Gr is the Grashoff number (dimensionless), and OG is used to mean the presence of other dimensionless numbers or relationships, a, b, c, d and e are experimental coefficients.

Even there are numerous correlations for h evaluation, they were developed for inorganic materials at most as the heated medium, and for Newtonian fluids as processing fluids. Since the studies of Peeples (1962a, 1962b) for fluid milk products, in the last 30 years some correlations have been proposed or fitted to handle of food processes in which either the food is heated or cooled as the processing fluid. Some of them are presented next.

- For heat transfer of fluid milk products, Peeples (1962a) established the constants affecting the corresponding Reynolds number for reconstituted skimmilk (12.5% solids non fat), fortified milk (10% fat, 12% solid non fat) and cream 25% (25% fat, 6% solids non fat), among others, while the power for Prandtl number was constant, and expressed as:

$$\begin{aligned} Nu &= 0.056Re^{0.708}Pr^{0.4} \\ Nu &= 0.170Re^{0.630}Pr^{0.4} \\ Nu &= 0.034Re^{0.804}Pr^{0.4} \end{aligned} \tag{5}$$

- And for the cooling process, Peeples (1962b) expressed similar equations for four milk fluids: skimmilk 8.8%, whole milk, milk 10% and cream 25%:

$$\begin{aligned} Nu &= 0.204Re^{0.586}Pr^{0.4} \\ Nu &= 0.118Re^{0.609}Pr^{0.4} \\ Nu &= 0.108Re^{0.632}Pr^{0.4} \\ Nu &= 0.260Re^{0.571}Re^{0.4} \end{aligned} \tag{6}$$

where: Nu is the Nusselt number (dimensionless), Re is the Reynolds number (dimensionless), and Pr is the Prandtl number (dimensionless).

- For heat transfer during canning implying non-Newtonian fluids, with their rheological properties, the next relationship was obtained by Anantheswaran and Rao (1985):

$$Nu = 1.41GRe^{0.482}GPr^{0.355} \tag{7}$$

where: GRe is the generalized Reynolds number (dimensionless), and GPr is the generalized Prandtl number (dimensionless).

- Rodríguez-Luna *et al.* (1987) proposed a relationship for non-Newtonian fluids (mango and papaya nectars), valid for Graetz numbers between 500 and 7000.

$$Nu = 1.75\left[Gz + 0.0083\left(Gr_w Pr_w\right)^{0.75}\right]^{0.33}\left[\frac{3n+1}{4n}\right]^{0.33}\left(\frac{K_b}{K_w}\right)^{0.14} \quad (8)$$

where: Gz is the Graetz number (dimensionless), Gr is the Grashoff number (dimensionless), Pr is the Prandtl number (dimensionless), both a Tw being the wall temperature (°C), n is the flow behavior index (dimensionless), and K is the consistency coefficient ($Pa \cdot s^n$) at Tb (K_b) as the bulk temperature (°C) and Tw (K_w), respectively.

- Rao and Anantheswaran (1988) mentioned a correlation for flame sterilization of viscous foods in cans 303 x 406 obtained by Teixeira in 1982:

$$Nu_D = 0.433\, Re_i^{0.456}\, Pr_i^{0.6}\, Re_e^{-0.68} \quad (9)$$

where: Nu is the Nusselt number for the can of D diameter (dimensionless), Re is the Reynolds number (dimensionless) for the fluid within the can (Re_i) and for the combustible gas (Re_e), and Pr is the Prandtl number within the can (dimensionless).

- A proposed equation for vapor condensation was modified by Zhang and Cavalieri (1991) for blanching of green beans:

$$h = 0.06\left[\frac{k_f^3 \rho_f^2 g \lambda_f}{\left(T_\infty - T_S\right)\eta_f D}\right] \quad (10)$$

where: k_f, ρ_f, λ_f and η_f are the properties (conductivity: W/m K, density kg/m^3, latent heat: kJ/kg and viscosity: Pa·s) of the steam as liquid phase, g is the acceleration of gravity (m^2/s), T_∞ is the steam temperature (°C), T_S is the surface temperature (°C), and D is the food item diameter (m).

- A simple correlation was obtained by Ratti and Crapiste (1995) for h during drying of foodstuffs:

$$Nu_D = 0.249\, Re_D^{0.64} \quad (11)$$

where: Nu_D is the Nusselt number and Re_D is the Reynolds number for drying process, both numbers are based on the diameter (D).

- Dincer (1997) developed a couple of correlations for cylindrical (cucumbers) and spherical (tomatoes and pears) food products being cooled in air (equation 12) or water (equation 13):

$$\left(Nu/Pr^{0.333}\right) = 1.264\,Re^{0.449} \text{ for } 1000 < Re < 70\,000 \quad (12)$$

$$\left(Nu/Pr^{0.333}\right) = 0.267\,Re^{0.432} \text{ for } 1000 < Re < 24\,000 \quad (13)$$

- A good correlation was obtained by Costa *et al.* (1999) for frying of French fries at 140°C:

$$h = 271 + 22808w \;\; (R^2 = 0.929) \quad (14)$$

where: w is the water loss rate (g/s).

- A model was proposed by Baik *et al.* (1999) for cakes baking inside an oven:

$$h_t = 0.0887T + 61.4V \quad (15)$$

where: h_t is the effective surface heat transfer coefficient (W/m^{2}K), T is the baking temperature (186-225°C), and V is the air velocity (0.027-0.043 m/s).

- The local h for a turbulent multiple jet impingement system for heating of cookies was related to the jets number utilized for the experiment (Nitin and Karwe, 2003) by the next correlations.

$$\begin{aligned} Nu &= 0.227Re^{0.595} \rightarrow for \cdot three \cdot jets \\ Nu &= 2.077Re^{0.387} \rightarrow for \cdot seven \cdot jets \\ Nu &= 1.680Re^{0.414} \rightarrow for \cdot eleven \cdot jets \end{aligned} \quad (16)$$

- The local h for cooling of a beef carcass exposed to air streams with different velocities, was deducted by Willix *et al.* (2006)

$$Nu = 0.076\,Re^{0.77} \text{ for } Tu \geq 2.5\% \quad (17)$$

$$Nu = 0.0074\,Re \text{ for } Tu \geq 20\% \quad (18)$$

where: Tu is the turbulence intensity (%).

Furthermore, in the covered reviews (Maesmans *et al.*, 1992; Barigou *et al.*, 1998; Zogzas *et al.*, 2002; and Krokida *et al.*, 2002), a wide range of h values: 8.1-7870 W/m^{2}K, has been reported. With the next particularities: 8.1-5472 W/m^{2}K for fluid to particle heat transfer coefficients for liquid foods containing particles; 55-1175 W/m^{2}K for particles immobilized in a flow with Reynolds of 123-2144, and 930-7870 W/m^{2}K for particles moving concurrently in a flow with Reynolds of 30-50000 these values were compiled by Maesmans *et al.* (1992).

Values of 55-7870 W/m^{2}K for particle heat transfer coefficients, were compiled by Barigou *et al.*, (1998) corresponding to Reynolds numbers from 0.006 to 50000. A h range of 83-155 W/m^{2}K, was computed by Zogzas *et al.* (2002) for Reynolds number from 520 to 2094. And, Krokida *et al.* (2002) evaluated the heat transfer factor J_H as a function of the flow regime to evaluate h, covering a J_H of 0.001-40 for Reynolds numbers of 10 to 2000000 in common food process operations.

All of them are expressed as a function of a diversity of variables, food processes, flow conditions, food properties of the solids and fluids, flow behavior, particle shapes; with different materials of solid items immersed in Newtonian and non Newtonian fluids. Finally, some data and experimental information from a good number of studies may be seen and analyzed in table 3. Whereas that different equations for specific food process operations, and their correspondent h values may be consulted in works of Zogzas *et al.* (2002) and Krokida *et al.* (2002).

5. Heat Transfer Coefficient between Fluids and Particles in Suspension

5.1. Studies on Convective Heat Transfer Coefficient for Fluid Foods and Particles

Over the past twenty years, strong research efforts around the world have been conducted to study and characterize a diversity of heating/cooling treatments in which the convective heat transfer coefficient for fluid foods and particles (h_{fp}), also known as surface heat transfer coefficient, has been evaluated and modeled.

In this specific situation in which food particles are heated or cooled within a liquid, the rate of heat transferred between particles and fluid, may be computed by the modified Newton´s law for cooling, that is equivalent to Eq. 3.

$$Q = h_{fp} A_p \left(T_{ps} - T_f\right) \tag{19}$$

where: Q is the heat flow (W), h_{fp} is the fluid-to-particle heat transfer coefficient (W/m^{2}K), A_p is the heat transfer are of the particle (m^2), T_{ps} is the particle surface temperature (°C), and T_f is the fluid temperature (°C).

In addition, for design considerations into processing liquid foods, there are a good number of important parameters associated with food particle heating. Some of the completed studies include heat exchange for non-Newtonian fluids and irregular particles (Alhamdan and Sastry, 1990; Bhamidipati and Singh, 1995), non-Newtonian fluids and regular items (Chandarana *et al.*, 1990; Awuah *et al.*, 1993; Awuah *et al.*, 1996; Baptista *et al.*, 1997; Ahmad *et al.* 2001; Chakrabandhu and Singh, 2002; Palazoglu, 2006); heating/cooling of food particles in Newtonian fluids (Chandarana *et al.*, 1990; Mankad *et al.*, 1997; Zareifard and Ramaswamy, 1997; Chen *et al.*, 1997; Dincer, 1997; Kelly, 1997; Nixon *et al.*, 1997; Hulbert *et al.*, 1997; Ahmad *et al.*, 2001; Palazoglu and Sandeep, 2002; Carson *et al.*, 2006);

frying processes of food items (Costa *et al.*, 1999; Sahin *et al.*, 1999; Budzaki and Seruga, 2005), aseptic processing in continuous flow (Awuah *et al.* 1995; Ramaswamy and Zareifard, 2000; Tessner *et al.*, 2001; Chakrabandhu and Singh, 2002; Palazoglu and Sandeep, 2002), industrial freezing (Amarante and Lanoisellé, 2005), and for static cans or during end-over-end agitation of cans and packed containers (Anantheswaran and Rao, 1985; Chang and Toledo, 1990; Sablani and Ramaswamy, 1996; Sablani *et al.*, 1997; Chen *et al.*, 1997; Sablani and Ramaswamy, 1997; Sablani and Ramaswamy, 1998; Varga & Oliveira, 2000); among others. Additional information on heat transfer between fluid and particulate foods is required; its quantification has remained as a challenging issue.

Works on natural and forced convection in food particles include those of Alhamdan and Sastry (1990) in which natural convection between CMC solutions and an irregular shape particle was investigated, finding h_{fp} values between 22 and 153 W/m^{2}K for cooling, and 75 to 310 W/m^{2}K for heating. Chandarana *et al.* (1990) studied the effect of the fluid behavior at UHT conditions, they reported values of 56-90 for particles in a non-Newtonian fluid, and 65-107 W/m^{2}K for particles heated in a Newtonian fluid. Awuah *et al.* (1993), evaluated the h_{fp} in CMC solution with food particles, finding h_{fp} values of 46-456 and 96-556 W/m^{2}K for cylinders of potatoes and carrots, respectively. Gadonna *et al.* (1996) developed predicting correlations for h_{fp} for particle and a conveying fluid inside a horizontal pipe, where high values (290-1587 W/m^{2}K) were obtained as a consequence of the turbulent flow during the forced convection. Zareifard and Ramaswamy (1997), proposed an experimental technique to evaluate h_{fp} under conditions imparting particle oscillatory motion while being heated, and they obtained a range of 430-840 W/m^{2}K for free particles, and 560-1370 W/m^{2}K for fixed particles. Ramaswamy and Zareifard (2000) carried out a set of experiments to evaluate h_{fp} and to know the influence of some factors, carrier fluid concentration (0-1% of CMC), flow rate (9-19 L/min), particle size (19-125.4 mm) and temperature (50-70°C); h_{fp} was in a range of 650 to 2260 W/m^{2}K, and increased with flow rate and particle size while decreased with CMC concentration and temperature. Tessner *et al.* (2001) used an ablation heat transfer method to determine h_{fp} of 1369 and 2749 W/m^{2}K for Reynolds numbers of 2700 to 17000. Ahmad *et al.* (2001) reported a range of 400 to 1350 W/m^{2}K for h_{fp} in heterogeneous pipe flow system using polypropylene particle, in which the liquid-to-particle heat transfer coefficient increased as a function of flow rate (0.25-1 L/s), particle concentration (0-20% v/v) and temperature (42.5-81.6°C). Chakrabandhu and Singh (2002) quantified the h_{fp} for spherical particles (green peas) in continuously flowing food suspensions, with values of 538 to 1450 in the coiled portion and 204 to 1224 W/m^{2}K in the straight tube of the holding section of an aseptic processing system, in which significant difference was found between sections. h_{fp} also increased as a function of flow rate (2 and 4 gal/min), particle concentration (0-30% v/v), and temperature (96-124°C).

Related to thermal processes, several of the studies on h_{fp} have been related to pasteurization and sterilization treatments. Chang and Toledo (1990) developed a sterilization system to evaluate h_{fp} for carrot cubes heated under UHT conditions, in which the coefficient varied from 359 to 1533 W/m^{2}K as a function of the studied variables. Awuah *et al.* (1995) reported h_{fp} values of 15 to 420 W/m^{2}K for particles in two shapes as stationary samples in the holding tube. Sablani and Ramaswamy (1997) developed an experimental work to evaluate h_{fp} during end-over-end agitation of cans and obtained 170-1165 W/m^{2}K with oil, while a range of 175-1550 W/m^{2}K was determined in water; h_{fp} was related inversely to particle diameter and increasingly to concentration. Kelly (1997) reported a range of 660 to

1440 W/m^2K for a study of aseptic processing of suspensions, determining the h_{fp} with aluminium particles under processing conditions in open channel flow. Palazoglu and Sandeep (2002) developed a computer program to determine the effect of h_{fp} on microbial and nutritional responses during aseptic processing. An increase in h_{fp}, taking values of 240 to 4800 W/m^2K, reduced the heating time, and did not improve the nutrient retention to reach a given lethality. Further, using the most conservative value of h_{fp}, a high quality food product was obtained by Pannu *et al.* (2003), who carried out a study to asses the effect of several parameters, flow direction and rate, food particle interaction, shape and size, as well as steam temperature, on the product temperature during packed bed heating of carrot and potato cubes and slices. Food center and surface temperatures were measured by these researchers under diverse process conditions, they reported h_{fp} values of 1500 and 2000 W/m^2K for potato and carrot cubes with low Biot numbers, 14.2 and 17.8, respectively.

5.2. Applied Methodology for h_{fp} Determination

Reviews developed by Maesmans *et al.* (1992) and Barigou *et al.* (1998), on the determination of fluid to particle heat transfer coefficient for food processes, mention the influence of important factors in this heat transfer phenomena. The main factors include natural or forced convective phenomena, fluid rheology, particle size and concentration, particle properties and particle displacement (translational and rotational), among the most important and studied.

To quantify the convective or surface heat transfer coefficient there are different approaches, the sample temperature as a function of time is indispensable for all of them. Barigou *et al.* (1998) mention six experimental approaches for h_{fp} determination at that time: i) biological or biochemical methods, ii) relative velocity method, iii) thermocouple method, iv) liquid crystal method, v) melting point indicator method, and vi) remote temperature sensor technique that offers the application for thermal process measurements but they have not been fully developed and applied to food processing operations.

Some of these methods are very sophisticated, employing numerical solution (Gadonna *et al.*, 1996; Chakrabandhu and Singh, 2002) or explicit finite difference technique (Palazoglu and Sandeep, 2002; Palazoglu, 2006), considering Biot numbers of 2 to 40, to elucidate the transient heat conduction. Other methods include the use of a time-temperature integrator combined with mathematical modeling (Maesmans *et al.*, 1992; Maesmans *et al.*, 1994); both approaches could be utilized where conventional measurement techniques are not possible. An alternative instrumental approach was utilized by Amarante and Lanoisellé (2005) to measure the heat transfer coefficients in freezing equipment, in which heat sensors were coupled to temperature recording devices with satisfactory results. A calorimetric technique was reported by Ramaswamy and Zarefreid (2000). Several more are based on energy balances between the gained heat for a phase change and the convective phenomenon (Tessner *et al.*, 2001; Ahmad *et al.*, 2001) that is very practical and manageable.

Other proposed approaches are simple, two of these are the rate method and the ratio approach, both methods require accurate thermophysical properties in order to get conservative h_{fp}. The first one is based on heating rate of one point of the food item, being consistent and conservative, while the second method is based on the ratio of temperature gradients at two locations in the food and simultaneous solving of two equations. Awuah *et al.* (1995) compared these two methodologies in the measuring of h_{fp} for four different materials, they found out that both methods were sensitive and matched well at low Biot numbers (< 10), but the rate method gave more consistency and less variation.

Most of the applied methods for h_{fp} evaluation, including those aforementioned and other, with some of their characteristics are presented in table 4, in a brief presentation. Some of them appear with different names but the fundamental basis is the same.

Table 4. Methods of evaluation of the fluid-to-particle heat transfer coefficient (h_{fp})

Method	Characteristics and observation of the measurement method	Reference
T-t Integrator method	Small device showing the T-t data, the response in thermal processing is usually described by first order kinetics, wireless measurements	Maesmans *et al.*, 1994
Rate method Ratio method	Based on heating rate at any given location, hfp ranged from 15 to 420 W/m^2K Based on the ratio of T gradients at two locations. Both consider transient T-t data	Awuah *et al.*, 1995
Transmitter particle	Based on particle center T measurements with a sensor system and energy balance, hfp values of 108-196 W/m^2K that were related to viscosity and T of the carrier fluid	Bhamidipati and Singh, 1995
Moving thermocouple	A thermocouple is attached to the moving particle, it yields hfp values compared to other methods, hfp ranged from 363 to 2010 W/m^2K, cost: 3000 US$	Balasubramaniam & Sastry, 1996
Liquid crystal	Chemical compound which changes its color, obtaining a color-temperature equation, limited by resolution capabilities of imagen system and video pictures	Balasubramaniam & Sastry, 1996; Gadonna *et al.*, 1996
Modified rate method	Based on the heating rate index estimated from the straight line of T vs t, obtained with a holding tube simulator, hfp values of 100-700 W/m^2K for Teflon and potato samples	Awuah *et al.*, 1996
FD Computer simulation	Solution of the partial differential equations of conduction in spherical coordinates in transient heating, hfp range of 23-145 W/m^2K, headspace was the most significant on hfp	Sablani & Ramaswamy, 1996
Combined method	Finite element modeling in combination with 2D MRI temperature mapping, hfp ranged from 117 to 389 W/m^2K for heating of carrot particles in water	Hulbert *et al.*, 1997
Lumped capacity	Energy balance under transient process conditions for Bi less than 0.1, T for the bulk fluid and particle center are needed, hfp ranged from 56 to 2612 W/m^2K	Baptista *et al.*, 1997

Table 4. (Continued)

Method	Characteristics and observation of the measurement method	Reference
Computer simulation	Computer- aided analytical method derived from PDE, thermocouple determinations of the particle center and fluid positions, hfp ranged from 165 to 2376 W/m^2K	Cheng *et al.*, 1997
Stationary particles	Stationary solid in a stream of flowing liquid, particle with different shapes, thermocouples to measure T at the particle surface and in the bulk liquid	Kelly, 1997; Nixon *et al.*, 1997; Mankad *et al.*, 1997
Calorimetric approach	Involves bulk and particule T measurements, a calibration curve for each particle, aluminium spherical particles were utilized, D: 19.05, 22.22 and 25.4 mmm	Ramaswamy & Zareifard, 2000
Ablation method	Melting of ice with the continuous removal of the melted phase. An apparent asymptotic trend as a function of the Re number was observed	Tessner *et al.*, 2001
Energy balance	Thermocouples to measure fluid, particle (cylinder of L=D= 26 mm) center and close to the surface, polypropylene particles (0-2 % w/v) in water	Ahmad *et al.*, 2001
Indirect method	Comparison of F for enzyme data to F for particle T profile, with tracer particles (8.4 mm) a video camera was use to record particle times, 0-30% (v/v) of green peas in 1.5 % CMC	Chakrabandhu & Singh, 2002
Numerical method	Explicit finite difference program to simulate heat transfer in spherical particles (5, 7.5, 10 mm), 10 nodes in the radial direction, Biot numbers: 8.3, 12.5 and 16.7.	Palazoglu & Sandeep, 2002
Evaluation of f_h	Determinations of T at center, bottom and top of particle and surface layer through heating by thermocouples, food cubes of uniform size: 12 mm, can 300 x 400	Pannu *et al.*, 2003
Thermo-couples	Determinations of T for particle and liquid, at different conditions: retort T (110-130°C) rotation speed (6-23 rpm), three can sizes, test liquids and particles, among others	Meng & Ramaswamy, 2007

T : temperature, t: time, FD: finite difference, MRI: Magnetic resonance image, PDE: Partial differential equations, D: diameter, F: lethality time; CMC: sodium carboximethylcellulose, f_h: negative inverse slope of the heating curve (heating rate index).

5.3. h_{fp} Empirical Correlations

Most of the revised works have established empirical correlations between the Nusselt number as a function of Reynolds, Grashof, Prandtl, and other dimensionless relationships, depending of the experimental conditions; those founded in the consulted literature are shown next:

- Alhamdan and Sastry (1990) studied the cooling and heating convection between non-Newtonian fluids and a mushroom shaped particle, developing two relations:

For cooling $$\log \mathrm{Nu} = 1.59 - 0.37\mathrm{Fo} \qquad (R^2 = 0.977) \tag{20}$$

For heating $$\log \mathrm{Nu} = 3.11 - 0.77\mathrm{Fo} \qquad (R^2 = 0.977) \tag{21}$$

where: Nu is the Nusselt number (dimensionless), including the h_{fp}, and Fo is the Fourier number (dimensionless)

- Chandarana *et al.* (1990) carried out a study in which silicone cubes were immersed in water and starch solution at low Reynolds magnitudes, and the resultant equations of the analysis were:

For water $$\mathrm{Nu} = 2.0 + 3.33\mathrm{x}10^{-2}\,\mathrm{Re}^{1.08} \qquad \mathrm{Re}: 287\text{-}881\ (R^2 = 0.89) \tag{22}$$

For starch $$\mathrm{Nu} = 2.0 + 2.82\mathrm{x}10^{-2}\,\mathrm{Re}^{1.6}\,\mathrm{Pr}^{0.89} \qquad \mathrm{Re}: 287\text{-}881\ (R^2 = 0.89) \tag{23}$$

where: Re is the Reynolds number (dimensionless), and Pr is the Prandtl number (dimensionless).

- Awuah *et al.* (1993) yielded a couple of relations to evaluate the surface heat transfer coefficient associated with heating of food particles in CMC solutions, with limits of $3 \times 10^3 < \mathrm{GGr}\,\mathrm{GPr} < \mathrm{x}\ 10^4$:

$$\mathrm{Nu} = 2.45(\mathrm{GGr} \cdot \mathrm{G\,Pr})^{0.108} \qquad \text{for carrot } (R^2 = 0.80) \tag{24}$$

$$\mathrm{Nu} = 2.02(\mathrm{GGr} \cdot \mathrm{G\,Pr})^{0.113} \qquad \text{for potato } (R^2 = 0.80) \tag{25}$$

where: GGr is the generalized Grashoff number (dimensionless), and GPr is the generalized Prandtl number (dimensionless).

- Gadonna *et al.* (1996) obtained some correlations for free particles and conveying fluid in a horizontal tube, as a function of Reynolds magnitude:

$$\mathrm{Nu} = 2 + 4.98\,\mathrm{Re}_f^{0.234}\left(\frac{d_p}{D}\right)^{1.441} \qquad \mathrm{Re_f}: 1400\text{-}2300\ (R^2 = 0.98) \tag{26}$$

$$\mathrm{Nu} = 2 + 19.31\,\mathrm{Re}_f^{0.279}\left(\frac{d_p}{D}\right)^{2.217} \qquad \mathrm{Re_f}: 2300\text{-}5000\ (R^2 = 0.96) \tag{27}$$

$$Nu = 2 + 0.933\, Re_p^{0.727} \qquad Re_p: 84\text{-}2300\ (R^2 = 0.95) \tag{28}$$

$$Nu = 2 + 0.228\, Re_p^{0.639} \left(\frac{d_p}{D}\right)^{0.514} \qquad Re_p: 84\text{-}2300\ (R^2 = 0.96) \tag{29}$$

$$Nu = 2 + 14.44 Fr_p^{0.777} \qquad Fr_p: 0.177\text{-}1.714\ (R^2 = 0.91) \tag{30}$$

$$Nu = 2 + 129.35 Fr_p^{0.712} \left(\frac{d_p}{D}\right)^{1.99} \qquad Fr_p: 0.177\text{-}1.042\ (R^2 = 0.96) \tag{31}$$

where: Re_f is the Reynolds number for the fluid (dimensionless), Re_p is the Reynolds number for the particle (dimensionless) including the d_p, d_p is the particle diameter (m), D is the tube diameter (m), and Fr_p is the Froude number for the particle (dimensionless).

- Holdsworth (1997) mentions two proposed equations by Fernandez *et al.* in 1988 and Merson and Stoforos in 1990. The first one was used for canned snap beans, and the second, for a rotating particle in a liquid:

$$Nu = 2.7 x 10^4\, Re^{0.294}\, Pr^{0.33}\, S_f^{6.98} \tag{32}$$

$$Nu = 2 + 0.4\, Re^{0.5}\, Pr^{0.4} \left(\frac{\mu}{\mu_s}\right)^{0.25} + 0.06\, Re^{0.66}\, Pr^{0.4} \left(\frac{\mu}{\mu_s}\right)^{0.25} \tag{33}$$

where: S_f is the shape factor ($6V_p\pi^{1/3}/A_p$, m), V_p is the particle volume, A_p is the particle area, μ is the liquid viscosity (Pa s), and μ_s is the suspension viscosity (Pa s).

- Baptista *et al.* (1997) modified the Fröszling equation to fit the experimental results for heating of still particles immersed in a moving fluid with h_{fp}: 56-2612 W/m^{2}K, and rotating particles in a stagnant fluid with h_{fp}: 67-1782 W/m^{2}K:

$$Nu = Nu_s + 0.17 G\, Re^{0.71}\, G\, Pr^{0.42} \left(\frac{d_p}{D}\right)^{0.28} \quad (R^2 = 0.955) \tag{34}$$

where: Nu_s is the Nusselt number for natural convection of a non-Newtonian fluid (dimensionless), d_p is the particle diameter (m), and D is the tube diameter (m).

- Chen *et al.* (1997) obtained a regressional analysis ($R^2 = 0.975$) of the experimental data for the surface heat transfer coefficient of spherical particles at low Reynolds numbers (<106):

$$\ln\left(Nu\, Re^{-2}\right) = 2.14 \ln\left(\frac{Pr^{0.55}\, Gr^{0.45}}{Re}\right) - 11.43 \tag{35}$$

where: Gr is the Grashoff number (dimensionless).

- Sablani *et al.* (1997) developed dimensionless correlations for convective heat transfer to liquid and particles in cans subjected to end-over-end rotation, for one particle (equations 36 to 38) and multiple particles (equations 39 and 40):

$$Nu = 2 + 1971 Fr^{0.31} Pr^{0.33} Ds^{0.34} \left(\frac{h_s}{H}\right)^{1.56} \left(\frac{D_r + D_c}{D_c}\right)^{0.21} \quad \text{for spheres } (R^2 = 0.83) \qquad (36)$$

$$Nu = 7.26\, Re^{0.42} Pr^{0.33} Ds^{1.12} \quad \text{for cylinders } (R^2 = 0.91) \qquad (37)$$

$$Nu = 8.03\, Re^{0.52} Pr^{0.33} Ds^{1.63} \quad \text{for cubic shape } (R^2 = 0.91) \qquad (38)$$

$$Nu = 0.167 Pe^{0.61} \left(\frac{k_p}{k_l}\right)^{0.48} \left(\frac{d_e}{D_c}\right)^{-0.70} \left(\frac{\varepsilon}{100-\varepsilon}\right)^{0.067} \Psi^{0.23} \quad \text{equivalent equations for} \qquad (39)$$

$$Nu = 0.718\, Re^{0.61} \left(\frac{k_p}{k_l}\right)^{1.98} \left(\frac{d_e}{D_c}\right)^{-0.70} \left(\frac{\varepsilon}{100-\varepsilon}\right)^{0.067} \Psi^{0.23} \quad \text{multiple particles } (R^2 = 0.96) \qquad (40)$$

where: Ds is the density simplex $[(\rho_p - \rho_f)/\rho_f)]$ including the particle density (ρ_p: kg/m^3) and fluid density (ρ_f: kg/m^3), h_s is the can headspace (m), H is the height of the can (m), D_r is the rotation diameter (m), D_c is the can diameter (m), k_p is the thermal conductivity for the particle (W/m°C), k_l is the thermal conductivity for the liquid (W/m°C), d_e is the particle equivalent diameter ($d_e = (6V_p/\pi)^{0.33}$, m), V_p is the particle volume (m^3), ε is the particle concentration (%, particle volume/can volume), and Ψ is the sphericity of the particle (A_{sp}/A_p, being A_{sp} the corresponding area of a sphere and A_p the particle area).

- Nixon *et al.* (1997) and Mankad *et al.* (1997) experimentally derived four correlations in a study of heat transfer within solid-liquid food mixtures:

$$Nu = 2 + 0.97\, Re_p^{0.45} Pr^{0.25} \quad \text{for a single particle} \qquad (41)$$

$$Nu = 2 + 1.215 \left(\frac{l}{d_p}\right)^{0.05} Re_p^{0.44} Pr^{0.20} \quad \text{for two particles} \qquad (42)$$

$$Nu = 0.50\, Re_{local}^{0.60} Pr^{0.28} \quad \text{for packed bed} \qquad (43)$$

$$Nu = 2.7 + 1.1\varepsilon^{1.1} Re_{local}^{0.46} Pr^{0.23} \quad \text{for expanded bed} \qquad (44)$$

where: Re_p is the Reynolds number for the particle (dimensionless), l is the distance between particles (m), Re_{local} is the Reynolds number for the local position, and ε is the void space in the bed (fraction).

- Lewis and Heppell (2000) cited a modified equation from Ranz and Marshall for non-Newtonian fluids with particles:

$$Nu = 2.0 + 0.6G\,Re_p\ G\,Pr^{0.33} \tag{45}$$

where: GRe_P is the generalized Reynolds number for the particle (dimensionless), and GPr is the generalized slip Prandtl number (dimensionless).

- Chakrabandhu and Singh (2002) developed a couple of dimensionless correlations for particulate flow in a coiled and straight holding tube.

$$Nu = 44.08G\,Re_{sb}^{0.13}\ G\,Pr_{sb}^{-0.18} \qquad \text{for the coil } (R^2 = 0.71) \tag{46}$$

$$Nu = 693.27G\,Re_{sb}^{0.28}\ G\,Pr_{sb}^{-0.61} \qquad \text{for the straight tube } (R^2 = 0.82) \tag{47}$$

where: GRe_{sb} is the generalized slip Reynolds number (dimensionless), and GPr_{sb} is the generalized slip Prandtl number (dimensionless).

- Recently, Meng and Ramaswamy (2007) developed a couple of correlations to fit h_{fp} for fluids during end-over-end rotation, as a function of the experimental variables. Their results were summarized in the next two relationships.

for non-Newtonian fluids ($R^2 = 0.87$)

$$Nu = 0.14G\,Re^{0.613}\ G\,Pr^{0.383}\left(\frac{\rho_p}{\rho_l}\right)^{0.518} Fr^{-0.213}\left(\frac{L_{can}}{D_{can}}\right)^{1.131}\left(\frac{V_p}{V_C}\right) \tag{48}$$

for Newtonian fluids ($R^2 = 0.92$)

$$Nu = 0.020\,Re^{0.53}\ Pr^{0.451}\left(\frac{\rho_p}{\rho_l}\right)^{0.446} Fr^{-0.113}\left(\frac{L_{can}}{D_{can}}\right)^{1.18} \tag{49}$$

where: ρ_p is the density of the particles (kg/m^3), ρ_l is the density of liquid (kg/m^3), L_{can} is the length of the can (m), D_{can} is the can diameter (m), V_p is the particle volume (m^3), and V_{can} is the volume of the can (m^3).

Additional relations for h_{fp} evaluation, may be found in works of Maesmans *et al.* (1992), Gadonna *et al.* (1996), Baptista *et al.*, (1997), Barigou *et al.*, (1998).

Nevertheless, there are several studies on h_{fp} evaluation and modeling, and reported data, still there is a lack of information on it. Little is known about the h_{fp} in food processing, and particularly the effect of particle load on the h_{fp}. On the other hand, there exist variations from one study to other, and the studied experimental conditions still are limited. Consequently more studies are needed.

6. Examples

6.1. Illustrative Example 1

Determine the heat flow by convection on a solid surface of a heat exchanger for two situations: i) flowing saturated steam (T = 100°C) which is condensing on the surface, and ii) for heated orange juice (T = 60°C) flowing at rate of 3 m/s in a tube of 2 in, when the metal surface is at 30°C.

Solution

i) Taking a conservative value for condensing steam: h of 30000 W/m^{2}K (Welty *et al.*, 1976) and using equation 3:

$$Q = 30000 \frac{W}{m^2K}(30-100)K = \mathbf{2{,}100{,}000 \frac{W}{m^2}}$$

Comment: the total heat flow it will depend of the heating area in which the steam is contacting.

ii) For the orange juice, considered as a Newtonian fluid, its physical properties are needed for h evaluation. Thus, using a correlation for turbulent flow in tubes (Welty *et al.*, 1976) by previous evaluation of Reynolds number, and taking the physical properties (density viscosity, specific heat and thermal conductivity) from Vélez-Ruiz and Soriano-Morales (2003):

$$Re_D = \frac{3\frac{m}{s}\cdot 1043\frac{kg}{m^3}\cdot\left(2.067in\frac{0.0254m}{in}\right)}{3.9x10^{-3}\frac{N\cdot s}{m^2}} = \frac{164.28\frac{kg}{m\cdot s}}{3.9x10^{-3}\frac{kg}{m\cdot s}} = 42123$$

$$Nu_D = 0.023(42123)^{0.8}\,Pr_D^{0.3} = 115.2\left(\frac{3.89\frac{kJ}{kg\cdot K}\left(\frac{1000J}{1kJ}\right)\cdot 3.9x10^{-3}\frac{kg}{m\cdot s}}{0.554\frac{W}{mK}}\right)^{0.3} = 310.9$$

$$h = 310.9\left(\frac{0.554\frac{W}{m\cdot K}}{2.067x.0254m} = 3280.4\frac{W}{m^2K}\right) \Rightarrow Q = 3280.4\frac{W}{m^2K}(30-60)K = \mathbf{98{,}412\frac{W}{m^2}}$$

And finally, if the surface of a tube of 16 ft long, in which the involved area of πDiL (3.1416 x 0.0525m x 4.88m = 0.08049 m^2), it implies a heat flow of the following magnitude:

$$Q = 98412\frac{W}{m^2}0.805m^2 = \mathbf{79{,}209.5W}$$

Comment: the total heat flow it will depend of the number of tubes in which the juice is flowing.

6.2. Illustrative Example 2

Determine the heat flow by convection in which a mango nectar (0.75 kg/s per plate and 45°C) is in contacting with the surface of a plate heat exchanger (T = 75°C). The total plates are 161, with 1.74 m x 0.55 m, creating an effective surface area per plate of 0.752 m^2 (l = 1.504 m, and a = 0.50 m), 45° of Chevron angle; plate pitch and thickness of 0.0035 and 0.0006 m, respectively, with an enlargement factor of 1.17 (Saunders, 1988).

Solution

For the mango nectar, considered as a non Newtonian fluid, its physical properties are needed for h evaluation. Thus, adapting a correlation from laminar flow in horizontal tubes (Rodriguez-Luna *et al.*, 1987) in which previous evaluation of Reynolds number is needed. In this case, the physical properties are taken from Rodriguez-Luna *et al.* (1987) and Vélez-Ruiz and Soriano-Morales (2003): $\rho = 1053$ kg/m^3, K = 2.5 Pa s^n and 0.55 Pa s^n at T_b (45°C) and T_w (75°C) respectively, n = 0.21, Cp = 3.85 kJ/kgK, k = 0.564 W/mK.

For the generalized Reynolds number, the velocity (v) is obtained from the flow rate (w), density (ρ) and flow area (A_f = ba) as v = w/ρA_f [= (0.75 kg/s)/(1053 kg/m^3)(0.0029 m x 0.50 m)]. Also a channel equivalent diameter was evaluated (Deq = 2b/η = 0.00496 m), being b the mean channel flow gap (b = pitch – thickness), and η the enlargement factor (1.17, Saunders, 1988; Vélez-Ruiz, 2008). The other numbers are computed with equations coming of table 1.

$$\mathrm{G\,Re} = \frac{1053\frac{\mathrm{kg}}{\mathrm{m}^3}\left(0.49\frac{\mathrm{m}}{\mathrm{s}}\right)^{2-0.21}(0.00496\mathrm{m})^{0.21}}{\left(2^{0.21-3}\right)\left(2.5\frac{\mathrm{N\cdot s}^{0.21}}{\mathrm{m}^2}\right)\left(\frac{3(0.21)+1}{0.21}\right)^{0.21}} = \frac{96.36\frac{\mathrm{kg}}{\mathrm{m\cdot s}^{1.79}}}{0.556\frac{\mathrm{kg}}{\mathrm{m\cdot s}^{1.79}}} = 173.4$$

$$\mathrm{Gz} = \frac{\left(3850\frac{\mathrm{J}}{\mathrm{kgK}}\right)\left(0.75\frac{\mathrm{kg}}{\mathrm{s}}\right)}{\left(0.564\frac{\mathrm{W}}{\mathrm{mK}}\right)(1.504\mathrm{m})} = 3404 \cdots \mathrm{and} \cdots \mathrm{Gr} = 12\cdot(\mathrm{sup\,posed})$$

$$\mathrm{G\,Pr} = \frac{\left(3850\frac{\mathrm{J}}{\mathrm{kgK}}\right)\left(2.5\frac{\mathrm{N\cdot s}^{0.21}}{\mathrm{m}^2}\right)(1.54)(0.578)}{4\left(0.564\frac{\mathrm{W}}{\mathrm{mK}}\right)\left(0.49\frac{\mathrm{m}}{\mathrm{s}}\right)^{0.79}(0.00496\mathrm{m})^{-0.79}} = \frac{8567.4\frac{\mathrm{J}}{\mathrm{kg}}\frac{\mathrm{kg}}{\mathrm{ms}^{1.79}}}{85.11\frac{\mathrm{W}}{\mathrm{m}}\frac{\mathrm{m}^{0.79}}{\mathrm{s}^{0.79}}\mathrm{m}^{-0.79}} = 100.7\frac{\frac{\mathrm{J}}{\mathrm{s}}}{\mathrm{W}}$$

With these adimensionless groups and the equation 8:

$$Nu = 1.75\left[3404 + 0.0083(12x100.7)^{0.75}\right]^{0.33}\left(\frac{3\cdot 0.21+1}{4\cdot 0.21}\right)^{0.33}\left(\frac{0.55Pa\cdot s^{.21}}{2.5Pa\cdot s^{0.21}}\right)^{0.14}$$

$$Nu = 1.75[15](1.25)(0.81) = 25.86 \rightarrow h = 26.58\frac{0.564\frac{W}{mK}}{0.00496m} = \mathbf{3059.2\frac{W}{m^2K}}$$

$$\rightarrow Q = 3059.2\frac{W}{m^2K}0.752m^2(45-75)K = \mathbf{69016W\ (per\ plate)}$$

Comment: the total heat flow it will depend of the number of plates in which the juice is flowing (in this example, with 161 plates, 80 are involved for mango nectar; therefore the total heat is 5.52 MW). Additionally to h evaluation; for the design of heat transfer equipment other parameters (h for the other process fluid, metal resistance and fouling effects), must be determined in order to obtain the overall heat transfer coefficient.

6.3. Illustrative Example 3

Determine the quantity of evaporated water during the frying of potato dough, taking some of the convective heat transfer coefficients reported by Budzaki. and Seruga, 2005) for three cases: i) a sphere of 2.5 cm diameter, in oil with a h of 210 W/m^2K at 160°C for the beginning of the process, ii) a cylinder of 2 cm (D) and 3 cm (l), and h of 450 W/m^2K at 170°C, and iii) a plate of 0.3 (a) x 2 (b) x 3 (l)cm and h of 600 W/m^2K at 190°C.

Solution

i) From the heat balance proposed by Hubbard and Farkas in 1999 (Budzaki and Seruga, 2005) in which the heat of water vaporization and the surface area of the samples are assumed to remain constant: $\frac{dm}{dt} = \frac{hA\Delta T}{\Delta H_{vap}}$, that is the applied equation by the authors to quantify h.

Thus, ΔH_{vap} = 2256.4 kJ/kg, and A = 3.1416 $(0.025m)^2$ = 1.9635 x 10^{-3} m^2.

$$\frac{dm}{dt} = \frac{210\frac{W}{m^2K}1.9635x10^{-3}m^2(160-100)K}{2256.4\frac{kJ}{kg}x\frac{1000J}{1kJ}} = \frac{24.74W}{2256400\frac{J}{kg}} = \mathbf{1.1x10^{-5}\frac{kg}{s} = 0.66\frac{g}{min}}$$

Comment: this is the quantity of evaporated water predicted for a ball of potato dough at the beginning of frying and assuming steady state conditions.

ii) Similarly, for the cylinder: A = $2\pi r^2 + \pi Dl$ = 2·3.14·$(0.01m)^2$ + 3.14·(.02m)·(0.03m) = 2.5 x 10^{-3} m^2):

$$\frac{dm}{dt}=\frac{450\frac{W}{m^2K}2.5x10^{-3}\,m^2\left((170-100)\right)K}{2256.4\frac{kJ}{kg}x\frac{1000J}{1kJ}}=\mathbf{3.5x10^{-5}\,\frac{kg}{s}}=\mathbf{2.1\frac{g}{min}}$$

Comment: this is the quantity of evaporated water predicted for frying of a cylinder of potato dough at a higher temperature and with a greater h, and also, assuming steady state conditions.

iii) For the plate: A = 2la + 2lb + 2ab = 2 (0.03 m x 0.003 m) + 2 (0.03 m x 0.02 m) + 2 (0.003 m x 0.02 m) = 1.5 x 10^{-3} m^2:

$$\frac{dm}{dt}=\frac{600\frac{W}{m^2K}1.5x10^{-3}\,m^2\left((190-100)\right)K}{2256.4\frac{kJ}{kg}x\frac{1000J}{1kJ}}=\mathbf{3.6x10^{-5}\,\frac{kg}{s}}=\mathbf{2.15\frac{g}{min}}$$

Comment: this is the quantity of evaporated water predicted for frying of a plate of potato dough at a higher temperature (with a higher h) and assuming steady state conditions. Major evaporation rates are reached with higher h and ΔT, as it is expected.

References

Ahmad M.N., Iweida J.A. and Magee T.R.A. 2001. Liquid-to particle heat transfer coefficients in heterogeneous pipe flow systems: The effect of loading. *Proceedings of the Eighth International Congress on Engineering and Food (ICEF 8)*. Technomic Pub. Co. Inc. Lancaster, PN, USA. Vol. I: 332-336.

Alhamdan A. and Sastry S.K. 1990. Natural convection heat transfer between non-Newtonian fluids and an irregular shaped particle. *Journal of Food Process Engineering,* 13: 113-124.

Amarante A. and Lanoisellé J.L. 2005. Heat transfer coefficient measurement in industrial freezing equipment by using heat flux sensors. Journal of Food Engineering, 66: 377-386.

Anantheswaran R.C. and Rao M.A. 1985. Heat transfer to model non-newtonian liquid foods in cans during end-over-end rotation. *Journal of Food Process Engineering,* 4: 21-35.

Awuah G.B., Ramaswamy H.S. and Simpson B.K. 1993. Surface heat transfer coefficients associated with heating of food particles in CMC solutions. *Journal of Food Process Engineering,* 16: 39-57.

Awuah G.B., Ramaswamy H.S. and Simpson B.K. 1995. Comparison of two methods for evaluating fluid-to-surface heat transfer coefficients. *Food Research International,* 28 (3), 261-271.

Awuah G.B., Ramaswamy H.S., Simpson B.K. and Smith J.P. 1996. Fluid-to-particle convective heat transfer coefficient as evaluated in an aseptic processing holding tube simulator. *Journal of Food Process Engineering,* 19: 241-267.

Balasubramaniam V.M. and Sastry S.K. 1996. Liquid-to-particle heat transfer in continuous tube flow: Comparison between experimental techniques. *International Journal of Food Science and Technology,* 31: 177-187.

Baik O.D., Grabowski S., Trigui M., Marcotte M. and Castaigne F. 1999. Heat transfer coefficients on cake baked in a tunnel type industrial oven. *Journal of Food Science,* 64 (4): 688-694.

Bhamidipati S. and Singh R.K. 1995. Determination of fluid-particle convective heat transfer coefficient. *Transactions of the ASAE,* 38 (3): 857-862.

Baptista P.N., Oliveira F.A.R., Oliveira J.C. and Sastry S.K. 1997. Dimensionless analysis of fluid-to-particle heat transfer coefficients. *Journal of Food Engineering,* 31: 199-218.

Barigou M., Mankad S. and Fryer P.J. 1998. Heat transfer in two-phase solid-liquid food flows: A review. *Transactions of the Institute of Chemical Engineers,* 76-C (March), 3-29.

Bialobrzewski I. 2006. Determination of the heat transfer coefficient by inverse problem formulation during celery root drying. *Journal of Food Engineering,* 74: 383-391.

Budzaki S. and Seruga B. 2005. Determination of convective heat transfer coefficient during frying of potato dough. *Journal of Food Engineering,* 66: 307-314.

Carson J.K., Willix J. and North M.F. 2006. Measurements of heat transfer coefficients within convection ovens. *Journal of Food Engineering,* 72: 293-301.

Chakrabandhu K. and Singh R.K. 2002. Fluid-to-particle heat transfer coefficients for continuous flow of suspensions in coiled tube and straight tube with bends. *LWT Food Science and Technology*, 35: 420-435

Chandarana D.I., Gavin A. and Wheaton F.W. 1990. Particle/fluid interface heat transfer under UHT conditions at low particle/fluid relative velocities. *Journal of Food Process Engineering,* 13: 191-206.

Chang S.Y. and Toledo R.T. 1990. Simultaneous determination of thermal diffusivity and heat transfer coefficient during sterilization of carrot dices in a packed bed. *Journal of Food Science,* 55 (1): 199-205.

Chen S.L., Yeh A.I. and Wu J.S.B. 1997. Effects of particle radius, fluid viscosity and relative velocity on the surface heat transfer coefficient of spherical particles at low Reynolds numbers. *Journal of Food Engineering,* 31: 473-484.

Costa R.M., Oliveira F.A.R., Delaney O. and Gemas V. 1999. Analysis of heat transfer coefficient during potato frying. *Journal of Food Engineering,* 39: 293-299.

Desrosier N. 1970. The Technology of Food Preservation. AVI Pub. Co. Inc., Westport, CN, USA.

Dincer I. 1997. New effective Nusselt-Reynolds correlations for food-cooling applications. *Journal of Food Engineering,* 39: 59-67.

Duquenoy A. and Nouafo J.L. 1997. Measurement of convection and radiant heat flux in a domestic oven. *Proceedings of the Seven International Congress on Engineering and Food (ICEF 7).* Sheffield Academic Press. Sheffield, UK. Sect. C: 25-28.

Farkas B.E. and Hubbard L.J. 2000. Analysis of convective heat transfer during immersion frying. *Drying Technology,* 18 (6): 1269-1285.

Fikiin A.G., Fikiin K.A. and Triphonov S.D. 1999. Equivalent thermophysical properties and surface heat transfer coefficient of fruit layers in trays during cooling. *Journal of Food Engineering,* 40: 7-13.

Gadonna J.P., Pain J.P. and Barigou M. 1996. Determination of the convective heat transfer coefficient between a free particle and a conveying fluid in a horizontal pipe. *Transactions of the Institute of Chemical Engineers, 74*-C (March): 27-39.

Ghisaberti L. and Kondjoyan A. 1999. Convective heat transfer coefficients between air flow and a short cylinder. Effect of air velocity and turbulence. Effect of body shape, dimensions and position in the flow. *Journal of Food Engineering,* 42: 33-44.

Holdsworth S.D.1997. Thermal Processing of Packaged Foods. Blackie Academic & Professional. London, UK.

Hulbert G.J., Litchfield J.B. and Smichdt S.J. 1997. Determination of convective heat transfer coefficients using 2D MRI temperature mapping and finite element model. *Journal of Food Engineering,* 34: 193-201.

Kelly B.P. 1997. Experimental and modeling studies related to aseptic processing of suspensions. *Proceedings of the Seven International Congress on Engineering and Food (ICEF 7).* Sheffield Academic Press. Sheffield, UK. Sect. C: 65-68.

Krokida M.K., Zogzas N.P. and Maroulis Z. B. 2002. Heat transfer coefficient in food processing: Compilation of literature data. *International Journal of Food Properties,* 5 (2): 435-450.

Landfield A. and Houska M. 2006. Prediction of heat and mass transfer during passage of the chicken through the chilling tunnel. *Journal of Food Engineering,* 72: 108-112.

Lewis M. and Heppel N. 2000. Continuous Thermal Processing of Foods. Pasteurization and UHT Sterilization. Aspen Publishers Inc. Maryland, USA.

Maesmans G., Hendrickx M., De Cordt S., Fransis A. and Tobback P. 1992. Fluid-to-particle heat transfer coefficient determination of heterogeneous foods: A review. *Journal of Food Processing and Preservation,* 16: 29-69.

Maesmans G.J., Hendrickx M.E., De Cordt S.V. and Tobback P. 1994. Feasibility of the use of a time-temperature integrator and a mathematical model to determine fluid-to-particle heat transfer coefficients. *Food Research International,* 27: 39-51.

Mankad S., Nixon K.M. and Fryer P.J. 1997. Measurements of liquid-particle heat transfer in systems of varied solids fraction. *Journal of Food Engineering,* 31: 9-33.

Meng Y. and Ramaswamy H.S. 2007. Dimensionless heat transfer correlations for high viscosity fluid-particle mixtures in cans during end-over-end rotation. *Journal of Food Engineering,* 80: 528-535.

Naveh D. and Kopelman I.J. 1980. Effect of some processing parameters on the heat transfer coefficients in a rotating autoclave. *Journal of Food Processing and Preservation,* 4: 67-77.

Nitin N. and Karwe M.V. 2003. Heat transfer coefficient for model cookies in a turbulent multiple jet impingement system. In *Transport Phenomena in Food Processing* (J Welti-Chanes, J.F. Vélez-Ruiz and G.V. Barbosa-Cánovas; Eds.). CRC Press. Boca Raton, USA. pp. 355-375.

Nixon K., Mankad S. and Fryer P.J. 1997. Heat transfer with solid-liquid food mixtures. *Proceedings of the Seven International Congress on Engineering and Food (ICEF 7).* Sheffield Academic Press. Sheffield, UK. Sect. C: 105-108.

Pajonk A.S., Saurel R., Blanc D., Laurent P. and Andrieu J. 2001. Heat transfer study and modeling during Emmental (Swiss cheese) ripening. *Proceedings of the Eighth International Congress on Engineering and Food (ICEF 8).* Technomic Pub. Co. Inc. Vol. I: 326-331.

Palazoglu K. 2006. Influence of convective heat transfer coefficient on the heating rate of materials with different thermal conductivities. *Journal of Food Engineering*, 73: 290-296.

Palazoglu T.K. and Sandeep K.P. 2002. Assessment of the effect of fluid-to-particle heat transfer coefficient on microbial and nutrient destruction during aseptic processing of particulate foods. *Journal of Food Science,* 67 (9): 3359-3364.

Pannu K.S. and Chinnan M.S. 2003. Relating food frying to daily oil abuse. Part I. Determination of surface heat transfer coefficients with metal balls. In *Transport Phenomena in Food Processing* (J Welti-Chanes, J.F. Vélez-Ruiz and G.V. Barbosa-Cánovas; Eds.). CRC Press. Boca Raton, USA. pp. 399-412.

Pannu K.S., Castaigne F. and Arul J. 2003. Thermal processing of particulate foods by steam injection. Part 1. Heating rate index for diced vegetables. In *Transport Phenomena in Food Processsing* (J Welti-Chanes, J.F. Vélez-Ruiz and G.V. Barbosa-Cánovas; Eds.). CRC Press. Boca Raton, USA. Pp. 309-326.

Peeples M.I. 1962a. Forced-convection heat transfer characteristics of fluid milk products. *Journal of Dairy Science,* 45: 303-310.

Peeples M.I. 1962b. Forced-convection heat transfer characteristics of fluid milk products during cooling. *Journal of Dairy Science*, 45: 1456-1458.

Rabiey L., Flick D. and Duquenoy A. 2007. 3D simulations of heat transfer and liquid flow during sterilization of large particles in a cylindrical vertical can. *Journal of Food Engineering,* 82: 409-417.

Ramaswamy H.S. and Zareifard M.R. 2000. Evaluation of factors influencing tube-flow fluid-to-particle heat transfer coefficient using a calorimetric technique. *Journal of Food Engineering,* 45: 127-138.

Rao M.A. and Anantheswaran R.C. 1988. Convective heat transfer to fluids in cans. *Advances in Food Research,* Vol. 32: 39-84.

Ratti C. and Crapiste G.H.1995. Determination of heat transfer coefficients during drying of foodstuffs. *Journal of Food Process Engineering,* 18: 41-53.

Rodríguez-Luna G., SeguraJaúregui J.S., Torres J. and Brito E. 1987. Heat transfer to non-Newtonian fluid foods under laminar flow conditions in horizontal tubes. *Journal of Food Science,* 52 (4): 975-979.

Rumsey T., Farkas D.F. and Hudson J.S. 1980. Measuring the steam heat transfer coefficient to vegetables. *Transactions of the ASAE,* 23 (4): 1048-1053.

Sablani S.S. and Ramaswamy H.S. 1996. Particle heat transfer coefficients under various retort operating conditions with end-over-end rotation. *Journal of Food Process Engineering,* 19: 403-424.

Sablani S.S. and Ramaswamy H.S. 1997. Heat transfer to particles in cans with end-over-end rotation: influence of particle size and concentration (% v/v). *Journal of Food Process Engineering,* 20: 265-283.

Sablani S.S. and Ramaswamy H.S. 1998. Multi-particle mixing behavior and its role in heat transfer during end-over-end agitation cans. *Journal of Food Engineering,* 38: 141-152.

Sablani S.S., Ramaswamy H.S. and Mujumdar A.S. 1997. Dimensionless correlations for convective heat transfer to liquid and particles in cans subjected to end-over-end rotation. *Journal of Food Engineering,* 34: 453-472.

Sahin S., Sastry S.K. and Bayindirli L. 1999. The determination of convective heat transfer coefficient during frying. *Journal of Food Engineering,* 39: 307-311.

Saunders E.A.D. 1988. Heat exchangers. Selection, design and construction. Longman Scientific & Technical, Harlow, England.

Simpson R., Jiménez M., Vega M., Romero A. and Costa M. 2001. Evaluation of commercial UHT milk and optimization of industrial process. *Proceedings of the Eighth International Congress on Engineering and Food (ICEF 8)*. Technomic Pub. Co. Inc. Lancaster, PN, USA. Vol. I: 708-712.

Sosa-Morales M.E., Orzuna-Espíritu R., and Vélez-Ruiz. J.F. 2006. Mass, thermal and quality aspects of deep-fat frying of pork meat. *Journal of Food Engineering.* 77: 731-738

Tessner W.A., Farkas B.E. and Sandeep K.P. 2001. Use of ablation to determine the convective heat transfer coefficient in two phase flow. *Journal of Food Process Engineering,* 24: 315-330.

Unklesbay K., Unklesbay N. and Biedrzycki.K. 1991. Convection heat transfer coefficient for a restructured pork/soy hull product. *Journal of Food Process Engineering,* 14: 197-208.

Varga S. and Oliveira J.C. 2000. Determination of heat transfer coefficient between bulk medium and packed containers in a batch retort. *Journal of Food Engineering,* 44: 191-198.

Vélez-Ruiz J.F. 2008. Notes of Transport Phenomena in Food Engineering. Universidad de las Américas, Puebla. Unpublished document. Puebla, Mexico.

Vélez-Ruiz J.F. and Soriano-Morales A. L. 2003. Evaluation of physical properties of foods by a computer program. *Revista Internacional Información Tecnológica*, 14 (4): 23-28. (In Spanish).

Vélez-Ruiz J.F. and Sosa-Morales M.E. 2003. Heat and mass transfer during the frying process of donuts. In *Transport Phenomena in Food Processing* (J Welti-Chanes, J.F. Vélez-Ruiz and G.V. Barbosa-Cánovas; Eds.). CRC Press. Boca Raton, USA. Pp. 429-444.

Verboven P., Scheerlinck N., De Baerdemaeker J. and Nicolai B.M. 1997. Variability of the local surface heat transfer coefficient and consequences for the heat processing of food products. *Proceedings of the Seven International Congress on Engineering and Food (ICEF 7).* Sheffield Academic Press. Sheffield, UK. Sect. C: 13-16.

Welty J.R., Wicks C.E. and Wilson R.E. 1976. Fundamentals of Momentum Heat and Mass Transfer. John Wiley and Sons, New York, NY, USA.

Wichchukit S., Zorrilla S.E. and Singh R.P. 2001. Contact heat transfer coefficient during double-sided cooking of hamburguer patties. *Journal of Food Processing and Preservation*, 25: 207-221.

Willix J., Harris M. B. and Carson J.K. 2006. Local surface heat transfer coefficients on a model beef side. *Journal of Food Engineering,* 74: 561-567.

Yanniotis S. and Kolokotsa D. 1997. Heat transfer to liquid films with phase change on a rotating disc. *Proceedings of the Seven International Congress on Engineering and Food (ICEF 7).* Sheffield Academic Press. Sheffield, UK. Sect. C: 1-4.

Zareifard M.R. and Ramaswami H.S. 1997. A new technique for evaluating fluid-to-particle heat transfer coefficients under tube-flow conditions involving particle oscillatory motion. *Journal of Food Process Engineering,* 20: 453-475.

Zhang Q. and Cavalieri R.P. 1991. Thermal model for steam blanching of green beans and determination of heat transfer coefficient. *Transactions of the ASAE*, 34 (1): 182-186.

Zogzas N.P., Krokida M.K., Michailidis P.A. and Maroulis Z. B. 2002. Literature data of heat transfer coefficients in food processing. *International Journal of Food Properties*, 5 (2): 391-402.

Zorrilla S.E. and Singh R.P. 2003. Heat transfer in double-sided cooking of meat patties considering two-dimensional geometry and radial shrinkage. *Journal of Food Engineering,* 57: 57-65.

Zorrilla S.E., Wichchukit S. and Singh R.P. 2001. Contact heat transfer coefficient during double-sided cooking of meat patties. *Proceedings of the Eighth International Congress on Engineering and Food (ICEF 8). T*echnomic Pub. Co. Inc. Lancaster, PN, USA. Vol. I: 321-325.

Zorrilla S.E., Wichchukit S. and Singh R.P. 2003. Transport phenomena during double-sided cooking of meat patties. In *Transport Phenomena in Food Processing* (J Welti, J. Vélez and G. Barbosa; Eds.). CRC Press. Boca Raton, USA. Pp. 295-308.

In: Food Processing and Engineering Topics
Editors: M. E. Sosa-Morales and J. F. Vélez-Ruiz

ISBN: 978-1-60741-788-0

Chapter 6

FOULING OF HEAT EXCHANGERS IN THE FOOD INDUSTRY

Jorge F. Vélez-Ruiz
Chemical and Food Engineering Department
Universidad de las Americas, Puebla, Mexico

1. INTRODUCTION

Plate and tubular heat exchangers are widely used in food industry, whatever is the production of the food item. They are more utilized to handle liquid foods, such as beer, concentrates, juices, milk fluids, nectars, soups, wine, and yogurt, just to mention a few. Food industry faces a common problem of fouling formation inside of heat exchanger equipments; it is a frequent concern in which the information is scarce.

Fouling may be defined as an accumulation or deposition of undesirable material on a heat transfer surface. It decreases the heat flow as well as fluid velocity, by reducing the heat transfer efficiency and increasing the pressure drop. In the food industry, fouling deposits can also act as sites for bacteria growing, causing product safety concerns. Fouling in the food industry is therefore an important economic factor.

It is well known that surfaces of the plate and tube heat exchangers do not remain clean after some time of operation. Scaling or deposits formed over the interior and/or exterior surfaces, may be analyzed as an "static" resistance, but the phenomenon is really a gradual build up of layers of damaged major food components, salts, minerals, dirty impurities, among others. Those formed deposits can severely affect the heat transfer equipment and process, particularly affecting the overall heat transfer coefficient. Two pictures of fouled surfaces may be appreciated in figure 1. When fouling becomes substantial, cleaning needs to be carried out, sometimes taking an important time that stops the normal production.

Fouling minimization requires engineering considerations in equipment and process designs and also in deciding operational conditions (fluid flow rate, fluid location, pressure, and temperature). This chapter will include the fundamentals of fouling, methods of evaluation, process variables, as well as available and experimental data of scale composition and magnitudes.

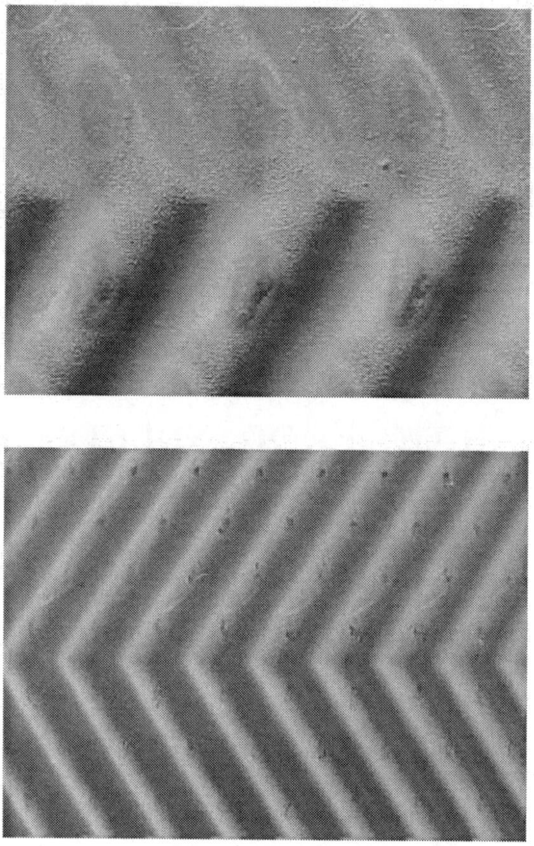

Figure 1. View of tow plates of a HE fouled with milk fluids (Tejeda-Urrutia *et al.*, 2003).

2. Fundamentals

Fouling has been broadly studied through the time, due to the economical and operational impact on the industry. Particularly, in food industry there is scarce information. Most of the works have been developed in milk processing for dairy products manufacturing, and very few data has been generated for other food fluids. Fouling of heated metallic surfaces of both heat exchanger types (plates and tubes) directly contributes toward increased production costs: capital costs, energy costs, maintenance costs, and shutdown costs (Vélez-Ruiz, 2008).

2.1. Costs of Fouling

Heat transfer equipment must be provided with additional surface area to compensate the reduction in the heat flow due to fouling. This increasing in exchanger size implies additional costs of installation, and larger pumps and motors due to pressure loss.

Energy costs are related to higher needs for pumping and power to reach the required operational flow. Higher energy demands and pressure losses are also involved in refrigerators, cooling towers, and condensers.

While maintenance costs arise from cleaning activities, dismantling and reassembling of heat transfer equipments, as well as on failures of the overall process performance. Additional costs are required for water treatment or chemical reagents for fouling treatment and cleaning. And finally, shutdown costs are involved with the value of lost production, which occurs when the heat transfer equipment is shut down due to fouling implications.

In according with Saunders (1988), the total costs in England, for the period of 1978-1980 due to the aforementioned cost aspects, were equivalent to 450 – 750 million of US dollars.

2.2. Heat Transfer

The main variables affecting the performance of a heat exchanger, in which two fluid streams are involved, include inlet and outlet temperatures of the cold and hot fluids, mass flow rates, available surface, thermal properties of the fluids and heat exchanger material, internal and external convective heat transfer coefficients, and the degree of scaling formed on the surfaces, inside and outside in the case of tubes, and hot and cool side in the case of plates. In the food industry, most of the processing fluids are liquid phases and gases are rarely handled. Thus, the rate of heat flow transferred through a heat exchanger may be computed by (Kern, 1997; Mc Cabe *et al.*, 2001; Vélez-Ruiz, 2008).

$$\dot{Q} = \dot{\omega} Cp \Delta T = UA\Delta T_{LM} \tag{1}$$

where: $\dot{Q}$ is the transferred heat (W), $\dot{\omega}$ is the mass flow rate (kg/s), it may be any of the two process fluids, the hot ($\dot{\omega}_h$) or the cold ($\dot{\omega}_c$), Cp is the specific heat capacity (kJ/kg°C), ΔT is the difference between the outlet and inlet temperatures, being positive for the fluid to be heated and negative for the fluid to be cooled, U is the overall heat transfer coefficient (W/m^2K), A is the surface area for heat transfer (m^2), and ΔT_{LM} is the logarithmic mean temperature difference or driving force of the whole process.

The logarithmic mean temperature difference is given by the relationship 2. The specific heat and other thermal properties that are needed for evaluation of the convective heat transfer coefficient must be obtained at average temperature, as the mean between the inlet and outlet temperatures for each fluid.

$$\Delta T_{LM} = \frac{(\Delta T)_{x=0} - (\Delta T)_{x=L}}{\ln \frac{(\Delta T)_{x=0}}{(\Delta T)_{x=L}}} \tag{2}$$

where: ΔT denotes the temperature difference between the hot and the cold fluids at a given location (°C), $(\Delta T)_{x=0}$ is for the inlet location (°C), and $(\Delta T)_{x=L}$ is for the outlet location (°C).

To evaluate the heat transfer in terms of the individual resistances, for a plate heat exchanger a simpler relation is used. For tube heat exchanger is necessary to consider an area of reference, in according to the following equations (Kern, 1997; Vélez-Ruiz, 2008).

$$\frac{1}{U} = \frac{1}{h_h} + \frac{\Delta x}{k} + \frac{1}{h_c} \tag{3}$$

$$U_e = \frac{1}{\dfrac{1}{h_e} + \dfrac{D_e \Delta x}{D_L k} + \dfrac{D_e}{D_i h_i}} \tag{4}$$

$$U_i = \frac{1}{\dfrac{1}{h_i} + \dfrac{D_i \Delta x}{D_L k} + \dfrac{D_i}{D_e h_e}} \tag{5}$$

where: U, U_e and U_i are the clean overall heat transfer coefficients (W/m^2K), U for plates, U_e and U_i for tubes heat exchanger considering the external or internal area as reference, respectively; h_h, h_c, h_e, h_i are the convective heat transfer coefficients (W/m^2K), with the specification for hot (h_h), cold (h_c), external (h_e) or internal (h_i) fluids, Δx is the metal thickness (m), k is the thermal conductivity of the exchanger material (W/mK), D_i is the inside diameter of the tube (m), D_e is the outside diameter of the tube (m), and D_L is the logarithmic mean diameter for the tube [$D_L = (D_e - D_i)/\ln (D_e/D_i)$] (m),

The evaluated U by some of the equations 3-5, is identified as clean overall coefficient because it does not include the effect of fouling. The effect of deposits is represented quantitatively by the fouling factor, which must be determined experimentally. Some data of the convective and overall heat transfer coefficients are included in tables 1 and 2, correspondly.

Table 1. Some reported values of the convective heat transfer coefficient (h) in heat exchanger equipments

Process Fluid(s)	h range (W/m^2K)	Equipment	Reference
Free convection, air	5-50	In general	Welty *et al.*, 1976
Forced convection, air	25-50	In general	
Forced convection, water	250-15000	In general	
Boiling water	2500-25000	In general	
Condensing water vapor	5000-100000	In general	Welty *et al.*, 1976
Tomato paste concentration	795-15000	Rotary evaporator	Chen *et al.*, 1979
Sugar solution concentration	1800-13000	Thin film evaporator	Stankiewicz & Rao, 1988
Milk heating (UHT)	5160-8170	Tubular exchanger	
Orange juice pasteurization	983- 6500	Plate heat exchanger	Kastanas *et al.*, 1995
Heating water	8387-24245	Plate heat exchanger	Kim *et al.*, 1999
Steam condensation (10 kPa)	8000-12000	Condenser	
Steam condensation (1 MPa)	15000-25000	Condenser	Peters *et al.*, 2004
Water vaporization (<500 kPa)	3000-10000	Vaporizer	
Water vaporization (0.5-10 MPa)	4000-15000	Vaporizer	

UHT: ultra high temperature.

Table 2. Common values of the overall heat transfer coefficient (U) in heat exchanger equipments

Process Fluid(s)	U (W/ m^2K)	Major resistance	Reference
Steam condenser	1000-5000	Coolant	Karlekar and Desmond, 1982
Freon 12 to water	300-850	Freon	
Gas to gas	10-40	Gas	Holman, 1986
Water to oil	110-350	Oil	
Water to water	850-2000	Water	
Feedwater heater	1100-8500	---	
Steam condenser	1100-5600	---	
Process water to other fluids in finned-tube, double tube	77-326 (gas)-(water)	Gas	Saunders, 1988
Evaporation of whole milk	1270-1530	Milk	Dodeja *et al.*, 1990
UHT of milk fluids	1730-2980	Milk	Kastanas *et al.*, 1995
Heating of WPC solution	560-640	WPC	Delplace *et al.*, 1997
Pasteurization of orange juice	942-3640	Juice	Kim *et al.*, 1999
Air to air	30	Air	Saravacos and Kostaropoulos, 2002
Steam or water to air	100	Air	
Water or steam to vegetable oil	800	Oil	
Water or steam to viscous food	1000	Food	
Water or steam to liquid food	1500	Food	
Steam to water	2500	Water	
Water heating (low flow)	1000-2600	Water	Tejeda-Urrutia *et al.*, 2003
Water heating (high flow)	2300-4000	Water	
Milk heating (low flow)	900-2400	Milk	
Milk heating (high flow)	1700-4000	Milk	
Milk pasteurization	880-1500	Milk	Jun and Puri, 2006

UHT: Ultra high temperature, WPC: Whey Protein Concentrate.

The presence of deposits on the heat transfer surface, increases the thermal resistance to heat flow. U clean may be related to the fouling resistance by equation 6. In the case of the plate heat exchanger (PHE), the fouling effect is related with the hot and the cold process fluids.

$$\frac{1}{U_D}=\frac{1}{U}+R_{fi}+R_{fe} \tag{6}$$

where: U_D is the design or dirty overall heat transfer coefficients (W/m^2K), U is the clean overall heat transfer coefficient (W/m^2K), and R is the fouling resistance (m^2K/W), including the internal effect of fouling (R_{fi}) and the external effect of fouling (R_{fe}).

With regard to equipments calculations and design, the food engineer must start with the use of simple factors, evaluated under clean process conditions, unless data of the fouled coefficients on operating conditions are available. The fouling factor concept assumes a uniform thickness and a constant performance of the formed deposit (Mc Cabe *et al.*, 2001). Some numerical values of fouling factors for chemical and food industries are given by several authors (table 3), corresponding to satisfactory processing under normal performance of the heat exchangers.

Table 3. Reported values for fouling factor (R_f) in heat exchanger equipments

Fluid involved in the fouling	Rf (m^2K/W)	Reference
Muddy or silty water (< 1 m/s at 51.7°C)	0.00050	Rohsenow and Hartnett, 1973
Muddy or silty water (< 1 m/s and >51.7°C)	0.00070	
Sea water below 1 m/s and 51.7°C	0.00010	
Sea water below 1 m/s and over 51.7°C	0.00020	
Industrial air	0.00050	Karlekar &Desmond, 1982
Refrigerating liquid	0.00020	
Distilled water	0.00009	Walker, 1982
Treated water for cooling tower	0.00018	
Vegetable oils	0.00053	
Alcohol vapor	0.00009	Holman, 1986
Refrigerating liquid	0.00020	
Sea water below 51.7°C	0.00009	
Treated boiler feedwater above 51.7°C)	0.00020	
City water	0.00058	Kern, 1997
River water	0.00058-0.00088	
Process water	0.00003	Thonon *et al.*, 1999
Solution of TiO2	0.00004-0.00013	
Solution of CaCO3 (1 m/s)	0.00002	
Solution of CaCO3 (0.8 m/s)	0.00008	
Solution of CaCO3 (0.4 m/s)	0.00012	
Clean steam	0.00005	Saravacos & Kostaropoulos, 2002
Distilled water	0.00010	
Inorganic gases	0.00150	
Polymer solutions	0.00100	
Sea water	0.00020	
Vegetable oil	0.00050	
Whole milk	0.00011-0.00030	Tejeda-Urrutia *et al.*, 2003

3. Mechanisms and Measurements of Fouling

Fouling deposits are formed as a result of adhesion of species from the liquid food to the heat transfer surface and cohesion between elements of such food item. In according with Kim and Lund (1998) deposition of milk components may be initiated with the formation of a monolayer, formed either by supersaturation or by a reaction of adsorption on the heat transfer metal. Once the monolayer is deposited, successive build-up of other layers results on the interaction between the formed monolayer and the bulk components of the process fluid.

From numerous works and observations on fouling of heat transfer equipment, some typical patterns have been identified, they may be observed in figure 2. In according with Sandu and Lund (1982), curve A, or linear fouling, describes a phenomenon in which the deposition rate predominates over the removal rate. Curve B, or asymptotic fouling, illustrates a response in which a balance between removal and deposition rates is developed. And an intermediate fouling trend is represented by the curve C or fouling falling rate. Finally, curves A*, B* and C* denote fouling behaviors in which an induction period (incubation or nucleation) is observed, and followed of similar trends of A, B, C kinetics.

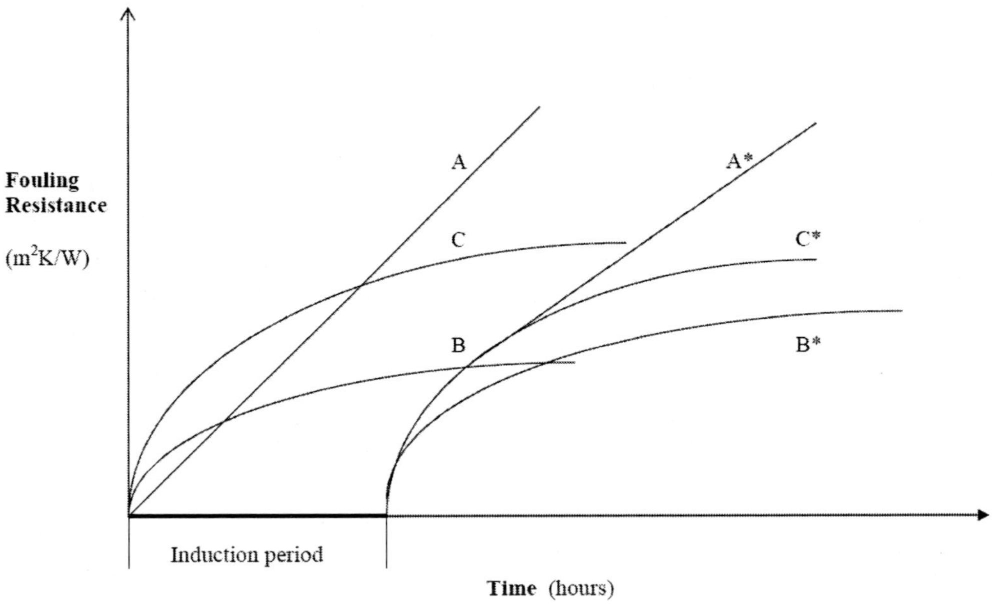

Figure 2. Fouling trends in heat transfer equipment (From Sandu and Lund, 1982).

3.1. General Mechanisms of Fouling

Saunders (1988) mentions six categories in which fouling may be formed, they are summarized next:

1) Chemical fouling, in which the deposits are generated from chemical reactions in the process fluid or at the heat transfer surface, the surface material itself is not a reactant. This mechanism is also known as reaction fouling (Paterson and Fryer, 1988). Denaturation of proteins, caramelization of sugars, formation of salts in food fluids, and polymerization of hydrocarbons are some examples.

Formation of aggregates for fouling of milk fluids is due to protein denaturation, and lately to aggregation of them. It has been experimentally corroborated that the denaturation behavior is strongly related to process temperature (Simmons *et al.*, 2007) and pH in the kineticsof fouling for β-lactoglobuline (Gotham *et al.*, 1992).

2) Corrosion fouling, that implies the reaction of the metallic surface, producing adherent corrosion materials. The corrosion products may promote in turn, the incorporation of other fouling materials.

3) Biological fouling, related with living organisms that adhere to the heat transfer surface. Bacteria, fungi, alga and slime may deposit or even live on that surface. Algae and various marine organisms grows on heat exchangers tubes, forming a film or coating (Schwartz, 1981).

Although most of the fouling films are chemical in nature, the term biofilm is being widespread and gaining diffusion to describe those deposits of biological origin in which microbial activity exists. In the case of bacterial adhesion, the surface energies of bacteria and surface have a significant effect; therefore that energy should be conserved on an optimum value that will favor an easy microorganism removal (Zhao and Liu, 2006).

4) Freezing fouling, that is developed as a result of the solidification of liquid materials or components under subcooling temperatures.

5) Particulate fouling, occurring when finely divided solids that are suspended in the process fluid are accumulated on the metallic surface. In case of settling by gravity, the process is identified as sedimentation fouling.

According to Simmons *et al.* (2007), in the aggregation phenomenon as a type of particulate fouling, the bulk reactions are responsible of that scaling as a first stage. And, a second step in fouling of milk fluids results from an interaction between denaturated proteins particles. Aggregation is a function of the applied shear fields and strength of formed layers.

6) Precipitation fouling, that is related to the precipitation of dissolved substances on the heat transfer surface. Calcium sulfate, lithium carbonate and magnesium silicate having inverse solubility may precipitate on that surface. Suspended solids of some cooling/heating fluids are usually found as sand, iron, silt or other solid particles. Then if fluid velocities are not high enough to keep them in suspension, those particles settle down causing scale problems (Schwartz, 1981). Aggregation of β-lactoglobuline has been identified by Gotham *et al.* (1992) as a simultaneous phenomenon, of denaturation and precipitation, in fouling formation.

In a more simple analysis of the fouling mechanisms, de Jong (1997) and Kastanas *et al.* (1995) established only two types, a first type known as protein fouling with relatively soft deposits and formed at temperatures between 70-100°C, being voluminous and white, as may be appreciated in figure 1. The second type, identified as mineral fouling is a hard and granulose deposit with a high mineral content, and formed at high temperatures, above 110°C, it is brittle and with brownish appearance.

3.2. Factors Involved in Fouling

There exist several factors influencing the fouling phenomenon, they are divided by Lewis an Heppell (2000) in two types, those related with the process and those related with the food product.

Process dependent variables include flow rate, shear rate/stress, pressure, temperature and turbulence; in this group the heat transfer equipment (plates or tubes, material type, corrugation angles, channels dimensions) may be also considered. Composition, particle type, physicochemical properties, solids concentration, and reaction velocities are considered as product dependent.

At the beginning in heat exchanger equipments for milks, the fouled material is predominantly formed by proteinic deposits, and then the deposit is mineral (Sahoo *et al.*, 2005). And when the fluid is based on tomato, the deposits include protein and pectin residues. Therefore, depending of the fluid to be handled it will be the fouling mechanism and the suitable methodology that must be applied to analyze the fouling phenomenon (Vélez-Ruiz, 2008).

A diversity of techniques to monitor this physicochemical phenomenon have been reported, such as weighting of accumulated mass, analysis of chemical composition of the scales and surface, use of column adsorption chromatography, measure of electrical conductivity/resistance of the solid deposit, recording of fluids temperature profiles, heat fluxes through the equipment, overall heat transfer coefficient, and pressure increases. Also the quantification of radial diffusion, scale thickness, thermal resistance, X-ray microanalysis, among others, have been utilized.

4. FOULING IN FOOD HEAT TRANSFER PROCESSES

Fouling of thermal process equipments by food industry is a severe concern, in which a diversity of components, organic and minerals, are deposited. Fouling occurs in continuous food processes involving heat transfer between liquids, particularly in ultrahigh temperatures treatments and evaporation. Scale is formed as a result of adhesion of species and cohesion between elements. Heat may induce reaction in the food fluid that causes formation of compounds depositing on the heat transfer surface.

Fouling can occur due to several reactions taking place within the food item, in milk, a denaturation phenomenon may be related with proteins, milkstone may be associated with inverse solubility salts in dairy products. In sugar refining, scales consist principally of calcium salts, metallic oxides and silica (Hartel, 1992). Fouling may also contain carbohydrates, fat, minerals, as well as dead or alive microorganisms.

In dairy industries, scaling consists of protein aggregates and minerals, being the β-lactoglobulin one of the major contributors, due to thermal denaturation of this protein (Guérin *et al.*, 2007). In pasteurization treatments, in the order of 85-110°C milk generates fouling highly proteinaceous; while at 110-114°C used in UHT processes, the deposit is mainly composed of Ca and P, due to the inverse solubility of their salts (Simmons *et al.*, 2007).

The fouling phenomenon is a dynamic process in which process conditions contribute to a higher or lower degree; the rate of fouling is a balance between the "growing force" or deposition and the "destructing force" or erosion (Jun and Puri, 2005). Whether the transport phenomena is considered, fouling may be shear (momentum), temperature and concentration dependent, as a function of the control driven force on fouling.

Fouling phenomenon often takes place in several stages: an induction period has been noted before significant deposits are observed, then a fouling period, that is the most important and mainly studied, and finally a post-fouling stage. The second stage as the most studied has been correlated to temperature and fluid velocity (Paterson and Fryer, 1988). In the last period, the fouling level is asymptotic and no further decreasing in the heat transfer rate is noticed.

There are two main trends in which the fouling has been faced in industrial work, one is the modification of the fluid process composition; and the other is the modification of surface characteristics of the heat exchangers. Modification of fluid nature by adding of chemical additives may be very effective, but this is not possible in most of the industrial food applications, and much less in the dairy industry (Rosmaninho *et al.*, 2007). In order to characterize and identify the fouling extension and its effect on process variables, a good number of methods to measure has been utilized, such as: i) temperature monitoring, ii) deposit formation, iii) overall heat transfer coefficient decreasing, iv) pressure increasing, v) application of fouling devices, vi) utilization of fouling sensors, vii) application of instrumental technique, and viii) use of indirect indicators (acidity-pH changes, alcohol susceptibility, denaturation degree, electric conductivity, heat stability, gelation, ammonia concentration and sedimentation, among others (Lewis and Heppell, 2000).

Regarding to surface modification, still it is in stage of research and not major modifications have been introduced for heat transfer equipment in the food industry; in general, the stainless steel continues being the best material.

Finally, very related with the fouling phenomenon is the cleaning process as a lately and critical stage in food manufacturing. Currently, a typical sequence of cleaning research investigation includes a device where fouling deposit is first completed, and then the fouled surfaces are cleaned; being the main difficulty in these studies, that is not possible to form a reproducible fouling scale. Due to this relation fouling-cleaning, a good number of research works face fouling as a previous part, when the final purpose is to produce cleaning-in-place routines, in which it is necessary an understanding of the interactions between food deposits and metallic surfaces.

5. Fouling Research in Food Processes

Given the economic impact of fouling in heat exchangers utilized in the food industry, it is not surprising that there is a considerable amount of research on the fouling and its implications; even though the fouling phenomenon is not completely understood. More of the investigations have utilized liquid milk as the studied material generating foulings; and as β-lactoglobulin is one of the main contributors in heat exchanger fouling in the food industry, whey protein concentrate (WPC) solution has been employed as a model fluid. However, other food items in which minor research has been conducted are creams, desserts, juices, milk based drinks, purees, sauces and spreads (Lewis and Heppell, 2000). With water on the other side, fouling problems may occur, mainly calcium or magnesium salts deposits related with hardness; and biofilms related with natural microorganisms.

5.1. Studies Related with Heat Transfer

Skudder *et al.* (1981) conducted a work to analyze the effect of potassium iodate added to milk on the fouling formation during an UHT treatment. They found an important effect of this salt in reducing protein deposits. This addition permitted the utilization of longer processing times for milk, before unacceptable pressures were recorded in the PHE.

Swartzel (1983) developed an investigation to find a relation between fouling and total thermal treatment given to the milk in a PHE with sedimentation rates of the packaged product. Both phenomena, fouling and sedimentation were positively correlated to thermal treatments.

An experimental study to analyze the effect of time on the fouling of PHE used for lowfat milk UHT sterilization was carried out by Lalande *et al.* (1984). The composition analysis allowed to distinguish two types of deposits: one in the preheating section with protein 50%, minerals 40%, and fat 1% (w/w), and another with minerals 75%, protein 15%, and fat 3%

(w/w). Fouling rate was higher during pasteurization of raw milk than during sterilization of pre-pasteurized milk.

A pilot plant PHE was utilized by Yoon and Lund (1994) to study the fouling response in which two operating methods were applied. Method B (control by increasing the inlet temperature of the heating medium) resulted better than the method A (control by increasing the flow rate of the heating medium) in maintaining constant temperatures for the process fluid, during the fouling period.

Delplace and Leuliet (1995) studied the fouling of a PHE in which WPC solutions were utilized as model fluid. Experimental data and predictions were very similar. For experimental analysis, measurements of U and dry mass were completed; whereas predictions were based on an empirical approach. Those predictions showed a fouling rate depending of concentration of denaturated proteins and the mixing intensity in the channels of the PHE.

An UHT plant was designed by Kastanas *et al.* (1995) to study the effect of fluid type and pH on the fouling behavior. The effect of factors on fouling was evaluated with the overall heat transfer coefficient, finding that the relations between the fouling response and pH of processed milk was non-linear. At pH above 7.0 fouling did not affect the U.

A comparative study was carried out by De Rafael and Calvo (1996) to analyze the scale formation of three types of milk (cow, ewe, goat) during heating. They observed that the differences in deposit concentration were not related to milk composition. Thus, they suggested two main factors involved in fouling: milk changes during heating and milk stability.

Delplace *et al.* (1994 and 1997) developed an experimental work to measure the unfolding of β-lactoglobuline also identified as protein heat denaturation in a PHE in a pilot plant, by using an immunodiffusion technique. They predicted the fouling effects by application of a reaction engineering approach with a numerical method. Experimental results were successfully compared with numerical calculations, in which the bulk temperature and the second order kinetics β-lactoglobuline were measured and predicted.

The effect of geometry and flow conditions on particulate fouling with small-scale and full-scale PHE including three corrugation angles (30, 45 and 60°) was studied by Thonon *et al.* (1999). They observed that the fouling resistance curves exhibited asymptotic trends influenced directly by velocity, and inversely by corrugation angle; the angle of 30° developed greater fouling in comparison with 60°.

Petermeier *et al.* (2002) developed a hybrid model of fouling in a tubular exchanger. Biological and chemical effects, as well as pressure drop and temperature profiles were and may be calculated by the model. By comparing measured and calculated values, they obtained differences of 10-20% for pressure, and 5-10% for temperature distribution.

Simmons *et al.* (2007) completed a study of fouling to know the effect of temperature and shear rate. They found that at temperatures lower than 75°C, the aggregates of WPC were small and weakly bonded, while at higher temperatures and shear rates, the formed aggregates were more rigid and denser.

Guérin *et al.* (2007) carried out an experiment to measure the electrical conductivity for deposits of WPC solutions in a PHE. They investigated the effect of calcium concentration (70-88 mg), Reynolds number (3200-5000) and temperature (60-96°C), founding that calcium ions and temperature favored the fouling; whereas flow rate, above a critical Reynolds magnitude, had a contrary influence. At higher temperatures the aggregation was due to covalent bonding.

5.2. Studies Related with Surface Modifications

Foster and Green (1990) conducted an investigation to analyze the fouling on a stainless steel surfaces at 140°C. They observed that deposits containing a higher protein content were easier of removing than other type of fouling. At 140°C, the deposit mass increased as a linear function of heating time, becoming more uneven in appearance.

The deposition phenomenon of milk proteins onto a stainless steel surface was followed up by Kim and Lund (1998). A dynamic and continuous flow adsorption set was adopted for adsorption studies, in which the column adsorption chromatography was the main element of the system. Based on adsorption characteristics of milk proteins onto the metallic surface, it has allowed to the researchers, the interpretation of a multiple layer deposition as the mechanism of fouling. Adsorption behavior of β-lactoglobuline and κ-casein was different, those differences were reflected on the adsorption curve as well as the amount of adsorption.

Zhao and Liu (2006) modified a stainless steel surface (by electroless plating of Ni-P and PTFE) in order to minimize bacterial adhesion. Once that they identified the characteristics of the coating, their initial experiments resulted in a reduction of 88-93% of attachment of *E. coli*, in comparison with the free surface of the steel.

A work in which the stainless steel was modified by coating, ions implantation, sputtering, and vapor deposition was carried out by Rosmaninho *et al.* (2007), in order to know the effect of such modifications on the fouling phenomenon. They found that calcium phosphate, milk model fluid and whey protein fouling was minimun in a surface auto-catalyzed with Ni-P-PTFE, being the most suitable. Whereas for bacterial adhesion, the most promising material corresponded to the ion TiC implanted surface.

5.3. Studies Related with Cleaning

Karlsson *et al.* (1996) modified a stainless steel surface in order to study the β-lactoglobuline fouling at five temperatures (25, 60, 73, 77 and 80°C) and its removal by an anionic surfactant. They concluded that the fouling and cleaning response of chromium oxide and stainless steel differs markedly from the behavior of the methylated silica. At low temperatures the silica was the material most easily cleaned, while at 73°C chromium and steel showed the higher cleanability.

Xin *et al.* (2002) investigated the dissolution kinetics and mechanical properties of fouling formed with WPC as part of a cleaning protocol. They found similar physical and chemical behaviors of WPC and whey protein deposits and concluded that dissolution is very important for cleaning purposes.

The fouling and cleaning processes on a metallic surface in an experimental unit was analyzed by Chen *et al.* (2004). Fouling was propitiated by recycling of the sample fluid and setting a high wall temperature. Measuring of fouling was done with an electrical resistance method, that has the potential to measure the fouling thickness in situ and in real time.

Liu *et al.* (2006) carried out a study to determine the adhesive and cohesive forces that are involved during fouling formation of WPC solution, finding an apparent cohesive/adhesive strength of 0-10 J/m^2 as a function of temperature. At room temperature, the adhesive strength was greater than the cohesive strength. A micromanipulation technique was used to obtain the whey protein scales.

Several investigations related to fouling in food industry and their characteristics are briefly cited in table 4.

Table 4. Summarized information on fouling research

Equipment	Method	Characteristics/Observations/ Comments	Measures & Reference
UHT plant	Experimental	Raw milk with addition of KIO_3 before UHT treatment Pressure increased as a function of processing time	Weight and composition Skudder *et al.*, 1981
UHT plant	Experimental	Raw whole milk at different temperatures and times Two categories of deposit were distinguished	Weight and composition Lalande *et al.*, 1984
UHT plant	Experimental	Raw milk at different temperatures (70, 80, 90, 100, 110°C, 90s) and different times (30, 60, 90 and 120 s at 90° C)	ΔP and mass deposits. Patil and Reuter, 1986
Stainless steel	Experimental	Reconstituted skimmed milk Reaction fouling	T. Paterson & Fryer, 1988
Stainless steel surface	Secondary ion mass spectrometry	Milk at different temperatures (0-140° C) and focused on 140° C Formation and adhesion of fouling were studied separately	Amount and thickness Foster and Green, 1990
Metalic surface	Differential Scanning Calorimeter	Bovine serum albumin and bovine β-lactoglobuline solutions The resultant trace was the sum of simultaneous denaturation and aggregational enthalpies Fouling decreased with increasing pH	Enthalpies and weight changes Gotham *et al.*, 1992
Three PHE	Immunodiffusion	WPC solution 1% (w/w) Only proteins denatured in the thermal boundary layer were involved in deposit formation	U, ΔP and mass deposits Delplace *et al.*, 1994
PHE (two sections)	Deposit analysis	Raw, clarified and whole milk Patterns of T distribution were not affected by the Reynols number, in the range of 1100-4400	T, and mass deposits Yoon and Lund, 1994
PHE	Immunodiffusion	WPC solution 1% (w/w) as fouling model fluid and several arrangements of a PHE	U and mass deposits. Delplace and Leuliet, 1995
Tubular exchanger	Thermal and hydraulic characterization	Skimmed milk at different pH, goat milk, and UF Reduction in pH significantly increases the rate of fouling	Temperature and pressure Kastanas *et al.*, 1995

Table 4. Continued

Equipment	Method	Characteristics/Observations/ Comments	Measures & Reference
Modified surface	Ellipsometry	β-lactoglobuline solutions Fouling formation and cleanability were studied at 25, 60, 73, 77 and 80° C	Adsorbed protein Karlsson *et al.*, 1996
Lab scale pasteurizer	Deposit analysis	Cow, ewe and goat milks adjusted at pH 6.7 Batch of 20 L of whole milk per run Milks were heated from 62° C to 80° C	Weighing and analysis De Rafael & Calvo, 1996
PHE (three sections)	Immunodiffusion	WPC solution 1% (w/w) composed by α-lactoalbumin, β--lactoglobulin, lactose, and minerals Mass deposits augmented with time	U and mass deposits Delplace *et al.*, 1997
Test tubes	UV spectroscopy	WPC solution Dissolution kinetics and mechanical characterization (dynamic properties) through the fouling-cleaning processes	Dissolution rate, Xin *et al.*, 2002
Stainless steel surface	Sputter coating (vapor deposition)	The layer was added, and several modified SS samples were characterized Surface modification can reduce CaSO4 scale, bacteria deposits and protein adsorption.	Composition, roughness and topography Santos *et al.*, 2004
Test section	Electric resistance	Test section simulating the heat transfer on a plate. Skim milk 10-30% Vertical and horizontal growing of deposits	Thickness Chen *et al.*, 2004
Stainless steel discs	Spinning disc measurement	Discs were fouled during heat transfer by skim milk and whey protein Skimmed milk 4% protein and WPC solution 5% protein Composition and cleaning of fouling was completed	Dry mass: 50-100 g/m^2 Thickness: 0.1-0.2 mm Morison & Larsen, 2005
PHE	Micromanipulation	T-shape probe to pull formed deposits and measuring of imposed force WPC solution 1.49% at 54 to 86° C Fouling runs lasted 2 h Composition and structure of fouling was analyzed	Forces and work Liu *et al.*, 2006
Stainless steel 304	Electroless plating	Coating of SS surface as a function of pH, T and PTFE concentration Coated surfaces with Ni-P-PTFE reduced *E. coli* attachment by 88 - 93% in comparison with SS 304	Thichness, composition, and morfhology Zhao & Liu, 2006
Couette apparatus	Deposition with step changes in T and γ	1.5% solutions of WPC35 with added minerals in different levels Fouling with different characteristics as a function of T and γ	Taylor number: 40-229 Simmons *et al.*, 2007
Stainless steel surface	Modification of surface by four techniques	The layer was added, several modified SS samples were characterized Fouling was affected by the surface, as well as the cleaning step	Rosmaninho *et al.*, 2007

Equipment	Method	Characteristics/Observations/ Comments	Measures & Reference

PHE	Electrical conductivity of deposits	WPC solutions 1% at 60-96° C Calcium was added at controlled level - Thickness measured by weighing and with a pneumatic lifting device Water was used as heating medium	Thickness and fouling rate Guérin *et al.*, 2007

UHT: Ultrahigh temperature, ΔP: pressure difference, T: temperature, PHE: Plate heat exchanger, WPC: whey protein concentrate, U: overall heat transfer coefficient, UF: ultrafiltered milk permeate,

SS: stainless steel, PTFE: polytetra-fluoroethlene, γ: shear rate.

5.4. Fouling Models

Similar to other food engineering challenges, the fouling formations have been modeled with different approaches, such as hydrodynamic, kinetic and thermodynamic in which analytical and numerical solutions have been proposed for one, two and three dimensions, and even some statistical solutions have been utilized. Furthermore, one of the interesting and fundamental facts, whatever is the modeling, the fouling formation involves simultaneous momentum and heat transfer phenomena, and also mass transport in less degree. The knowledge of this phenomenon and its proposed mechanisms will contribute to improve and control process conditions with a main objective: fouling reduction in the food industry.

Such considerations would no doubt involve the interactions between the compositional changes as well as the effect of fluid motion. The process is very complex and is difficult to model. However, without a comprehensive modeling of the process, which can be backed up by practical observations, optimization of the design and operational conditions is not possible or at least, would not be realistic. There are many intrinsic issues that can not be observed when many assumptions are made in order to simply the mathematics involved. And although an exact mathematical modeling of the fouling behavior seems impossible, any model with good background will be very useful to face the industrial problem.

One of the first efforts for fouling modeling was proposed by Sandu and Lund (1982), who considered momentum, heat and mass transfer fundamentals, as well as reaction elements to design a computational program based on a numerical procedure. It was a very good tool to model and elucidate some fouling characteristics.

To control the fouling challenge, one of the simplest and straightforward approach, is to utilize a model based on the physical and chemical fundamentals that are implied in the fouling phenomenon (de Jong, 1997). Thus, this model may be applied to optimize the operational conditions for the heat transfer equipment.

Petermeier *et al.* (2002) tested a hybrid approach, getting differences of 5-20% for pressure drop and temperature. They modeled analytical equations of fluid flow and heat transfer with fuzzy logic systems, valid for tubular heat exchanger over a wide range of operational restrictions.

In a review paper by Jun and Puri (2005) a general classification of fouling models grouped them in two main divisions: a first group of models based on hydrodynamic and thermodynamic fundamentals, and a group with integrated fouling dynamics models.

A computer model that allowed the control of milk temperature of heat transfer equipment affected by fouling, was developed by Nema and Datta (2005). The accurate control of temperature by increasing of the steam temperature is one of the ways to maintain the continuity of the process before the equipment is stopped for cleaning.

Similarly, Sahoo *et al.* (2005) proposed a mathematical model oriented to predict the fouling phenomena. Fouling behavior in a triple heat exchanger was simulated by application of fluid dynamics and heat transfer fundamentals. They identified that fouling deposit was stabilized after 105 min, fouling was function of the interface temperature (fouling-bulk fluid), and also the shear stress of the fluid on the surface equipment.

Jun and Puri (2006) implemented and validated a 2D fouling model using a PHE, the model was based on coupled transport phenomena equations and protein reactions. In it, the fouling performance was predicted, showing strong correlation between predicted and operating conditions. Basic ideas and some particularities of the proposed models for fouling are included in table 5.

Table 5. Proposed models for fouling and related aspects

Exchanger	Model	Characteristics/Observations/ Comments	Reference
None	Computation	Theoretical, based on momentum, heat and mass transfer, and reactions	Sandu & Lund, 1982
Stainless Steel	Reaction approach	Reconstituted skimmed milk Fouling rate predicted by the model Fouling rate related to surface temperature by an Arrhenius relation	Paterson & Fryer, 1988
PHE (three sections)	Numerical and kinetic	β lactoglobulin denaturated during the thermal treatment was predicted by a numerical determination of temperature profile and a kinetic denaturation rate	Delplace *et al.*, 1994
PHE (two sections)	Empirical	Empirical model to predict the temperature profile A quadratic model, under predicted T in the lightly fouled region, and overpredicted T in the heavy region	Yoon & Lund, 1994
PHE (three sections)	Numerical approach	Empirical model developed to predict the dry mass of deposits An accurate prediction of U as a function of time was obtained	Delplace & Leuliet, 1995
Heating and Cooling	Computer simulation	Combination of β.lactoglobulin reaction kinetics and transport phenomena Modelling worked very well in the heating regime	Chen *et al.*, 2001
UHT plant (with 3 heating sections)	Hibrid	Combination of deterministic DE with cognitive elements Both the fouling model and simulated plant can be parametrized Computer data was compared with experimental results	Petermeier *et al.*, 2002
Tubular (triple)	Computer simulation	Software based on iterative solution for estimation of U and fouling Control of the steam temperature to avoid process failures	Nema & Datta, 2005

Exchanger	Model	Characteristics/Observations/ Comments	Reference

PHE	CFD simulation	2D fundamental equations System of differential equations for a plate combining flow, heat transfer and local transport of β-lactoglobulin	De Bonis & Ruocco, 2006
PHE	2D dynamic	Adapted from a previous one Based on flow and energy equations in 2D Evaluation of mass deposit by plate and the total obtained by integration A reaction scheme for proteins was included U: 880-1506 W/m^2K	Jun & Puri, 2006

PHE: Plate heat exchanger, T: temperature, U: overall heat transfer coefficient, UHT: Ultrahigh temperature, DE: Differential equations, D: dimension, CFD: Computational fluid dynamics.

In spite of all investigations and efforts, still there are two main aspects of enormous interest for fouling in the food industry: one is how to predict the mass of deposit after certain process time, for the different parts of an industrial heat exchanger; and the other one is, how to evaluate the fouling resistance from a supposed quantity of fouling previously determined.

5.5. Plant Modifications

In addition to direct modifications of process parameters, such as greater surface area, minor flows and higher energy requirements, attempts have been carried out to face fouling problems. They include the application of advanced CIP (clean in place) techniques, the design of scale resistant equipment, the installation of control devices on processing parameters, and the utilization of scraped surface heat exchangers, as the most importants (Nema and Datta, 2005).

6. Final Remarks

Fouling in food industry is more important that in other industries, due to the presence of hundreds of components as part of the liquid foods, and also due to the thermal susceptibility of several of them. As it was mentioned, fouling has many implications in process operation; and mainly, a marked influence on food safety and quality when microbiological flora may survive.

Whether fouling is used for previous evaluation, that is to say for design, it results in an increasing of the needed heat transfer area. On the other hand, when it is developed during the heat treatment, it results in a drop of the product outlet temperature and/or a fall in the overall heat transfer coefficient, in higher energy requirements; situations that are more common for tube heat exchangers. As consequence of the scale forming, the annular section of double heat exchangers, channels of plate heat exchangers, and tubes of shell heat exchangers, will become narrower; and depending upon the pumping scheme, it can have a fall in fluid pressure or a drop in flow rates, demanding a greater pump power, situation that is more common for plate heat exchangers. Whatever that is the operational restriction, fouling implies several economic concerns.

Optimum design of any kind of heat exchanger for food processing is very important to enhance its performance and better utilization. Process optimization is possible if fouling behavior is considered, prevented if such is possible, or predicted accurately.

7. Examples

7.1. Illustrative Example 1

Consider a shell and tube (1 in, 16 BGW) used to preheat a flow of city water inside commercial steel tubes. The inside film heat transfer coefficient is 7200 W/m^{2}K, while the outside convective coefficient, in which treated water is flowing, has a h of 5700 W/m^{2}K (table 1). Evaluate the overall heat transfer coefficient for: i) the initial process (clean equipment), and, ii) after a given time of working in which a fouling phenomenon is expected, and also, iii) determine the percentage of decreasing in the heat transfer coefficient.

Solution

i) To evaluate the clean overall heat transfer coefficient, the equation 4 is applied. Needed data are: De = 1 in, Di = 0.87in, k_{steel} = 36.3 W/mK, (Karlekar and Desmond, 1982; Kern, 1997). Also the mean diameter should be computed:

$$D_L = \left[\frac{1-0.87}{\ln\dfrac{1}{0.87}}\right] = 0.933\text{in}$$, to be substituted in the aforementioned equation:

$$U_e = \frac{1}{\dfrac{1}{5700\dfrac{W}{m^2K}} + \dfrac{1.0\text{in}}{0.933\text{in}}\left(\dfrac{0.065\text{in}\cdot\dfrac{0.0254\text{m}}{\text{in}}}{36.3\dfrac{W}{mK}}\right) + \dfrac{1.0\text{in}}{0.87\text{in}\left(7500\dfrac{W}{m^2K}\right)}} = \mathbf{2649.4\frac{W}{m^2K}}$$

ii) For the fouled equipment, the fouling factors should be known, they are taken from table 3, 0.00018 m^{2}K/W for treated water and 0.00058 m^{2}K/W for city water. Then by applying of equation 6:

$$\frac{1}{U_D} = \frac{1}{2649.4\dfrac{W}{m^2K}} + 0.00018\frac{m^2K}{W} + 0.00058\frac{m^2K}{W} \therefore U_D = \mathbf{879.2\frac{W}{m^2K}}$$

iii) Finally, the percentage of heat decreasing will be directly related to the overall heat transfer coefficient, therefore: $\Delta U = \dfrac{2649.4 - 879.2}{2649.4} x100 = 66.82\%$

Conclusion: When the effect of fouling is considered, it implies severe changes in design parameters for the heat transfer equipment, thus the heat rate will decrease 66.8%, and consequently the needed heat transfer area will increase proportionally

7.2. Illustrative Example 2

A PHE of stainless steel has been designed for pasteurization of orange juice; consider a clean overall heat transfer of 3000 W/m^{2}K (table 2). Determine the effect of fouling on the overall coefficient (U) and heat transfer area (A), for a process in which the juice temperatures are 45 for inlet and 75°C for outlet with a flow rate of 216 000 kg/h; whereas the heating water has an inlet temperature of 90 and an outlet temperature of 60°C in parallel arrangement.

Solution

The magnitude of the heat rate may be evaluated by equation 1:

$$Q = \left(216000 \frac{kg}{h} \frac{1h}{3600s}\right)\left(3.9 \frac{kJ}{kg°C}\right)(75 - 45)°C = 7{,}020kW$$

i) For the clean overall coefficient, in which the driving force is 30°C (being the same for both fluids, the mean temperature difference does not make sense for these process conditions), the needed area is:

$$A = \frac{7020000W}{3000 \frac{W}{m^2K}(30K)} = 78m^2$$

ii) For the dirty equipment, overall coefficient decreases and affects the needed area. Thus, similarly to the previous example, now taking 0.00003 m^{2}K/W for process water (table 3), and 0.00060 m^{2}K/W for liquid food (supposed)

$$\frac{1}{U_D} = \frac{1}{3000 \frac{W}{m^2K}} + 0.00003 \frac{m^2K}{W} + 0.00060 \frac{m^2K}{W} \therefore U_D = \mathbf{1038.1 \frac{W}{m^2K}}$$

Comment: if the fouling factor for orange juice is lower than the supposed value for this example, the effect on U will be lower.

and finally:

$$A = \frac{7{,}020{,}000 \cdot W}{1038.1 \frac{W}{m^2 K}(30K)} = 225.4 m^2$$

Conclusion: The consideration of fouling, it causes a decreasing of 65.4% on the overall heat transfer coefficient and increases for the heat transfer area of 189%.

REFERENCES

Chen C.S., Lima V.M. and Marsaioli A. 1979. A heat transfer correlation for rotary steam-coil vacuum evaporation of tomato paste. *Journal of Food Science,* 44 (1): 200-203.

Chen X.D., Chen J. and Farid M.M. 2001. A computational investigation of whey protein based fouling on heat exchangers and holding tube. The effect of loading. *Proceedings of the Eighth International Congress on Engineering and Food* (*ICEF 8*). Technomic Pub. Co. Inc. Lancaster, PE, USA. Vol. II: 1821-1826.

Chen X.D., Li D.X.Y., Lin S.X.Q. and Özkan N. 2004. On-line fouling/cleaning detection by measuring electric resistance-equipment development and application to milk fouling detection and chemical cleaning monitoring. *Journal of Food Engineering,* 61: 181-189.

De Bonis M.V. and Ruocco G. 2006. Fouling of heat exchangers in the dairy industry by coupling flow and kinetics modeling. *Proceedings of the COMSOL Users Conference.* Milano, Italy. 5 pages.

de Jong P. 1997. Impact and control of fouling in milk processing. *Trends in Food Science and Technology,* 8 (Dec): 401-405

Delplace F. and Leuliet J.C. 1995. Modelling fouling of a plate heat exchanger with different flow arrangements by whey protein solutions. *Transactions of the Institute of Chemical Engineering,* 73 (C): 112-120.

Delplace F., Leuliet J.C. and Tissier J.P. 1994. Fouling experiments of a plate heat exchanger by whey protein solutions. *Transactions of the Institute of Chemical Engineering,* 72 (C): 163-169.

Delplace F., Leuliet J.C. and Levieux D. 1997. A reaction engineering approach to the analysis of fouling by whey proteins of a six-channels-per-pass plate heat exchanger. *Journal of Food Engineering,* 34: 91-108.

De Rafael D. and Calvo M.M. 1996. Deposit formation during heat treatment of cows', goats′ and ewes′ milks. *Journal of Dairy Research,* 63: 635-638.

Dodeja A.K., Sarma S.C. and Abichandani H. 1990. Heat transfer during evaporation of milk to high solids in thin film scraped surface heat exchanger. *Journal of Food Process Engineering,* 12: 211-225.

Foster C.L. and Green M.L. 1990. A model heat exchange apparatus for the investigation of fouling of stainless steel surfaces by milk II. Deposition of fouling material at 140°C, its adhesion and depth profiling. *Journal of Dairy Research,* 57: 339-348.

Gotham S.M., Fryer P.J. and Pritchard A.M. 1992. β-Lactoglobuline denaturation and aggregation reactions and fouling deposit formation: a DSC study. *International Journal of Food Science and Technology,* 27: 313-327.

Guérin R., Ronse G., Bouvier L., Debreyne P. and Delplace G. 2007. Structure and rate of growth of whey protein deposit from *in situ* electrical conductivity during fouling in a plate heat exchanger. *Chemical Engineering Science,* 62: 1948-1957.

Hartel R.W. 1992. Evaporation and Freeze Concentration. In *Handbook of Food Engineering* (D.R. Heldman and D.B. Lund, Eds). Marcel Dekker, Inc. New York, NY., USA

Holman J.P. 1986. Heat Transfer. Mc Graw-Hill Book Company. New York, NY, USA.

Jun S. and Puri V.M. 2005. Fouling models for heat exchangers in dairy processing: A review. *Journal of Food Process Engineering,* 28: 1-34.

Jun S. and Puri V.M. 2006. A 2D dynamic model for fouling performance of plate heat exchangers. *Journal of Food Engineering,* 75: 364-374.

Karlekar B.V. and Desmond R.M. 1982. Heat Transfer. West Publishing Co., St. Paul MN., USA.

Karlsson C.A.C., Wahlgren M.C. and Trägardh A.C. 1996. β-Lactoglobulin fouling and its removal upon rinsing and by SDS as influenced by surface characteristics, temperature and adsorption time. *Journal of Food Engineering,* 30: 43-60.

Kastanas P., Lewis M.J. and Grandison A.S. 1995. Design and development of a miniature UHT plant for fouling studies and its evaluation using milk adjusted to different pH values, goats′ milk and milk permeate. *Transactions of the Institute of Chemical Engineering,* 73 (C): 83-91.

Kern D. 1997. Heat Transfer Process Operations. Ed. CECSA, México, DF, MEXICO. (In Spanish).

Kim J.C. and Lund D.B. 1998. Milk protein/stainless steel interaction relevant to the initial state of fouling in thermal processing. *Journal of Food Process Engineering,* 21: 369-386.

Kim H.B., Tadini C.C. and Singh R.K. 1999. Heat transfer in a plate heat exchangers during pasteurization of orange juice. *Journal of Food Engineering,* 42: 79-84.

Lalande M., Tissier J.P. and Corrieu G. 1984. Fouling of a plate heat exchanger used in ultra-high-temperature sterilization of milk. *Journal of Dairy Research,* 51: 557-568.

Liu W., Christyan G.K, Zhang Z. and Fryer P.J. 2006. Direct measurement of the force required to disrupt and remove fouling deposits of whey protein concentrate. *International Dairy Journal,* 16: 164-172.

Lewis M. and Heppel N. 2000. Continuous Thermal Processing of Foods. Pasteurization and UHT Sterilization. Aspen Publishers Inc., Maryland, USA. pp. 331-368.

McCabe, W.L., Smith, J.C. and Harriot P. 2001. Unit Operations of Chemical Engineering. 6th. Edition, McGraw Hill, Tokyo, Japan. pp. 315-335.

Morison K.R. and Larsen S. 2005. Spinning disc measurement of two-stage cleaning of heat transfer fouling deposits of milk. *Journal of Food Process Engineering,* 28: 539-551.

Nema P.K. and Datta A.K 2005. A computer based solution to check the drop in milk outlet temperature due to fouling in a tubular heat exchanger. *Journal of Food Engineering,* 71: 133-142.

Paterson W.R. and Fryer P.J. 1988. A reaction engineering approach to the analysis of fouling. *Chemical Engineering Science,* 43 (7): 1714-1717.

Patil G.R. and Reuter H. 1986. Deposit formation in UHT plants. Effect of forewarming in indirectly heated plants. *Miclhwissenschaft,* 41 (6): 337-339.

Petermeier H., Benning R., Delgado A., Kulozik U., Hinrichs J. and Becker T. 2002. Hybrid model of the fouling process in tubular heat exchangers for the dairy industry. *Journal of Food Engineering,* 55: 9-17.

Peters M.S., Timmerhaus K.D. and West R.E. 2004. Plant Design and Economics for Chemical Engineers. Ed. Mc Graw-Hill. Singapore.

Rohsenow W.M. and Hartnett J.P. 1973. Handbook of Heat Transfer. Mc Graw-Hill Book Company. New York, NY, US. 18: 1-113.

Rosmaninho R., Santos O., Nylander T., Paulsson M., Beuf M., Benezech T., Yiantsios S., Andritsos N., Karabelas A., Rizzo G., Múller-Steinhagen H. and Melo L.F. 2007. Modified stainless steel surfaces targeted to reduce fouling-surface. Evaluation. of fouling by milk components. *Journal of Food Engineering,* 80: 1176-1187.

Sahoo P.K., Ansari I.A. and Datta A.K. 2005. Milk fouling simulation in helical triple heat exchangers. *Journal of Food Engineering,* 69: 237-244.

Sandu C. and Lund D. 1982. Fouling of heat transfer equipment by food fluids: computational models. *The American Institute of Chemical Engineers*, 78 (218): 12-30.

Santos O., Nylander T., Rosmaninho R., Rizzo G., Yiantsios S., Andritsos N., Karabelas A., Múller-Steinhagen H., Melo L., Boulangé L., Gabet C., Braem A., Trágardh C. and Paulsson M. 2004. Modified stainless steel surfaces targeted to reduce fouling-surface characterization. *Journal of Food Engineering,* 64: 63-79.

Saravacos G.D. and Kostaropoulos A.E. 2002. Handbook of Food Process Equipment. Kluwer Academic/Plenum Publishers. New York, NY, USA.

Saunders E.A.D. 1988. Heat Exchangers. Selection, Design and Construction. Longman Scientific & Technical, Harlow, England, UK.

Schwartz M. 1981. Four types of heat exchangers failures. *Plant Engineering,* October 15: 145-150.

Simmons M.J.H., Jayaraman P. and Fryer P.J. 2007. The effect of temperature and shear rate upon the aggregation of whey protein and its implications for milk fouling. *Journal of Food Engineering,* 79: 517-528.

Skudder P., Thomas E.L., Pavey J.A. and Perkin A.G. 1981. Effects of adding potassium iodate to milk before UHT treatment. I. Reduction in the amount of deposit on the heated surfaces. *Journal of Dairy Research,* 48: 99-113.

Stankiewicz K. and Rao M.A. 1988. Heat transfer in thin-film wiped-surface evaporation of model liquid foods. *Journal of Food Process Engineering,* 10: 113-131.

Swartzel K.R. 1983. The role of heat exchanger fouling in the formation of sediment in aseptically processed and packaged milk. *Journal of Food Processing and Preservation,* 7: 247-257.

Tejeda-Urrutia M., Macedo-Ramìrez C.R. and Vélez-Ruiz J.F. 2003. Basic study of fouling formation in a plate heat exchanger. Memorias del XXIV Encuentro Nacional de AMIDIQ (Mexican Academy of Research and Teaching in Chemical Engineering): 1009-1013 (In Spanish).

Thonon B., Grandgeorge S. and Jallut C. 1999. Effect of geometry and flow conditions on particulate fouling in plate heat exchangers. *Heat Transfer Engineering,* 20 (3): 12-23.

Vélez-Ruiz J.F. 2008. Notes of Food Engineering II. Universidad de las Américas, Puebla. Puebla, Mexico. Unpublished Material.

Walker G. 1982. Industrial Heat Exchangers. Mc Graw Hill. New York, NY, USA.

Welty J.R., Wicks C.E. and Wilson R.E. 1976. Fundamentals of Momentum Heat and Mass Transfer. John Wiley and Sons, New York, NY, USA.

Xin H., Chen X.D. and Özkan N. 2002. Whey protein-based gel as a model material for studying initial cleaning mechanisms of milk fouling. *Journal of Food Science,* 67 (7): 2702-2711.

Yoon J. and Lund D.B. 1994. Comparison of two operating methods of a plate heat exchanger under constant heat flux condition and their effect on the temperature profile during milk fouling. *Journal of Food Process Engineering,* 17: 243-262.

Zhao Q. and Liu. 2006. Modification of stainless steel surfaces by electroless Ni-P and small amount of PTFE to minimize bacterial adhesion. *Journal of Food Engineering*, 72: 266-272.

In: Food Processing and Engineering Topics
Editors: M. E. Sosa-Morales and J. F. Vélez-Ruiz
ISBN: 978-1-60741-788-0

Chapter 7

EVAPORATION OF LIQUID FOODS

Jorge F. Vélez-Ruiz
Chemical and Food Engineering Department
Universidad de las Américas, Puebla, Mexico

1. INTRODUCTION

Many liquid foods are concentrated to decrease water content implying some considerations, such as enhancing of shelf life, inducing consistency and flavor changes, reduction of volume and weight, and diminishing of energy consumption for drying. The concentration of liquid foods by evaporation is usually utilized for liquid foods such as milk, fruit juices, vegetable purees, soups and syrups; whereas other concentration processes are applied to a limited extent (Vélez-Ruiz, 2008). Products like cheese whey, fermentation wastes, molasse residues, sea water among others, also can be concentrated by evaporation (APV, 2003).

Food evaporation is the unit operation in which a liquid food is heated by means of high quality steam, reaching the boiling point of the solution. Then the generated vapor, in equilibrium with the liquid food is separated from the rest of the solution. Consequently, the remnant food increases its solids content and a preservative effect is gotten due to a reduction of the water activity.

Therefore, the main purpose of the concentration process by evaporation may be one of the next:

1) To reduce the weight and volume from liquid foods with a high water content,

2) To reduce needs and costs of materials and space during the packaging, storage and distribution of the fresh liquid foods,

3) To increase the stability of the fruit juices, milk, soups, syrups, and vegetable purees by reduction of their water activity,

4) To increase the consistency or viscosity of the liquid foods,

5) To improve the acceptability and convenience by consumers, or

6) To facilitate and economize the food dehydration of foods such as, coffee, juices, milk, purees and sugar cane.

Evaporation of food items has been utilized and researched extensively, but there is little information in which all the engineering aspects of this important food process operation have been collected. Most of the liquid food evaporation processes are carried out in tubular heat exchangers: single effects are employed for small scale operations, scraped surface is used when the food material is very viscous or heat sensitive, and multiple effect evaporators are the commonly process equipment in which the concentration of milk, fruit juices and purees is supported. And recently, the design and utilization of plate heat exchangers for evaporation of liquid foods have increased. Thus, an overall approach of this food process operation is presented in this chapter.

1.1. Performance of a Typical Evaporator

By using steam in the evaporation process, a turbulent film is generated on the heat transfer surface and it becomes the most efficient means for the evaporation phenomena, including conduction and convection heat transfer. During the process, the change of phase takes place in the vicinity of the surface while the vapor generated forms a core within the walls, thus the solution, also identified as liquor, is forced to form a thin film in which the large volume in a confined space causes high velocity. Most of the equipments have been designed and manufactured based on the film concept as the fundamental phenomenon of the evaporation, main. Hall and Hedrick (1971) established that for a good heat transfer of a falling film, the Reynolds number should exceed a magnitude of 2000.

The performance and design of industrial equipments are based on mass and heat balances and transport phenomena that have been well understood, although aspects as the modeling and evaluation of heat transfer convective coefficient has been recently studied and researched. On the other side, physical properties of foods, such as boiling point increasing, viscosity increasing, rheological properties changes, and thermal properties evolution through the whole process are not completely known, and therefore they are being investigated.

1.2. Liquid Food Characteristics and Evaporators Types

The performance and design of evaporation equipments is strongly affected by the character and properties of the liquid food to be concentrated. The most important properties of the food item that should be considered are:

i) Thermal sensibility of foods. Milk as an example may suffer protein denaturation or vitamins damage; whereas fruit juices may loss part of its nutritional value or flavor due to vitamins or color damage, and aroma volatilization. Short processing times and vacuum are needed in order to avoid these inconveniences.

ii) Viscosity and rheological changes due to the concentration phenomenon. The viscosity of Newtonians fluids increases as the solids content of the concentrate increases, like syrups and sugar solutions; other food items modify their rheological nature from Newtonian to non-Newtonian behavior, thus liquids such as milk, non-clarified juices, and purees may develop this rheological-structural change, causing lower heat transfer coefficients and higher energy needs.

iii) Boiling point and density augment are also a consequence of the solids increasing, the boiling point elevation as a colligative property causes more steam consumption.

iv) Fouling of the heat transfer surface may be the result of thermal damage, viscosity and density increasing, or components deposition, that implies a reduction in the overall heating coefficient and major needs for cleaning.

v) Foaming of some food components.

v) Other liquid foods properties, such specific heat and thermal conductivity may be altered. Because of this variation in the liquor properties, many evaporator designs have been developed by the manufactures, and the selection of the evaporator type depends of the mentioned characteristics of the liquid food.

Any equipment for evaporation process include three zones: the heating surface, the boiling zone and the separator; furthermore, the existing evaporator types may be grouped in four main categories (Kessler, 1981; Hartel, 1992, Mc Cabe *et al.*, 2001; Singh and Heldman, 2001; APV, 2003; Vélez-Ruiz, 2008), they are briefly described next.

1) Batch-type pan evaporator that is one of the simplest and oldest designed equipments for the food industry in which the quality of the concentrated product was low due to very long residence times as a consequence of a small heating surface.

2) Tubes evaporators with natural and forced flow, being of three types:

2.1 Short tubes of horizontal and vertical arrangement. Short tubes are placed in the heating zone in horizontal or vertical position and were widely used in the food industry for many years.

2.2 Long tubes are the most popular type of evaporator. Long tubes of diameter 1-2 in for upward flow, and 2-10 in for downward flow, with 3-15 m length are used to create the so known rising and falling films evaporators. These designs are the most popular and commercial equipments for the food industry.

2.3 Agitated thin-film evaporator for viscous fluids in which the heat transfer is enhanced by a sweeping device or mechanical agitator. It has a very low processing capacity and it has been used for the tomato industry.

3) Plates evaporator that operates in a similar way to the heat exchangers. Liquor flow is a combination of rising and falling films. This is the newest and more expensive equipment, but offers several advantages: good heat transfer rates, low residence times, flexibility of operation.

4) Other configurations. Some evaporators may be constructed with different characteristic to the three aforementioned groups. For instance, a thin film, spinning cone evaporator is commercialized by FT Industrial (2004) as a single effect with hollow conical discs with certain advantages such as short contact time, low thermal impact, compact installation and processing flexibility.

In the concentration of liquid food products, the unit is generally made of energy saving evaporation equipment, whose multistage arrangement leads to decrease specific steam consumption. Multi effect equipments may be operated in several configurations: forward flow, in which the fresh feed enters to the first evaporator where the steam is supplied; in reverse flow, also the steam is supplied to the first effect where the concentrated food or product leaves out of the evaporation unit; in parallel flow, the feed is split among all the evaporation units; combination of these three ways of fees flow may be used. Most liquid food concentration is carried out in the forward flow arrangement, parallel flow has been used in the industry of cane sugar, whereas the reverse flow can be applied for not too sensitive liquid foods (Schwartzberg, 1977).

2. FUNDAMENTALS

Most evaporators for liquid foods operate under vacuum conditions with a boiling temperature in the range of 70 to 45°C as a function of the reached vacuum at the correspondent evaporator stage. This decreasing in the boiling point of food items minimizes the thermal degradation of food components. Although the vapor generated in each unit is superheated, when it enters to the next evaporator, the condensation inside the heat transfer zone happens at a temperature that corresponds to the pressure of the unit where it was generated.

In each stage of the evaporator, the concentrated food stays a time that is governed by the vacuum, the desired concentration and the heat transfer characteristics. The boiling liquid flows up or down along the tube or plate depending of the specific arrangement and evaporator type. On the outside, the steam condenses on the external surface of the tubes and plates, delivering the heat for water evaporation. Temperature difference between steam and boiling liquid is smaller at the top of the tube or plate than at the bottom due to the pressure losses along the tube.

Mass and energy balances as the basis of the evaporation process is well understood, but the heat transfer and pressure drop in evaporator tubes and plates are complex phenomena (Bouman *et al.*, 1993). Transport phenomena related to momentum and heat transfers through the evaporation process are very complicated as a consequence of the continuous changing in water content, solids increasing, viscosity augment, boiling temperature increasing and rheologic nature of the concentrated food.

2.1. Engineering and Design

Mass and Energy Balances

While it is possible to reduce the energy needs for concentration of foods, by adding more evaporative effects, the requirements of heat transfer area increases as a function of the number of effects, and consequently the cost of the evaporation equipment increases.

As a first and quick approximation in calculating energy requirements for an evaporation equipment of n effects, a water removal mass at each effect and a steam consumption of 1/n (kg/h) may be considered. A better estimation is provided by an overall energy balance as it will be discussed later. In feed forward units, the amount of evaporated water tends to increase successively due to the flashing effect induced by the superheated steam, as a consequence of the pass of one effect with higher boiling temperature to the next evaporator

Table 1. Proposed correlations and data for heat transfer coefficients (h and U)*

Equipment characteristics	Food item	Correlation	Comment & Reference
Evaporators	Milk	$Nu = 0.11(Re^a)^{0.69} Pr^{0.4}$ a = 0.964 for skim milk, 0.964 for whole milk	a (empirical constant) Hall & Hedrick, 1971
Rotary steam-coil vacuum	Tomato paste (20-50%)	$Nu = 0.60 X_s^{-2.51} Re^{0.62} Pr^{0.33} \left(\frac{\mu}{\mu_W}\right)^{0.14}$ h: 795 – 15000 W/m² K	X_s (mass fraction of solids) Chen *et al.*, 1979
Falling film	Liquid foods	$Nu = 0.042 Re^{0.17} Pr^{0.53}$, $Re > 10^4$	Muddawar & El-Masri,1986
Upward liquid film	Pineapple juice (1[illegible]-62 °Bx)	$Nu = 8.5 Re^{0.20} Pr^{0.33} S^{0.67}$ (R = 0.972)	S (slip ratio), Bourgois & LeMaguer, 1987
Thin film wiped surface (simple effect)	Sugar solution (10-45%)	$Nu = 0.048 Re_x^{0.59} Pr^{1.05} Fr^{0.12} \left(\frac{\eta}{\eta_W}\right)^{-2.93}$ (R^2=0.96) (h: 1800 – 12000 W/m² K)	Transition regime, M_f: 58-87 kg/h; N: 150-500 rpm Stankiewicz & Rao, 1988
Thin film wiped surface (simple effect)	Sugar solution (10-45%)	$Nu = 4.14 Re^{0.26} Pr^{0.33} Fr^{0.03} \left(\frac{\eta}{\eta_W}\right)^{-0.75}$ (R^2=0.63) (h: 11100 – 13000 W/m² K)	Turbulent regime, M_f: 87-133 kg/h Stankiewicz & Rao, 1988
Thin film scraped surface	Whole milk (19-70%)	$Nu = 6615 G Re^{0.133} Pr^{0.076} \left(\frac{\Delta T}{T_S}\right)^{0.284}$ U: 1270 – 1530 W/m² K	ΔT (temperature difference); Ts (steam temperature) °C Dodeja *et al.*, 1990
Falling film (multiple)	Skim and whole milk	$h = 2288 - 349\Delta T + 1.8.\Delta T\mu + 82\Delta TM -$ ($R^2 = 0.950$)	ΔT (temperature difference) °C; μ (viscosity) cP; M (steam mean momentum) kg/s. Jebson & Iyer, 1991

Table 1. (Continued)

Equipment characteristics	Food item	Correlation	Comment & Reference
Horizontal thin film scraped surface	Cream, milk and water	A multiple correlation for U was proposed	U: 600 to 1800 W/m^2 K Abichandani & Sarma, 1991
Falling film (one effect)	Skim milk Whole milk	$h = 0.77q^{0.69}\eta^{-0.41}$ (R = 0.911) $h = 6.05q^{0.47}m^{0.26}\eta^{-0.44}$ (R = 0.912)	q (heat flux): 0.6-12.7 W/m^2, m (wetting rate): 0.06-0.50 kg/m s, η (viscosity): Pa s. Bouman *et al.*, 1993
Thin film scraped surface	Tomato pulp (5.9-19.%)	$U = 411.4 + 113.4T - 149.3P$ $-1.6T^2 + 2.4TP - 0.3P^2$ ($R^2 = 0.984$)	T (steam temperature); P (feed pulp flow rate) Sangrame *et al.*, 2000
Plates (one to four effects)	Skim milk (9.0-48%), Sugar brine, (14-87,%) Waste brine (1.9-35%)	$Nu = 1.25x10^{-4}Pe_x^{0.80}Kp^{0.91}Pr^{-0.12}\left(\frac{L}{d_e}\right)^{0.1}$ $Nu = 3.79x10^{-2}Pe_x^{0.33}Kp^{0.62}Pr^{-0.12}$	q (heat flux) < 15-20 or q ≥ 22-25 kW/m^2, Pr: 2.3-842, Kp (pressure criterion), L (plate height): 1 m; d_e (equi-valent diameter): 4–32 mm Hoffman, 2004
Falling film (three effects)	Sucrose solution (9.3 – 36%)	$h = 29804\,Re_L^{-0.265}\,Pr_L^{0.159}$ ($R^2 = 0.988$) First, h: 4358 - 5401; Second, h: 4823 - 6168; Third effect: 1908 - 2995 W/m^2K	Re: 22.6-2613; Pr: 3.5-200; L (liquid) Prost *et al.*, 2006

* h: convective heat transfer coefficient; U: overall heat transfer coefficient; Nu: Nusselt number, Pe: Peclet number; Pr: Prandtl number; Re; Reynolds number; GRe: generalized Reynolds number; $\mu = \eta$: viscosity; $\mu_w = \eta_w$: viscosity at Tw (wall temperature).

Table 2. Reported magnitudes for the overall heat transfer coefficient (U) for evaporation of liquid foods

Food item	Equipment	U (W/m^2 K)	Reference
Corn syrup, dextrose	Falling film - one effect	568 – 1200	Schwartzberg, 1977
Skim milk	Rising film – one pass	980 – 2150	Schwartzberg, 1977
Sugar	Calandria, Falling film – one effect	1475 - 3400, 475 - 2725	Schwartzberg, 1977
Tomato paste	Forced circulation	720 – 3000	Schwartzberg, 1977
Tomato paste	Rotary steam-coil vacuum	704 - 4355	Chen *et al.*, 1979
Apple juice	Falling film - multiple effect (6, forward)	492 – 1540	Moresi, 1985
Apple juice	Falling film - multiple effect (6, backward)	874 – 1135	Moresi, 1985
Skim and whole milk	Falling film - multiple effect	300 – 3200	Jebson and Iyer, 1991
Tomato pulp	Thin film scraped surface	626 – 911	Sangrame *et al.*, 2000

within a lower boiling point. For a reverse flow unit the opposite situation occurs, the amount of evaporated water decreases from effect to effect, because additional steam condensation must be completed to heat up the liquid food to successive higher boiling temperatures.

Heat Transfer Coefficients

Both heat transfer coefficients, convective and overall, markedly vary through the evaporation process as a function of several factors, such as food product, concentration, evaporator type, boiling temperature, and temperature difference. Mc Cabe *et al.* (2001) and Ibarz and Barbosa (2003) mention some values for the overall heat transfer coefficient, a range of 1000 to 5000 W/m^2K have been cited for long-tube vertical evaporators, a range of 800 to 3000 W/m^2K for short-tube evaporators, and a range of 600 to 2000 W/m^2K for agitated-film evaporators as a function of the solution viscosity. Kessler (1981) also reported overall coefficients for skim milk of magnitude 1000 to 2600 W/m^2K, whereas for whole milk the coefficients are of 900 to 2200 W/m^2K, being the higher values of 2000-2600 for the first evaporator, and minor values (900-1200) for the third effect. Some reported data from studies on food evaporation for both heat coefficients, convective and overall including a diversity of liquid foods and equipments, are shown in tables 1 and 2.

2.2. Multiple Effect Operation

Evaporators design is generally taken to mean the determination of the heat transfer area and steam consumption required to reach a given concentration of the liquid food. Typical pictures of evaporation equipments are shown in figure 1.

a

Figure 1. Continued on next page

b

c

Figure 1. Continued on next page

d

Figure 1. Evaporation Equipment: a) Overall view, b) Shell with internal tubes.c) GEA Niro 5-effect falling film evaporator for grape juice concentration with thermal vapor recompression, d) 5 GEA Wiegand plate evaporator, directly heated, with plate condenser, aroma recovery and concentrate cooling system. Approximate evaporation rate: 13,000 kg/h.

Whereas a set of evaporator specifications is presented in figure 2 for three effects, for a forward feeding whose performance may be summarized in four steps:

1) The liquid food or "binary" mixture to be separated is introduced to the first effect as a feed stream. The energy required to evaporate the water or solvent for the first effect, is supplied by the latent heat of vaporization coming up from the steam upon condensing, the steam flows normally at the external side of the heat exchanging zone and it is removed as condensate.
2) The hydraulic pump forces the liquid food up through the tubes or plates at a relatively high velocity of 2 to 6 m/s (Holland, 1975) and the evaporation phenomenon occurs in each effect, generating n condensates and n evaporated waters. In this case, n is equal to 3.
3) After the liquid-vapor mixture in equilibrium at the boiling point leaves the tubes or plates, it strakes a deflector that separates the major part of the liquid; then, concentrated liquid or liquor is conducted to the next evaporator taking advantage of a higher vacuum pressure.
4) The vapor generated in the previous effect can be pumped to the next unit as the heating medium in the chest to supply the needed energy for the liquid food, or well is withdrawn to the thermo-compressor or to the condenser, such it happens in the last evaporator. For multiple effect operation in which $n > 3$, the sequence is the same, the vapor generated in the previous unit is supplied to the next evaporator and so on.

2.3. Calculation Procedure

Usually in designing a multiple system, the expected prediction results are the amount of evaporated water or capacity, the amount of needed steam or consumption, the approximate boiling temperatures in the different effects, and the area of heating surface required at each unit. Although the procedure is developed for a triple effect evaporator, the application to a single evaporator is completely similar and simpler.

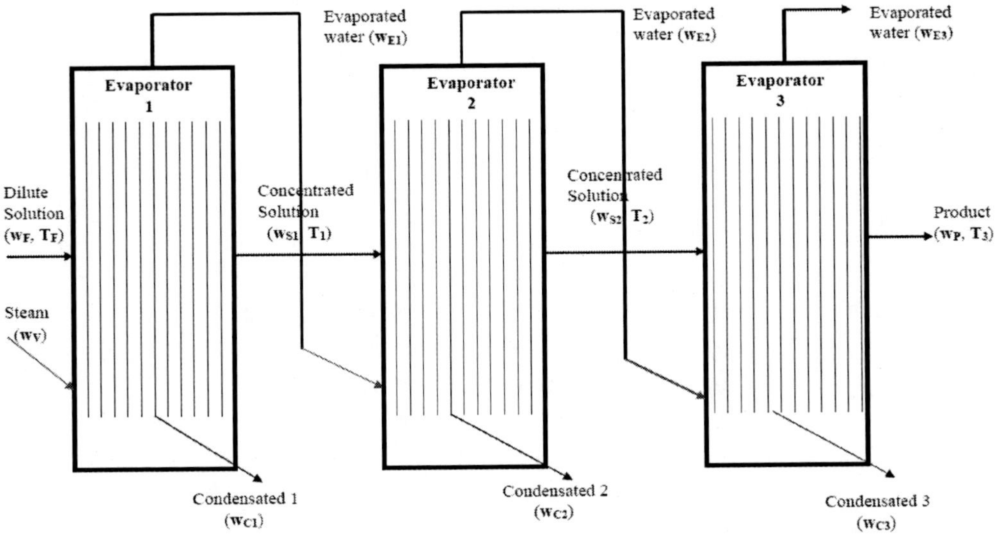

Figure 2. Three-effect evaporator, including streams specification.

To find out the desired information, the application of material and enthalpy balances as well as the heat transfer equation is obligatory, some assumptions should be considered:

i) Liquid solution (w_F, w_{S1}, w_{S2}, w_P, Figure 2) in the evaporator are composed of non-volatile solutes and water, it is well mixed.
ii) The available temperature difference is only reduced by the boiling point rise of each evaporation unit.
iii) Incondensable gases do not affect the overall pressure and the heat transfer effectiveness in the effect.
iv) Vapor phase is in equilibrium with the liquid phase, therefore the liquor and the generated vapor have the same temperature (T_n: T_1, T_2, T_3)
v) The supplied energy for the heating and boiling process is obtained only from the condensation; it means that steam is saturated as its condensate is a saturated liquid. The enthalpy of both phases, as well as the enthalpy of phase change may be found in thermodynamic tables for water
vi) The overall heat transfer for each evaporator is constant through the whole evaporation process.

2.3.1. Mass Balance

Taking as a basis the three effect evaporator schematized by Figure 2, the setting out of the performance and design equations are developed below.

A total mass balance for the overall system and for the liquor phase as well, are:

$$w_F + w_V = w_{C1} + w_{C2} + w_{C3} + w_{E3} + w_P \tag{1a}$$

$$w_F = w_{E1} + w_{E2} + w_{E3} + w_P \tag{1b}$$

where: w is mass flow (kg/s or kg/h) for each one of the streams (F: feed, V: steam, C: condensate at effect 1, 2 and 3, E: generated vapor at every effect (1,2,3), P: concentrated from the third unit, that is the product in this particular case).

In equation 1, w_v is the steam consumption, and $w_{C2} + w_{C3} + w_{E3}$ is the capacity of the triple effect system. Furthermore, taking the flux identity: $w_v = w_{C1}$ (assumption number v, in the past section), and doing a balance for the water and solid content:

$$\begin{aligned} w_F(x_{WF}) &= w_{C2}(1.0) + w_{C3}(1.0) + w_{E3}(1.0) + w_P(x_{WP}) \\ w_F(x_{SF}) &= w_P(x_{SP}) \end{aligned} \tag{2}$$

At the first effect, a balance for the water and solid content only for the liquor; whereas for the vapor phase $w_v = w_{C1}$:

$$\begin{aligned} w_F(x_{WF}) &= w_{E1}(1.0) + w_{S1}(x_{WS1}) \\ w_F(x_{WF}) &= w_{S1}(x_{SS1}) \end{aligned} \tag{3}$$

Similarly for the second unit, in which the vapor phase $w_{E1} = w_{C2}$:

$$\begin{aligned} w_{S1}(x_{WS1}) &= w_{E2}(1.0) + w_{S2}(x_{WS2}) \\ w_{S1}(x_{SS1}) &= w_{S2}(x_{SS2}) \end{aligned} \tag{4}$$

And finally, for the third evaporator (vapor phase $w_{E2} = w_{C3}$):

$$\begin{aligned} w_{S2}(x_{WS2}) &= w_{E3}(1.0) + w_P(x_{WP}) \\ w_{S2}(x_{SS2}) &= w_P(x_{SP}) \end{aligned} \tag{5}$$

where: x is the mass fraction of solids (dimensionless), for two components: solids (S) and water (W); and S1, S2 and S3 or P (being S3 = P) are the concentrated solutions or liquors coming up from each evaporator and expressed as mass flow (kg/s or kg/h).

2.3.2. Boiling Point Elevation

As a general consideration:

$$\Delta T_b = T_{solution} - T_{water} \quad \text{(at the same pressure)} \tag{6}$$

where: ΔT_b is the boiling point rise (°C), $T_{solution}$ is the boiling point of the correspondent solution (°C), and T_{water} is the boiling point of water (°C) for the same evaporator operating conditions (same pressure)

To evaluate the boiling point rise, the simplest case is to consider ΔT_b as insignificant; in the case of foods, a common correlation establishes that ΔT_b is a function of the total solids, as it is expected.

If the boiling point of the fluid to be concentrated is unknown, there are two options, the first one is to employ that boiling points of the liquid foods or solutions from Dühring diagrams, such those included in Figure 3 for sucrose and tamarindo juice (Ibarz and Barbosa-Cánovas, 2003). In the second option may be utilized some of the proposed correlations, like those included in table 3 from experimental work.

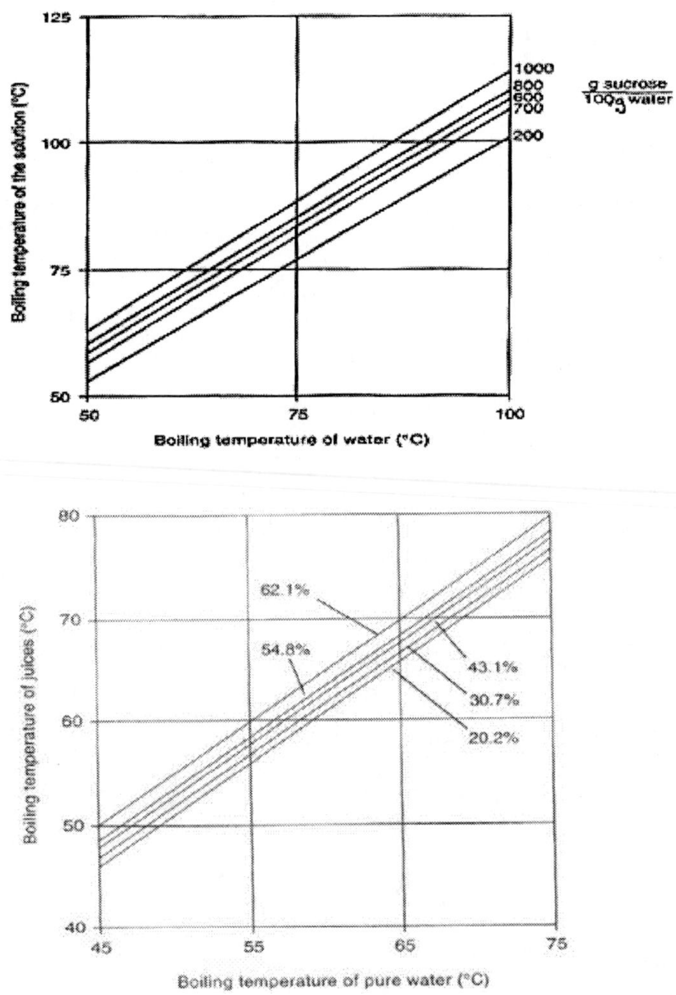

Figure 3. a) Duhring diagram for sucrose solutions; b) Duhring diagram for tamarindo juice

Most of the few reported diagrams have been developed for fruit juices (Moresi and Spinosi, 1980, 1984; Crapiste & Lozano, 1988; Ilangantileke *et al.*, 1991; Camiro-Cabrera *et al.*, 2007; Tellis-Romero *et al.*, 2007); whereas Dühring graphics for inorganic solutions appear at some books (Holland, 1975; McCabe *et al.*, 2001; Ibarz and Barbosa-Cánovas, 2003).

Table 3. Correlations for boiling point rise and related aspects

Food material	Equations/Correlations	Observations/ Reference
Orange juice (10-65°Bx)	$\Delta T_b = 3.2y - 2.4y^2 + 14y^3$	y (weight fraction of sucrose): 0.10-0.65 Moresi & Spinosi, 1980
Grape juice (0-73.1°Bx)	$\ln p = A - \frac{B}{T}$ A and B: characteristic constants: $A = 20.64 - 0.384y + 1.297y^2 - 2.045y^3$ $B = 5212 - 1375y$ $\Delta T_b = \frac{RT_w^2}{\Delta H_v} x$	p (vapor pressure): 100-760 mm Hg; T (absolute tempe rature): 324-385 K; y (weight fractions of equivalent of sucrose): 0-0.73; R (gas constant), T_w (water boiling point); ΔH_V (latent heat of vaporiza tion); x (solute molar fraction) Moresi & Spinosi, 1984
Sucrose, Reducing sugars Juices	$\Delta T_b = 0.031C^{0.094}P^{0.136}\exp(-0.0533C)$ $\Delta T_b = 0.023C^{0.588}P^{0.119}\exp(-0.0359C)$ $\Delta T_b = 0.014C^{0.749}P^{0.106}\exp(-0.0339C)$	C (concentration): °Bx; P (pressure): mbar Crapiste & Lozano, 1988
Tangerine juice (35-67°Bx)	$\Delta T_b = 0.0499C^{0.029}P^{0.113}\exp(-0.03889C + 0.00065C^2)$	C (concentration): °Bx; P (pressure): mbar Ilangantileke *et al.*, 1991
Grapefruit juice (9.3–60.6°Bx)	$BP_{GJ} = m_0 + m_1 BP_W$ $R^2 = 0.998\text{-}0.999$ m_0 and m_1 given at each concentration: m_0 = 0.173, 0.962, and 3.265°C; m_1= 1.001, 1.008, and 1.022 for 9.3, 33.6 and 56°Bx, respectively	In the range of 9.3-29°Bx, the rise in boiling point was nearly independent of pressure. Tellis-Romero *et al.*, 2007

ΔT_b = boiling point elevation = BP.

2.3.3. Energy/enthalpy Balance

Similar to the mass balance, the energy balance may be applied to the overall system and at each one of the effects, in order to get the design parameters.

Being the next, the overall enthalpy balance:

$$w_F \Delta h_F + w_V \Delta h_V = w_{C1} \Delta h_{C1} + w_{C2} \Delta h_{C2} + w_{C3} \Delta h_{C3} + w_{E3} \Delta h_{E3} + w_P \Delta h_P \quad (7)$$

where w is mass flow (kg/s), and Δh is the specific enthalpy (kJ/kg) for the correspondent stream.

In this case, the difference of energies between the steam ($w_V\Delta h_V$) and the first condensate ($w_{C1}\Delta h_{C1}$) is the enthalpy of condensation due to the phase change ($w_V\Delta h_{LV}$ = $w_V\lambda_{LV}$), in which $\Delta h_{LV} = \lambda_{LV}$ Is the latent heat of evaporation/condensation. Therefore, the equation may be partially modified:

$$w_F \Delta h_F + w_V \Delta h_{LV} = w_{C2} \Delta h_{C2} + w_{C3} \Delta h_{C3} + w_{E3} \Delta h_{E3} + w_P \Delta h_P \quad (8)$$

With these considerations, the energy balances for the three evaporators are:

First

$$w_F \Delta h_F + w_V \Delta h_V = w_{C1} \Delta h_{C1} + w_{E1} \Delta h_{E1} + w_{S1} \Delta h_{S1} \quad (9)$$

Second

$$w_{S1} \Delta h_{S1} + w_{E1} \Delta h_{E1} = w_{C2} \Delta h_{C2} + w_{E2} \Delta h_{E2} + w_{S2} \Delta h_{S2} \quad (10)$$

Third

$$w_{S2} \Delta h_{S2} + w_{E2} \Delta h_{E2} = w_{C3} \Delta h_{C3} + w_{E3} \Delta h_{E3} + w_P \Delta h_P \quad (11)$$

These three equations may be simplified taking the same consideration, that is to say, that the heating medium or vapor only supplies the condensing latent heat in each effect. On the other side, considering the heat flux as the energy needed for the liquor phase or the supplied energy from the steam, the energy equation for the three effects are given next:

First

$$\begin{aligned} w_F \Delta h_F + Q_1 &= w_{E1} \Delta h_{E1} + w_{S1} \Delta h_{S1} \\ w_V \Delta h_V &= w_{C1} \Delta h_{C1} + Q_1 \end{aligned} \quad (12)$$

Second

$$\begin{aligned} w_{S1} \Delta h_{S1} + Q_2 &= w_{E2} \Delta h_{E2} + w_{S2} \Delta h_{S2} \\ w_{E1} \Delta h_{E1} &= w_{C2} \Delta h_{C2} + Q_2 \end{aligned} \quad (13)$$

Third

$$w_{S2}\Delta h_{S2} + Q_3 = w_{E3}\Delta h_{E3} + w_{S3}\Delta h_{S3}$$
$$w_{E2}\Delta h_{E2} = w_{C3}\Delta h_{C3} + Q_3 \tag{14}$$

where: Q is the transferred heat in each one of the effects (Q_1, Q_2 and Q_3: kW), and P = S3..

2.3.4. Heat Transfer Area

The rate of heat transfer is required for designing and sizing the heat transfer area of evaporators. That heating area is evaluated from the following equation:

$$Q = UA\Delta T \tag{15}$$

where Q is the heat flux (kW), transferred from the condensing steam to the liquid food; U is the overall heat transfer coefficient (W/m^{2}K), including heat conduction and convection, as well as the fouling resistances, ΔT is the temperature difference or driven force for each effect (°C), evaluated as the difference between the steam temperature and the correspondent boiling point of the effect (°C); and A is the heat transfer area for each evaporation unit (m^2), that is the main parameter in calculations for evaporators design. Usually, U and A are based on the external surface of the heating tubes, that is not the case of plate evaporators.

Furthermore, considering each one of the aforementioned parameters:

$$U = \frac{1}{\frac{1}{he} + R_{fe} + \frac{\Delta x D_e}{kD_{ML}} + R_{fi} + \frac{De}{h_i D_i}} \tag{16}$$

where he and hi, are the external and internal heat transfer coefficients for convection (W/m^{2}K), respectively; R_{fe} and R_{fi} are the external and internal resistances for the correspondent fouling (m^{2}K/W); Δx and k are the tube thickness (m) and thermal conductivity (W/mK), forming the conduction resistance, De and Di are the external and internal tube diameters (m), and D_{ML} is the log mean diameter (m), representing the logarithmic mean of De and Di [$D_{ML} = (D_e - D_i)/\ln(D_e/D_i)$]. For evaporator plates, the diameter is not considered.

Additionally, the heat transfer area A (m^2) may be evaluated for tubes or plates heat exchanger, from the relationship

$$A = n(\pi D_e L) = n(L_e a) \tag{17}$$

Being n the number of tubes or plates, L is the length of the tubes (m); Le and a are the effective length (m) and effective width of the plates (m), respectively.

The thermal potential or driven force (ΔT) at each evaporator unit is:

$$\Delta T = T_v - T_n \tag{18}$$

where Tv is the steam temperature (°C), that it would be modified for multiple evaporators; Tn is the boiling point of the liquid food in the correspondent effect (°C).

Therefore, the needed heat fluxes for each one of the evaporators are:

First:

$$Q_1 = U_1 A_1 \Delta T_1 ; \tag{19}$$

Second:

$$Q_2 = U_2 A_2 \Delta T_2 \tag{20}$$

Third:

$$Q_3 = U_3 A_3 \Delta T_3 \tag{21}$$

That includes the proper data and specific conditions of operation at every effect, being $\Delta T_1 = T_v - T_1$; $\Delta T_2 = T_1 - T_1$; and $\Delta T_3 = T_2 - T_3$.

2.4. Solving Criteria

For a triple-effect equipment in which the fundamental equations are applied, there will be seven equations (1b, 9, 10, 11, 19, 20, and 21), with seven unknowns variables (Mc Cabe *et al.*, 2001; Vélez-Ruiz, 2008): the consumption (steam for the first effect, w_V), the mass flow leaving from each effect (w_{S1}, w_{S2}, and w_p), the boiling temperatures for the first and second effects (T_1 and T_2), and the heating surface per evaporator, assumed as identical for the three evaporators (A).

To solve the general set of equations associated with an evaporation system, several methods may be used, two of the most common are to solve by some generalized nonlinear numerical procedure, such the Newton-Raphson method (Holland, 1975), or well, by linearizing the equations and using integrated iteration loops (Ribeiro and Caño Andrade, 2002), among others.

One simple, iterative and practical solution, based importantly on the assumption of equal heating areas for the three effects, is summarized next in four steps (Mc Cabe *et al.*, 2001; Vélez-Ruiz, 2008):

i) Suppose the boiling points for effects 1 and 2 (T_n), taking as basis the overall heat transfer coefficients of the three effects, and the overall ΔT.
ii) Develop the enthalpy balances and mass balances for the three effects, that are reduced to a simultaneous system of 3 x 3 equations, to evaluate the flow rates.
iii) Solve the equations for Q_n and evaluate the surface area of each evaporator

iv) Compute the medium area to know if the heating area is approximately the same for each effect. If the area is different (> 10%) another iteration will be convenient.

With this procedure, the number of iterations will be less than three and an error less than 10% will be obtained. This procedure will be exemplified lately in the correspondent section. Ibarz and Barbosa (2003) mention other iterative procedures for solving of evaporator equations, when the boiling point rise is neglected or when it is important of evaluation.

3. Complementary Aspects of Evaporation

3.1. Physical Properties

Physical properties of liquid foods are required for adequate design and analysis of the evaporation equipment and process. Some of these properties may be evaluated easily by additive relationships, such as the density, specific heat and thermal conductivity, while others like viscosity, apparent viscosity, flow behavior index, and consistency coefficient cannot be predicted with precision. And although several researches have been conducted to know the physical properties of foods, still there is scarce information. Particularly the engineering calculations for evaporation need those involved physical properties as a function of concentration and temperature, some of the existing correlations and data for boiling point elevation, density, rheological and thermal properties for liquid foods of different origin as a function of the concentration or temperature, are presented in tables 3 to 6, correspondently.

3.2. Process Limitations

One of the more severe problems in the concentration by evaporation of the liquid foods is the fouling on hot surface. Other problems come from the thermal degradation of nutrients; both limitations are minimized in an agitated thin film evaporator. Recently, a thin-film spinning cone evaporator (centritherm) also represents an alternative for this type of foods (FT Industrial, 2004)

Aroma retention for coffee, fruit juice and tea extract is also a process limitation. Aroma are volatile and removal of water by evaporation results in losses of them. These food materials contain aromas that are complex aqueous mixtures of organic compounds; a typical low boiling component in the coffee aroma is methyl mercaptan (6°C), whereas a typical high boiling component in grapes is methyl anthranilate (225°C). These problem of aroma losses have been faced by two alternatives, mixing fresh and concentrated juices, or well, using an aroma recovery unit (Thijssen, 1970; Karlsson and Trägardh, 1998).

New technologies based on membrane separation are matching with the evaporation process in order to eliminate or compensate some of the evaporation disadvantages.

Table 4. Density (ρ) of liquid foods, solutions, concentrates, and related aspects

Food material	Equations/Correlations	Observations/ Reference
Tomato paste (20 – 50%)	$\rho = 62.4(0.44X_s - 0.997)$	ρ (lb/ft^3); Xs (mass fraction of solids) Chen *et al.*, 1979
Orange juice (10-65 °Bx)	$\rho = -0.9944 + 0.307y - 0.282y^2$	y (weight fraction of sucrose): 0.10-0.65 Moresi & Spinosi, 1980
Grape juice (0-73.1°Bx)	$\frac{1}{\rho} = \left(\frac{y}{\rho_{GS}} - \frac{(1-y)}{\rho_w}\right)$ $\rho_{GS} = 1.608$	ρ (g/cm^3); y (weight fraction of equivalent sucrose): 0-0.73 ρ_{GS}: density of pseudo-sucrose component, ρ_w: water density. Moresi & Spinosi, 1984
Apple juice (14 - 39°Bx)	$\rho = 0.83 + 0.35\exp(0.01X) - 0.000564T$ ($R^2 = 0.990$)	ρ (g/cm^3): 1.02-1.17 ; T (absolute temperature): 20-80°C; X (concentration, °Bx). Bayindirli, 1992
Grape juice (19 - 35°Bx)	$\rho = 0.74 + 0.43\exp(0.01X) - 0.000555T$ ($R^2 = 0.997$)	ρ (g/cm^3): 1.07-1.18 ; T (absolute temperature): 20-80°C; X (concentration, °Bx) Bayindirli, 1993
Fruit juice and puree (10 – 60 °Bx)	$\rho = 1006.6 - 0.5155T + 4.1951C + 0.0135C^2$ for peach juice ($R^2 = 0.999$) $\rho = 1025.4 - 0.3289T + 3.2819C + 0.0178C^2$ for peach juice ($R^2 = 0.999$) $\rho = 1025.4 - 0.3289T + 3.2819C + 0.0178C^2$ for peach juice ($R^2 = 0.999$)	ρ (kg/m^3); T (temperature): 5-75°C; C (soluble solids concentration, °Bx) Ramos & Ibarz, 1998
Orange juice (0.34-0.73 w/w)	ρ range: 1077 – 1275 kg/m^3	T: 0.5-62°C Tellis-Romero *et al.*, 1998

Table 4. (Continued).

Food material	Equations/Correlations	Observations/ Reference
Crab apple juice (17 – 70 °Bx)	$\rho = -0.0054C + 0.9671$ ($R^2 = 0.995$)	ρ (g/cm^3, at 25°C), C (solids concentration, °Bx) Cepeda & Villarán, 1999
Guava juice (9 - 11°Bx)	$\rho = 1047 - 0.5293\,T$ at 9°Bx $\rho = 1047.9 - 0.45T$ at 11°Bx	ρ (kg/m^3):1002-1012 at 9°Bx ρ (kg/m^3):1011-1019 at 11°Bx T (temperature): 65-85°C. Zainal *et al.*, 2000
Banana puree (10 – 80 °Bx)	$\rho = 1053.4 - 1.4421T + 0.2256C + 0.0864C^2$ $\rho = 1047 - 0.5293\,T$	ρ (kg/m^3); T (temperature): 5-75°C; C (soluble solids concentration, °Bx) Tsen and King, 2002
Soluble coffee (0 – 0.5 of w)	$\frac{1}{\rho} = \frac{w}{\rho_C} + \frac{1-w}{\rho_W}$ $\rho_C = 1654\text{-}1.79T\text{-}0.0063T^2$; $\rho_W = 1000\text{-}0.036T\text{-}0.004T^2$	ρ (kg/m^3); w (coffee mass fraction: kg dry coffee/kg solution); c (solid coffee); T (temperature): 25-95°C Sobolík *et al.*, 2002
Clear grape juice (22.9 – 70.6°Bx)	ρ range: 1097.3 – 1358.4 kg/ m^3 densities for 20-80°C	Zuritz *et al.*, 2005
Concentrated and reconstituted milk (44 – 60%)	$\rho = 988.1 + 2.8754\,X$ concentrated (R^2=0.981) $\rho = 944.0 + 3.0999\,X$ reconstituted (R^2=0.925)	ρ (kg/m^3); X (solids concentration, % w/w) Enriquez *et al.*,, 2009

Table 5. Rheological properties and related aspects

Food material	Equations/Correlations/Instrument	Observations/ Reference
Tomato paste (20 – 50%)	$\mu_a = 3.62X_s^{2.4}\exp\left(\frac{1710}{T+460}\right)$	μ_a (lb/ft h); Xs (solids mass fraction); T (temperature, °F) Chen *et al.*, 1979
Orange juice (10-65°Bx)	$\mu = \alpha T^{\beta}$ α and β: characteristic constants. Höppler Falling Sphere Viscometer $\alpha = 34.668 - 0.2024y + 0.0162y^2 - 2.045y^3$ $\beta = -6.1055 + 0.0396y - 0.00268y^2$	μ (cP); y (weight fraction of sucrose): 0.10-0.65 Moresi & Spinosi, 1980
Grape juice (0-73.1°Bx)	$\mu_a = f(y, T)$ $K = \exp\left(C + \frac{D}{T}\right)$ $C = -13.38 + 16.93y - 95.74y^2 + 248.5y^3 - 217.6y^4$ $D = 1903 - 1802y + 22960y^2 - 68310y^3 + 66330y^4$ $n = 1 - 0.252y$	μ_a: 0.600-1200 mPa s (depending of concentration and temperature); C and D: characteristic constants, y (weight fractions of equivalent sucrose): 0-0.73 ; Haake Rotational Viscometer (RV12) Moresi & Spinosi, 1984
Apple juice (14-39°Bx)	$\mu = \mu_W \exp\left(\frac{[-0.24 + 917.92/T]X}{100 - [2.03 - 0.00267T]X}\right)$ μ_W (viscosity of water at same temperature) Ostwald-Cannon Kenske Viscometer	μ_a: 0.5-7.2 cP; T(absolute tempera ture): 20-80°C; X (°Bx). Bayindirli, 1992
Grape juice (19 - 35°Bx)	$\mu = \mu_W \exp\left(\frac{[-0.24 + 1821.45/T]X}{100 - [0.86 + 0.000441T]X}\right)$ μ_W (viscosity of water at same temperature) Ostwald-Cannon Kenske Viscometer	μ_a: 0.5-4 cP; T(absolute tempera ture): 20-80°C; X (°Bx). Bayindirli, 1993
Orange juice	K range: 0.064-4.27 mPa s^n n: 0.675 – 0.852	Haake rotovisco viscometer (RV-12). Hernandez *et al.*, 1995

Table 5. (Continued)

Food material	Equations/Correlations/Instrument	Observations/ Reference
Concentrated milk	$\ln K = -9.863 + \frac{2557}{T} + 0.1426X$ ($R^2 = 0.960$) Physica UM rheometer (MC20)	K (mPa s^n); X (solids concen tration, % w/w); T (absolute temperature): 20-80°C Vélez-Ruiz & Barbosa-Cánovas, 1997
Date concentrates (40 – 70 °Bx)	K ranged: 2.00-37.1 mPa s^n for variety Khudary K ranged: 1.34-13.3 mPa s^n for variety Sufri K ranged: 3.30-35.9 mPa s^n for variety Serri Viscotester Haake (VT181)	n: 0.631-0.988; T: 25-95°C n: 0.788- 0.940; T: 25-95°C n: 0.606- 0.996 ; T: 25-95°C Hobani, 1998
Cherry juices (22 – 74 °Bx)	$\eta = 1.12x10^{-6} \exp\left(0.094C + \frac{3523}{T}\right)$ ($R^2 = 0.864$) Rotovisco (RV12)	η: 1.36-6850 mPa s; T (abso lute temperature): 5-70°C; C (solids concentration,°Bx). Giner *et al.*, 1997
Concentrated milk (12.6-42.6%)	$\ln K = -18.22 + \frac{2973}{T} + (0.1278X)$ ($R^2 = 0.980$) Physica UM Rheometer (MC20)	K (Pa s^n); X (concentration, % w/w); T (absolute temperature): 5-25°C Vélez-Ruiz & Barbosa-Cánovas, 1998
Clarified apple juice (17 – 70 °Bx)	$\mu = \exp\left(-53.46 + \frac{9458.6}{T} + 0.29C\right)$ ($R^2 = 0.995$) Rotatory Viscometer (Mettler RM 180)	μ: 0.003-3300 Pa s; T (absolute temperature): 10-60°C; C (°Bx). Cepeda and Villarán, 1999
Concentrated milk (12.5-46.7%)	$K = 0.224 \exp(0.148X)$ ($R^2 = 0.990$) $n = 1.11 - 0.0072X$, or $n = 1.13 \exp(-0.0079X)$ Physica UM Rheometer (MC20)	K (mPa s^n); X (solids concentration, % w/w) Vélez-Ruiz & Barbosa-Cánovas, 2000
Tomato concentrate (8.7–17 °Bx)	K range: 1.45-5.08 Pa s^n n: 0.22 – 0.17	Capilar Viscometry and In-line US Viscometer Dogan *et al.*, 2003

Table 5. (Continued)

Food material	Equations/Correlations/Instrument	Observations/ Reference
Clear grape juice (22-70.6°Bx)	μ range: 0.6 -18 Pa s Brokfield Viscometer (LVDV-II)	viscosities for 20-80°C. Zuritz *et al.*, 2005
Sour pomergranate juice (45 – 73 °Bx)	$\mu = 1.25\text{x}10^{-11}\exp\left(0.148C + \frac{4398}{T}\right)$ ($R^2 = 0.931$) RheoStress Rheometer (Haake RS1)	μ: 0.014-14.04 Pa s; T (temperature): 5-60°C; C (concentration, °Bx). Kaya & Sözer, 2005
Blueberry and raspberry juices (10 – 65 °Bx)	$\mu = 0.056\exp\left(\frac{[7C^2 + 9055]}{8.314\,T} - 0.0019\,C^2\right)$ for blueberry $\mu = 0.0206\exp\left(\frac{[6.3C^2 + 11196]}{8.314\,T} - 0.0017\,C^2\right)$ for raspberry TA Rotatory Rheometer (AR 2000)	μ: 1.2-150 mPa s; C (°Bx); T: 20-60°C. μ: 1.2-200 mPa s; C (°Bx); T: 20-60°C Nindo *et al.*, 2005
Tomato concentrate (5 – 24.°Bx)	K ranged: 5-94.3 Pa s^n, variety Heinz K ranged: 2.4-99.5 Pa s^n, variety Orsetti K ranged: 2.6-55.3 Pa s^n, variety AB Viscotester Haake (VT24) and In-line US Viscometer	n: 0.315 – 0.214 at 60°C n: 0.394 – 0.149 at 60°C n: 0.357 – 0.147 at 60°C. Choi *et al.*, 2006
Mango juice (7.6 - 26°%)	$K = aC^b$ a and b: characteristic constants ($R^2 = 0.990$) Brokfield Viscometer (LVDV-II)	K (Pa s^n), a = 0.0002, b = 3.9-3.99 (30-40°C); a = 0.0001, b= 4.12 (50°C); a = 0.00006, b = 4.29 (60°C); a = 0.00004, b = 4.39 (70°C). Dak *et al.*, 2007
Prickly pear (42 – 58°Bx)	$K = 26.24 - 1.263X + 0.208T + 0.015X^2 - 0.005XT$ $n = -0.949 + 0.082X - 0.001X^2$ Brookfield DVI	K: 89-3188 mPa s^n; n: 0.50-0.83. X (°Bx), T (temperature) Rodriguez-Martinez *et al.*, 2007

μ: viscosity; μ_a: apparent viscosity; K: consistency coefficient; n: flow behavior index (dimensionless).

Table 6. Thermal properties of solutions, concentrates, and related aspects

Food material	Equations/Correlations/Instrument	Observations/ Reference
Tomato paste (20 – 50%)	$Cp = 1 - 0.67X_s$ $k = \frac{(5.75 - 4.8X_s)[993 + 500X_s - 0.317(T - 32)]}{17300[1.1 - 0.002(T - 32)]}$	Cp (Btu/lb °F); Xs (mass fraction of solids); k (BTU/hft°F); T (temperature, °F) Chen *et al.*, 1979
Orange juice (10-65°Bx)	$Cp = 4.186 - 2.679y$ Thermal Analyzer Dupont 990	Cp (J/g K); y (weight fraction of sucrose): 0.10-0.65. Moresi & Spinosi, 1980
Tomato juice (4.8 – 80%)	$Cp = Cp_S X_S + Cp_W X_s$; $k = k_S X_S + k_W X_s$ $Cp_W = 4.188 + 0.001029T$ $Cp_S = 1.727 - 0.000745T + 0.00000959T^2$ $k_W = 0.6192 + 0.00053T_s$; $k_W = 0.6192 + 0.00053T_s$	Cp (kJ/kg °C); k (W/m K); S (total solids); W (water); T (temperature): 20–150°C Thermal conductivity probe Choi and Okos, 1983
Grape juice (0-73.1°Bx)	$Cp = yCp_S + (1 - y)Cp_w$ $Cp_S = 1.21$ Thermal Analyzer Dupont 990	Cp (J/g K), y (weight fraction of equivalent sucrose): 0-0.73; S: pseudo-sucrose component; W: water. Moresi & Spinosi, 1984
Whole milk (37 – 72.4%)	$k = (0.59 + 0.0012T)(1 - 0.0078X)$ Steady state, parallel-plate apparatus	k (W/m k); T(temperature, °C): 40-90; X (total solids, %) More and Prasad, 1988
Reconstituted milk (40-70%)	$Cp = 3833.2 - 28.019X - 1.759T - 0.11X^2 + 0.234XT - 0.05T^2$ $k = 0.5656 - 0.0027X + 0.00091T - 0.000021X^2 + 0.0000079XT - 0.0000014T^2$	X (concentration, %); T (temperature, °C): 35-65 Reddy and Datta, 1994
Orange juice (0.34-0.73 w/w)	$Cp = 1424.34 + 2673.19X_W + 2.446X_W$ $k_W = 0.0797 + 0.5238X_W + 0.00058T_s$	Cp (J/g °C); X_W (water content, w/w) ; T (temperature): 5-75°C; k (W/m °C). Two coaxial copper cylinders Tellis-Romero *et al.*, 1998
Guava juice (9 - 11°Bx)	$k = 0.522 - 0.00002T$ at 9°Bx $k = 0.512 - 0.00002T$ at 11°Bx	k (W/m °C); T (temperature): 65-85°C Zainal *et al.*, 2000
Soluble coffee (0 – 0.5 of w)	$k_W = 10^{-3}(565.1 + 1.8T - 0.0058T^2)(1 - 0.54w)$	k (W/m°C); w (coffee mass fraction: kg dry coffee/kg solution). Sobolík *et al.*, 2002

Cp: Specific heat; k: Thermal conductivity.

3.3. Additional Equipments

Utilized equipments to complete the operation of an evaporation unit include preheaters, pumps, compressors, vapor condensers, and controls, as the most important (Brown *et al.*, 1965; Mc Cabe *et al.*, 2001; Singh and Heldman, 2001; Ibarz and Barbosa Cánovas, 2003; Vélez-Ruiz, 2008).

Preheaters

Preheating of a liquid food to be concentrated precedes the evaporation process, tubular and plates heat exchangers are normally utilized for this pretreatment. Either double tube, shell and tubes or plates are common, and the employed equipment depends of the flow quantity to be processed, among others important process considerations.

Pumps

Pumping is needed when reverse feeding evaporation is handled or well, reciprocating pumps together with steam jet ejector are commonly used to produce vacuum. The concentrate leaving from the last effect should be moved out with a positive pump, and also the condensate water may be removed from the condenser with a centrifugal pump.

Condensers

Some times, when the vapor is not recycled to the process, the condensation of the vapor generated in the last effect is needed. The condenser is also a tubular heat exchanger able of cool down the vapor phase and transform it to liquid phase by removal of the latent heat. Simultaneously, the vacuum is maintained because the volume occupied by the liquid is considerably lower that the vapor volume. Occasionally a jet condenser as a direct condenser can be employed instead of the tubular condenser, thus the sprayed water and the condensate are mixed and may be reutilized.

Compressors

Vapor recompression increases the efficiency in energy utilization of an evaporation system. Energy is added to the generated vapor by means of two alternative units: a mechanical compressor or a steam jet (thermo compressor). The objective of vapor recompression is to increase the pressure and thereby the temperature of the generated vapor. In the first equipment the compression is accomplished mechanically, by using a compressor. In the second case, the released vapor is compressed by acting on it with a high-pressure steam.

3.4. Experimental Sets

Studies and experiments on evaporation and related properties have utilized a diversity of equipments: laboratory rotary evaporators (Chen *et al.*, 1979; Bayindirli, 1992; Bouman *et al.*, 1993; Enriquez *et al.*, 2009), pilot plant scale (Stankiewicz and Rao, 1988; Dodeja *et al.*, 1990; Vélez-Ruiz and Barbosa-Cánovas, 1997; 1998; Hobani, 1998), and industrial units (Zuritz *et al.*, 2005;), with several arrangements. Falling and climbing evaporators have been utilized more frequently, either with simple effect (Prost *et al.*, 2006) or multiple effects

(Jebson and Iyer, 1991; Hernandez *et al.*, 1995; Tellis-Romero *et al.*, 1998; Zuritz *et al.*, 2005;) and few studies exist on plate (Hoffman, 2004) and scraped surface evaporators (Abichandani and Sarma 1991; Sangrame *et al.*, 2000)

Morever, for studies of the physical properties affected by the evaporation process, the liquid foods have been reconstituted (Moresi and Spinosi, 1984; Tsen and King, 2002; Dak *et al.*, 2007), reformulated (Zainal *et al.*, 2000), and by majority, they have been concentrated on vacuum (Moresi and Spinosi, 1980; Stankiewicz and Rao, 1988; Bayindirli, 1993; Vélez-Ruiz and Barbosa-Cánovas, 1997, 2000; Choi *et al.*, 2006).

Heat treatment have implied the application of mathematical analysis and models for evaluation of heat transfer coefficients (Chen *et al.*, 1979; Stankiewicz and Rao, 1988; Jebson and Iyer, 1991; Prost *et al.*, 2006). Rheological properties have been measured either with viscometers (Moresi and Spinosi, 1980, 1984; Bayindirli 1992, 1993; Cepeda and Villarán, 1999; Sobolík *et al.*, 2002; Zuritz *et al.*, 2005; Enriquez *et al.*, 2009) or with rheometers (Hernandez *et al.*, 1995; Giner *et al.*, 1997; Vélez-Ruiz and Barbosa-Cánovas, 1997, 1998, 2000; Kaya and Sözer, 2005; Nindo *et al.*, 2005;), and even with in-line instruments (Dogan *et al.*, 2003; Choi *et al.*, 2006); whereas the boiling point determination have been measured by a pyrex lab set (Moresi and Spinosi, 1980; Camiro *et al.*, 2007), density by means of picnometers, and thermal properties with different thermal devices, cylindrical (Tellis-Romero *et al.*, 1998), parallel plates (More and Prasad, 1988), a line heat source probe (Choi and Okos, 1983), or a thermoanalyzer (Moresi and Spinosi, 1980, 1984).

Finally, the solids content determination has been carried out by measuring of the soluble solids with refractometers (Dogan *et al.*, 2003; Kaya and Sözer, 2005), or well the total solids by moisture determination in a vacuum oven (Dogan *et al.*, 2003; Rodríguez-Martínez *et al.*, 2007; Enriquez *et al.*, 2009).

3.5. Computer Simulation

Nowadays, with the presence of advanced computers and with the development of commercial simulators, the evaluation of an evaporator performance and its design may be faced with more precision, more controlled, more detailed, more quickly, and of course without experimentation. Some works have incorporated simulation techniques: Chuaprasert *et al.* (1999) utilized a data reconciliation method in Aspenplus to simulate an agitated evaporation by adjusting process measurements and satisfying the mass and energy balances, they found very good results and recognized the short time of the simulation. Cadet *et al.* (1999) also applied a simulation modeling and nonlinear predictive control on cane sugar evaporation with five effects, with very good and promising fittings for the sugar industry. Chawankul *et al.* (2001) developed a heat exchange-two flash model with AspenPlus software to simulate the concentration of orange juice in agitated thin film evaporators, finding good agreement among the experimental and simulated data. Ribeiro and Caño Andrade (2002) developed a steady state model for a plate evaporator with successful simulation of six industrial units. Miranda and Simpson (2005) modeled also a multiple effect evaporator for simulation and control purposes. They developed a dynamic model with differential equations and parameters sensitivity methodology in which the simulation showed acceptable behavior in the comparison of an industrial unit for tomato juice and the simulator results.

Similarly, the application of artificial intelligence for industrial production is a fact; thus, Benne *et al.* (2000) demonstrated the reliability of artificial neural networks for modeling and control of an industrial evaporation process for saccharose.

4. Examples of Application

As illustrative way of application of the aforementioned fundamentals, two examples are shown and solved below with academic purposes.

4.1. Illustrative Example 1

A single evaporation unit must be evaluated to concentrate 48000 kg/h with 9% (by weight) of a sodium hydroxide solution. The dilute solution at 39°C is to be fed to an evaporator to be concentrated up to 45% by utilizing saturated steam at 100°C. On the basis of an overall heat transfer coefficient of 1200 W/m^{2}K and an operation pressure of 13.5 kPa that is kept inside the equipment, determine the performance and/or design parameters.

Solution

The design parameters include, the capacity (rate of evaporated water), the consumption (rate of needed vapor), the economy (ratio between the capacity and the conpsumtion), and the heat transfer area.

i) Mass Balance for a single effect (equations 1a and 1b)

$$\text{Global}: 48000 + w_v = w_E + w_P + w_c$$

$$\text{Liquor}: 48000 = w_E + w_P$$

$$\text{Solids}:$$

$$(0.09)\left(48000\frac{kg}{h}\right) = w_E(0) + w_P(0.45) \therefore w_P = 9600\frac{kg}{h}(\text{of} \cdot \text{product})$$

$$\text{From}-\text{de}-\text{liquor}-\text{balance}: w_E = 48000 - 9600\frac{kg}{h} = 38400\frac{kg}{h}$$

Therefore, the quantity of product is 9600 kg/h, while the capacity of the evaporation system is 38400 kg/h.

ii) Boiling Point Rise (BPR)

Data for water (P = 13.5 kPa), it corresponds a saturation temperature of 51.48°C (by interpolation), based on thermodynamic tables (Cengel and Boles, 2006). While, for NaOH solution at 45%, the boiling point comes from Dühring diagram: ~ 87°C (Brown *et al.*, 1965), thus:

$$BPR = T_{bsol} - T_{bwater} \Rightarrow T_{bsol} = 87°C; T_{bwater} = 51.5°C \therefore BPR = 87 - 51.5°C = 35.5°C$$

iii) Energy Balance (equation 12)

For a T_{base} of 0°C, and by using enthalpy-concentration diagrams or tables for NaOH (Brown *et al.*, 1965), as well as thermodynamic tables for water (Cengel and Boles, 2006),. Needed data for the energy balance are: Δh for NaOH 9% at 39°C (Δh_F): 151 kJ/kg; Δh for the evaporated water (sat. vapor at 87°C, (Δh_E): 2654 kJ/kg; Δh for NaOH 45% at 87°C (Δh_P): 442 kJ/kg; and Δh for the vapor (condensation heat, for saturate steam at 100°C, (Δh_V - Δh_C): 2256.4 kJ/kg.

Liquor:

$$Q + \left(48000\frac{kg}{h}\right)\left(151\frac{kJ}{kg}\right) = \left(38400\frac{kg}{h}\right)\left(2654\frac{kJ}{kg}\right) + \left(9600\frac{kg}{h}\right)\left(442\frac{kJ}{kg}\right)$$

$$Q = 98{,}908{,}800\frac{kJ}{h}$$

$$\text{Vapor}: w_V \Delta h_V = w_C \Delta h_C + Q \therefore Q = w_V (\Delta h_V - \Delta h_C)$$

$$w_V = \frac{Q}{\Delta h_V} = \frac{98{,}908{,}800\frac{kJ}{h}}{2256.4\frac{kJ}{kg}} = 43834.78\frac{kg}{h} \rightarrow \text{that} - \text{is} - \text{the} - \textbf{consumption}$$

$$\text{Economy} = \frac{\text{Capacity}}{\text{Consumption}} = \frac{38400\frac{kg}{h}}{43834.78\frac{kg}{h}} = 0.876$$

iv) Heat transfer area (equation 15)

With the existing information, the last and very important step is the evaluation of the required area for the heating-evaporation process and equipment design.

$$A = \frac{Q}{U\Delta T} \rightarrow \text{the} - \text{driven} - \text{force} - \text{is}: T_{vapor} - T_{boiling} = (100 - 87)°C$$

$$A = \frac{\left(98{,}908{,}800\frac{kJ}{h}\right)\left(\frac{1000J}{1kJ}\right)}{\left(1200\frac{W}{m^2K}\right)\left(\frac{1\frac{J}{s}}{1W}\right)\left(\frac{3600s}{1h}\right)(13K)} = 1761.2m^2$$

Conclusion: From the developed calculations, it can be summarized that, for the concentration of 48000 kg/h of NaOH from 9 to 45% (by weight) entering at 39°C, the evaporator unit needs to evaporate 36400 kg/h of water (that is the effect capacity) to produce

9600 kg/h of concentrated. This evaporation unit implies a heating area of 1761.2 m^2 and a supply of 43835 kg/h of saturate steam (consumption) at 100°C.

Comment: The required heating area is big due to three main aspects: the first is a big quantity of feeding; secondly, a low feeding temperature; and third, is the small difference or few degrees for the driven force as a consequence of the boiling point elevation.

4.2. Illustrative Example 2

It is desired to design a triple effect evaporation system to concentrate the solids of grape juice from 6 to 51°Bx. The feed flow is 30000 kg/h entering to the third effect at 24°C, forced convection is used. Saturated steam at 115°C is available for heating requirements of the first unit. If the third effect is to be operated with a vacuum pressure of 4 in Hg, and overall heat transfer coefficients of 1800 (U_1), 1650 (U_2), and 1500 W/m^2K (U_3), are expected, determine the design parameters for this *backward triple evaporator,* neglecting the BPR.

Solution

Taking as reference the parameters included in Figure 1, with w_F in the third effect and w_P from the first, the algorithm for multiple effect evaporation is followed.

i) Mass Balance (equations 1a and 1b)

Liquor mass balances:

$$w_F = w_{E3} + w_{C3} + w_{C2} + w_P \rightarrow (1)$$

$$30000\frac{kg}{h}(0.06) = w_P(0.51) \therefore w_P = 3529.41\frac{kg}{h}$$

$$\text{capacity} = w_{E1} + w_{E2} + w_{E3} = 30000 - 3529.41 = 26470.59\frac{kg}{h}$$

ii) Boiling Temperatures and Energy Balances

To proceed with the energy balances on the three effects, and knowing the boiling point of water at 4 in Hg is 53°C (Cengel and Boles, 2006)..

Tn must be assumed based on an overall $\Delta T = (115 - 53)°C = 62°C = \Delta T_1 + \Delta T_2 + \Delta T_3$, and following the aforementioned steps. Boiling point assumption based on equal heat transfer surfaces:

$$Q_1 = Q_2 = Q_3 \therefore U_1A_1\Delta T_1 = U_2A_2\Delta T_2 = U_3A_3\Delta T_3$$

$$U_1\Delta T_1 = U_2\Delta T_2 = U_3\Delta T_3 \therefore \Delta T_2 = \frac{1800\frac{W}{m^2K}}{1650\frac{W}{m^2K}}\Delta T_1 = 1.091\Delta T_1;$$

$$\text{and}\cdot\text{similarly} \rightarrow \Delta T_3 = 1.20\Delta T_1 \therefore \Delta T_1 + 1.09\Delta T_1 + 1.20\Delta T_1 = \Delta T = 62°C$$

$$\Delta T_1 = 18.84 \rightarrow 18.8°C; \Delta T_2 = 20.54 \rightarrow 20.5°C; \Delta T_3 = 22.61 \rightarrow 22.7°C$$

generating that the boiling point correspondent to each effect becomes: $\Delta T_1 = 115 - T_1$; thus $T_1 = 96.2°C$; $\Delta T_2 = 96.2 - T_2$; thus $T_2 = 75.7°C$; and $\Delta T_2 = 75.7 - T_2$; thus $T_2 = 53°C$. The enthalpies corresponding to these boiling points are taken from Cengel and Boles (2006)..

Therefore, the energy balances, taking as T_{base} the boiling temperature at each effect, assuming the Cp for the flow to each effect are 3.990, 3.644 and 3.345 kJ/kg°C (Vélez-Ruiz, 2008), respectively, and considering the aforementioned assumptions (section 2.3):

First effect balances:

$$w_{S2} = w_{E1} + w_P; w_v = w_{C1} \rightarrow (2)$$

$$w_{S2}\Delta h_{S2} + Q_1 = w_{E1}\Delta h_{E1} + w_P\Delta h_P \rightarrow (3)$$

$$Q_1 = 2271\frac{kJ}{kg}w_{E1} - w_{S2}\left(3.345\frac{kJ}{kg°C}\right)(75.7 - 96.2)°C = w_V\left((2216\frac{kJ}{kg}\right)$$

$$\therefore 2216w_V = 2271w_{E1} + 68.6w_{S2} \rightarrow (3*)$$

Second effect Balance

$$w_{S3} = w_{E2} + w_{S2}; w_{E1} = w_{C2} \rightarrow (4)$$

$$w_{S3}\Delta h_{S3} + Q_2 = w_{E2}\Delta h_{E2} + w_{S2}\Delta h_{S2} \rightarrow (5)$$

$$Q_2 = 2320.6\frac{kJ}{kg}w_{E2} - w_{S3}\left(3.644\frac{kJ}{kg°C}\right)(53.0 - 75.7)°C = w_{E1}\left((2271\frac{kJ}{kg}\right)$$

$$\therefore 2271w_{E1} = 2320.6w_{E2} + 82.72w_{S3} \rightarrow (5*)$$

Third effect balance

$$w_F = w_{E3} + w_{S3}; w_{E2} = w_{C3} \rightarrow (6)$$

$$w_F\Delta h_F + Q_3 = w_{E3}\Delta h_{E3} + w_{S3}\Delta h_{S3} \rightarrow (7)$$

$$Q_3 = 2375\frac{kJ}{kg}w_{E3} - w_F\left(3.990\frac{kJ}{kg°C}\right)(24 - 53)°C = w_{E2}\left((2320.6\frac{kJ}{kg}\right)$$

$$\therefore 2320.6w_{E2} = 2375w_{E3} + 3471300\frac{kJ}{h} \rightarrow (7*)$$

Analyzing the equations 3*, 5* and 7*, there are 6 unknowns (w_V, w_{E1}, w_{S2}, w_{E2}, w_{S3}, and w_{E3}), that may be solved by considering the mass balances (eqns 2, 4 and 6), generating the next set of equations:

From the capacity, and equations 6 and 5*:

$$w_{E1} = 26470.59 - w_{E2} - w_{E3}$$
$$w_{S3} = 30000 - w_{E3}$$
$$2271(26470.59 - w_{E2} - w_{E3}) = 2320.6w_{E2} + 82.72(30000 - w_{E3})$$
$$w_{E3} = 26337.17 - 2.09827w_{E2} \rightarrow (8)$$

Now, equations 7* and 8 include the same two unknowns, solving them simultaneously:

$$2320.6w_{E2} = 2375(26337.17 - 2.098w_{E2}) + 3471300$$
$$w_{E2} = 9039.18\frac{kg}{h} \therefore eq(8) \rightarrow w_{E3} = 7370.18\frac{kg}{h}$$
$$and - susbtituting - in - capacity \rightarrow w_{E1} = 10060.88\frac{kg}{h}$$

Finally, by substitution of w_{E1} and w_{S2} (eq. 2: 3429.41+10060.889) kg/h in equation 3*:

$$2216w_V = 2271(10060.88) + 68.6(13490.3) \rightarrow w_V = 10728.2\frac{kg}{h}$$

iii) Heat transfer area

The heating surface for each evaporation unit are computed next:

$$A_1 = \frac{\left(2216\frac{kJ}{kg}\right)\left(10728.2\frac{kg}{h}\right)}{\left(1800\frac{W}{m^2\,°C}\right)(115 - 96.2)°C} = 702.35\frac{kJm^2}{Wh}\left(\frac{10^3\,W}{1kJ}\right)\left(\frac{1h}{3600s}\right) = 195.1m^2$$

$$A_2 = \frac{\left(2271\frac{kJ}{kg}\right)\left(10060.88\frac{kg}{h}\right)}{\left(1650\frac{W}{m^2\,°C}\right)(96.2 - 75.7)°C(3.6)} = 187.6m^2; A_3 = \frac{2320.6x9039.18}{1500x22.7x3.6} = 171.1m^2$$

iv) Second iteration

In order to get a better accuracy, and even though there exist a small difference between areas (< 15%), a second iteration may be completed. It should start from the evaluation of the mean area and lately, the new difference of temperatures is re-evaluated in the next way:

$$A_m = \frac{(195.1m^2)(18.8°C) + (187.6m^2)(20.5°C) + (171.1m^2)(22.7°C)}{62°C} = 183.83m^2$$

$$\Delta T_{12} = \frac{18.8x195.1}{183.83} = 20; \Delta T_{22} = \frac{20.5x187.6}{183.83} = 20.9; \Delta T_{32} = \frac{22.7x171.1}{183.83} = 21.1°C;$$

Now the boiling temperatures would be modified to 95.0, 74.1. and 53, respectively. As it can be observed only there is modification on T_1 and T_2, being 53°C a constant at the third effect

Conclusion: With these new boiling temperatures, the balances suffer some minor changes. After following the same methodology and solving the equations, it will generate accurate heating surfaces for the three effects with minimal differences between them. Therefore after the second iteration, the final results are: consumption (w_V) = 10676.33 kg/h; economy (26470.59/10676.33) = 2.48; evaporated water by effect: w_{E1} = 10026.41, w_{E2} = 9052.37, and w_{E3} = 7391.81 kg/h; and heat transfer areas: A_1 = 182.6, A_2 = 183.3, and A_3 = 184.5 m^2.

Comment: The required heating area is small in comparison to the first example, due to a smaller feeding flow, and to a neglected boiling point elevation. Comparing the first and the second iteration, there are small adjustments on the solved balances that conduct to a more accurate solution, taking the heating area as the decision criterion.

5. Final Remarks

Even though the knowledge and industrial application of evaporation have many years, there are several engineering aspects that have been studied in a limited way. It is very important to apply the fundamental knowledge of this process operation at the different sectors of the food industry. The requirements of energy for evaporation are very important (one to five of percentage of the total energy of a country). Therefore, it is remarkable from the ecological and economical viewpoints that evaporators operate at their maximum capacity and efficiency.

Evaporation of food items has been used and researched mediumly, but engineering information, covering aspects such as the boiling point elevation, physical properties and heat transfer coefficients, still are scarce. Only few works were found on these aspects, with exception of rheology in which more information is available.

The information of this important food process operation has been collected in this chapter to offer the different parts of it. Fundamentals of the unit operation, experimental information of researches, as well as examples of application and a variety of references have been compiled, with the goal of helping to students, food industrial workers, and investigators.

REFERENCES

Abichandani H. and Sarma S.C. 1991. Evaporation in a horizontal thin film scraped surface heat exchanger. *Journal of Food Process Engineering,* 14: 173-187.

APV 2003. Evaporator Handbook. Commercial brochure. APV Company. New York, USA.

Bayindirli L. 1992. Mathematical analysis of variation of density and viscosity of apple juice with temperature and concentration. *Journal of Food Processing and Preservation,* 16: 23-28.

Bayindirli L. 1993. Density and viscosity of grape juice as a function of concentration and temperature. *Journal of Food Processing and Preservation,* 17: 147-151.

Benne M., Grondin-Pérez B., Chabriat J.P. and Hervé P. 2000. Artificial neural networks for modeling and predictive control of an industrial evaporation process. *Journal of Food Engineering,* 46 (4): 227-234.

Bourgois J. and LeMaguer M. 1987. Heat transfer correlation for upward liquid heat transfer with phase change: Applications in the optimization and design of evaporators. *Journal of Food Engineering,* 6: 291-300.

Bouman S., Waalewijn R., De Jong P. and Van der Linden H.J.L.J. 1993. Design of falling–film evaporators in the dairy industry. *Journal of the Society of Dairy Technology,* 46 (3): 100-106.

Brown G.G., Foust A.S., Katz D.L., Schneidewind R., White R.R., Wood W.P., Brown G.M., Brownell L.E., Martin J.J., William G.B., Banchero J.T. and York J.L. 1965. Basic Operation in Chemical EngineeringEd. Marín, Spain. Pp. 499-517. (In Spanish).

Cadet C., Touré Y., Gilles G. and Chabriat J.P. 1999. Knowledge and nonlinear predictive control of evaporators in cane sugar production plants. *Journal of Food Engineering,* 40: 59-70.

Camiro-Cabrera, M., Torres-Pacheco, R.A., Enriquez-Fernandez, B.E., Vélez-Ruiz, J.F. and Sosa-Morales, M.E. 2007. *Physical properties of concentrated mango and orange juices* (008-23). IFT Annual Meeting, Chicago, IL, USA.

Cengel Y.A. and Boles M.A. 2006. Thermodynamics. An Engineering Approach. Mc Graw-Hill. Singapure. Pp. 890-893.

Cepeda E. and Villarán M.C. 1999. Density and viscosity of *Malus florinda* juice as a function of concentration and temperature. *Journal of Food Engineering,* 41: 103-107.

Chawankul N., Chuaprasert S., Douglas P. and Luewisutthichat W. 2001. Simulation of an agitated thin film evaporator for concentrating orange juice using AspenPlus^{TM}. *Journal of Food Engineering,* 47: 247-253.

Chen C.S., Lima V.M. and Marsaioli A. 1979. A heat transfer correlation for rotary steam-coil vacuum evaporation of tomato paste. *Journal of Food Science*, 44 (1): 200-203.

Choi Y. and Okos M.R. 1983. The thermal properties of tomato juice concentrates. *Transactions of the ASAE,* 26 (1): 305-311.

Choi Y.J., Milczarek R.R., Fleck C.E., Garvey T.C., McCarthy K.L. and McCarthy M.J. 2006. In-line monitoring of tomato concentrate physical properties during evaporation. *Journal of Food Process Engineering,* 29: 615-632.

Chuaprasert S., Douglas P. and Nguyen M. 1999. Data reconciliation of an agitated thin film evaporator using Aspen plus. *Journal of Food Engineering,* 39: 261-267.

Crapiste G.H. and Lozano J.E. 1988. Effect of concentration and pressure on the boiling point rise of apple juice and related sugar solutions. *Journal of Food Science,* 53 (3): 865-868.

Dak M., Verma R.C. and Jaaffrey. 2007. Effect of temperature and concentration on rheological properties of "Kesar" mango juice. *Journal of Food Engineering*, 80: 1011-1015.

Dodeja A.K., Sarma S.C. and Abichandani H. 1990. Heat transfer during evaporation of milk to high solids in thin film scraped surface heat exchanger. *Journal of Food Process Engineering,* 12: 211-225.

Dogan N., McCarthy M.J. and Powell R.L. 2003. Comparison of in-line consistency measurement of tomato concentrates using ultrasonics and capillary methods. *Journal of Food Process Engineering,* 25: 571-587.

Enríquez-Fernández B.E, Sosa-Morales M.E. and Vélez-Ruiz J.F. 2009. Effect of Reconstituted Milk on Properties of Spray Dried Milk. IADC 2009. *Inter-American Drying Conference.* Aug. 23-27, 2009, Montreal, Canada.

FT Industrial. 2004. Centritherm. Commercial brochure. FT Industrial Company. Australia.

Giner J., Ibarz A., Garza S. And Xhian-Quan S. 1997. Rheology of clarified cherry juices. *Journal of Food Engineering,* 30: 147-154.

Hall C.W. and Hedrick T.I. 1971. Drying of Milk and Milk Products. The AVI Pub. Co. Westport, CT, USA. pp. 17-46.

Hartel R.W. 1992. Evaporation and Freeze Concentration. In *Handbook of Food Engineering* (D.R. Heldman and D.B. Lund, Eds). Marcel Dekker, Inc. New York, NY, USA

Hernandez E., Chen C.S., Johnson J. and Carter R.D. 1995. Viscosity changes in orange juice after ultrafiltration and evaporation. *Journal of Food Engineering,* 25: 387-396.

Hobani A.I. 1998. Rheological behavior of date concentrates. *Journal of Food Engineering,* 36: 349-357.

Hoffman P. 2004. Plate evaporators in food industry-theory and practice. *Journal of Food Engineering*, 61: 515-520.

Holland C.D. 1975. Evaporation. In *Fundamentals and Modeling of Separation.* Ed Prentice-Hall. Englewood Cliffs, NJ. USA. Pp. 5-42

Ibarz A. and Barbosa-Cánovas G.V. 2003. Unit Operations in Food Engineeering. CRC Press, Boca Raton, FL, USA. Pp. 625-670.

Ilangantileke S.G., Ruba A.B. and Joglekar H.A. 1991. Boiling point rise of concentrated Thai tangerine juices. *Journal of Food Engineering,* 15: 235-243.

Jebson R.S. and Iyer M. 1991. Performances of falling film evaporators. *Journal Dairy Research.* 58: 29-38.

Karlsson H.O.E. and Trägardh G. 1998. Aroma recovery during beverage processing. *Journal of Food Engineering,* 34: 159-178.

Kaya A. and Sözer N. 2005. Rheological behavior of sour pomergranate juice concentrates (*Punica granatum* L.). *International Journal of Food Science and Technology*, 40: 223-227.

Kessler H.G. 1981. Food Engineering and Dairy Technology. V.A. Kessler. Germany. Pp. 208-238.

McCabe, W.L., Smith, J.C. and Harriot P. 2001. Evaporation. In *Unit Operations of Chemical Engineering.* McGraw Hill, Tokyo, Japan. Pp. 473-503.

Miranda V. and Simpson R. 2005. Modeling and simulation of an industrial multiple effect evaporator: tomato concentrate. *Journal of Food Engineering*, 66: 203-210.

More G.R. and Prasad S. 1988. V.R.N. Thermal conductivity of concentrated whole milk. *Journal of Food Process Engineering,* 10: 105-112.

Moresi M. 1985. Design and optimization of falling-film evaporators. In *Developments in Food Preservation,* Vol. 3 (S. Thorne, Ed.). Elsevier Applied Science Publishers, Barking UK. Pp. 183- 244.

Moresi M. and Spinosi M. 1980. Engineering factors in the production of concentrated fruit juices. I. Fluid physical properties of orange juices. *Journal of Food Technology,* 15: 265-276.

Moresi M. and Spinosi M. 1984. Engineering factors in the production of concentrated fruit juices. II. Fluid physical properties of grape juices. *Journal of Food Technology,* 19: 519-533.

Muddawar I.A. and El-Masri M. 1986. Momentum and heat transfer across freely falling turbulent liquid films. *International Journal of Multiphase Flow,* 12 (5): 771-790. Cited by Hernandez E., Chen C.S., Johnson J. and Carter R.D. 1995. Viscosity changes in orange juice after ultrafiltration and evaporation. *Journal of Food Engineering,* 25: 387-396.

Nindo C.I., Tang J. Powers J.R. and Singh P. 2005. Viscosity of blueberry and raspberry juices for processing applications. *Journal of Food Engineering,* 69: 343-350.

Prost J.S., González M.T. and Urbicain M.J.. 2006. Determination and correlation of heat transfer coefficients in a falling film evaporator. *Journal of Food Engineering,* 73: 320-326.

Ramos A.M. and Ibarz A. 1998. Density of juice and fruit puree as a function of soluble solids content and temperature. *Journal of Food Engineering,* 35: 57-63.

Reddy C.S. and Datta A.K. 1994. Thermophysical properties of concentrated reconstituted milk during processing. *Journal of Food Engineering,* 21: 31-40.

Ribeiro C.P. and Caño Andrade M.H. 2002. A heat transfer model for the steady-state simulation of climbing-falling-film plate evaporators. *Journal of Food Engineering,* 54: 309-320.

Rodríguez-Martinez V., Sosa-Morales M.E. and Vélez-Ruiz J.F. 2007. Physicochemical and flow characterization of prickly pear concentrates. CIBIA VI. *VI Congreso Ibero Americano de Ingeniería de Alimentos* November 5-8, Ambato, Ecuador. (In Spanish)

Sangrame G., Bhagavathu D., Thakare H., Ali S. and Das H. 2000. Performance evaluation of a thin film scraped surface evaporator for concentration of tomato pulp. *Journal of Food Engineering,* 43 (4): 205-211.

Schwartzberg H.G. 1977. Energy requirements for liquid food concentration. *Food Technology*, March: 67 – 76.

Singh R.P. and Heldman D.R. 2001. Introduction to Food Engineering. Academic Press. Glascow, Great Britain, UK. Pp. 625-670.

Sobolík V., Zitný R., Tovcigrecko V., Delgado M. and Allaf K. 2002. Viscosity and electrical conductivity of concentrated solutions of soluble coffee. *Journal of Food Engineering,* 51: 93-98.

Stankiewicz K. and Rao M.A. 1988. Heat transfer in thin-film wiped-surface evaporation of model liquid foods. *Journal of Food Process Engineering,* 10: 113-131.

Tellis-Romero J., Telis V.R.N., Gabas A.L.and Yamashita F. 1998. Thermophysical properties of Brazilian orange juice as affected by temperature and water content. *Journal of Food Engineering,* 38: 27-40.

Tellis-Romero J., Cantú-Lozano D., Telis V.R.N. and Gabas A.L. 2007. Thermal evaporation: Representation of rise in boiling point of grapefruit juice. *Food Science and Technology International,* 13 (3): 225-229.

Thijssen H.A.C. 1970. Concentration processes for liquid foods containing volatile flavors and aromas. *Journal of Food Technology,* 5: 211-229.

Tsen J.H. and King V.A.E. 2002, Density of banana puree as a function of soluble solids concentration and temperature. *Journal of Food Engineering,* 55: 305-308.

Vélez -Ruiz J.F. 2008. Notes of Food Engineering II. Universidad de las Américas, Puebla. Puebla, Mexico. Unpublished document.

Vélez-Ruiz J.F. and Barbosa-Cánovas G.V. 1997. Effects of concentration and temperature on the rheology of concentrated milk. *Transactions of the ASAE,* 40 (4): 1113-1118.

Vélez-Ruiz J.F. and Barbosa-Cánovas G.V. 1998. Rheological properties of concentrated milk as a function of concentration, temperature and storage time. *Journal of Food Engineering,* 35: 177-190.

Vélez-Ruiz J.F. and Barbosa-Cánovas G. V. 2000. Flow and structural characteristics of concentrated milk. *Journal of Texture Studies.* 31: 315-333.

Zainal B.S., Rahman R.A. Ariff A.B., Saari B.N. and Asbi B.A. 2000. Effects of temperature on the physical properties of pink guava juice at two different concentrations. *Journal of Food Engineering,* 43: 55-59.

Zuritz C.A., Muñoz E., Mathey H.H., Pérez E.H., Gascón A., Rubio L.A., Carullo C.A., Chernikoff R.E. and Cabeza M.S. 2005. Density, viscosity and coefficient of expansion of clear grape juice at different soluble solid concentrations and temperatures. *Journal of Food Engineering,* 71: 143-149.

PART III: TOPICS INVOLVED IN MASS TRANSFER DURING FOOD PROCESSING

In: Food Processing and Engineering Topics
Editors: M. E. Sosa-Morales and J. F. Vélez-Ruiz
ISBN: 978-1-60741-788-0

Chapter 8

EFFECT OF FRYING AND COMBINED TREATMENTS ON FRIED FOODS

María Elena Sosa-Morales and Jorge F. Vélez-Ruiz
Chemical and Food Engineering Department
Universidad de las Americas, Puebla., Mexico

1. INTRODUCTION

The world market for fried foods is continuously growing and it represents great quantities of money. Important products/snacks such as chicken nuggets, French fries, fried chicken, meat balls and beef burgers, regional items, toasted tortillas, among others, have a very high demand and consumption. These and other food products include a frying treatment as main part of the manufacturing process, in which oil or fat is incorporated to the product. On the other side, consumer preferences for low and free fat food items is a present driven force to modify them in the food industry.

In recent years, some research efforts have been focused in how to reduce oil uptake of fried items. Thus, some treatments and innovations such as dehydration, edible films, thermal treatments, vacuum application, and vapor injection in combination with immersion frying have been tested, in order to get fried foods of high quality with lower fat content.

The chapter covers the fundamentals and characteristics of the frying process, basic modeling, physical properties of fried items, and some obtained modifications of fried foods when one of the aforementioned treatments is applied. Experimental and reported data also have been included.

2. FRYING PROCESS OPERATION

2.1. Fundamental Aspects

Frying process operation is a traditional way of world wide food preparation, it produces desirable sensory characteristics, including aroma, flavor, and physical attributes. In this process, in addition to proteins and carbohydrates characteristic transformations, fat reacts with both food constituents, to generate desirable sensory properties (Weiss, 1983). Fried foods are very attractive, offering characteristics than increase their palatability. This process has evolved as a food process operation in which a great diversity of studies have contributed to a better understanding.

Frying, from the engineering point of view, is a unique unit operation in Food Industry. It is characterized by a simultaneous heat and mass transfer phenomena, heat is transferred from the oil to the food and moisture is evaporated from the foodstuff. During the frying process, physical, chemical and sensory properties are modified due to application of high temperatures (Krokida *et al.*, 2001a; Sosa-Morales and Vélez-Ruiz, 2008). General equations have been utilized for modeling of heat and mass transport during the frying process, considering four assumptions (Moreira *et al.*, 1995a): 1) the initial temperature and moisture are uniform, 2) shrinkage is negligible, 3) thermal and moisture diffusivities are constant, and 4) the oil is spontaneously absorbed into the food during the cooling-down stage.

$$\frac{\partial T}{\partial t} = \alpha\left(\frac{\partial^2 T}{\partial z^2}\right) \qquad (1)$$

$$\frac{\partial C_w}{\partial t} = D\left(\frac{\partial^2 C_w}{\partial z^2}\right) \qquad (2)$$

where: T is the food temperature (°C), t is the process time (s), α is the thermal diffusivity of food item (m^2/s), z is the distance across the thickness of the food (m), D is the diffusivity coefficient (m^2/s) and Cw is the moisture content (% dry basis).

These equations (second law of Fourier and Fick, respectively) must be solved by using adequate boundary conditions, in order to fit the two properties evolution through the whole frying process. Their solution and simulation have been the reason of many studies. More details about modeling may be consulted in several publications (Rice and Gamble, 1989; Moreira *et al.*, 1995b; Farkas *et al.*, 1996a, 1996b; Costa and Oliveira, 1999; Yamsaengsung and Moreira, 2002; Yildiz *et al.*, 2007; Sosa-Morales and Vélez-Ruiz, 2008)

The most popular and used frying method is by immersion, but there are some variations (pan frying, double side frying) in which the food may be partially immersed such in the case of donuts and regional food items, and sometimes the food is cooked on pans where a fine film of oil is present. Whatever is the frying method, there are three distinctive characteristics in this process operation: i) a high temperature of the frying medium (110-200°C) that implies a short cooking time; ii) an internal food temperature that does not exceed 100°C; and iii) the leaching of water-soluble compounds is minimal (Saguy and Pinthus, 1995).

In according with Singh (1995) and Farkas *et al.* (1996a), the immersion frying is characterized by four stages: an initial heating, followed of a surface boiling, a falling velocity stage and a bubble final point. Initial heating as a short period of few seconds, it is characterized by a heating up of the food item to reach the water boiling temperature. Natural convection is the transport mechanism, in which the sensible heat is involved. In this period, there is not surface evaporation. Surface boiling begins with the evaporation of water and an intense bubble formation. Natural convection is replaced by forced convection, and a strong turbulence is developed in the oil. The convective heat transfer coefficient increases rapidly, and the crust formation begins outside the food. This stage is analogous to food evaporation.

The decreasing rate period is the longest; it is characterized by a notable surface evaporation and food heating, where the water boiling temperature within the food is reached. Simultaneously, several biochemical changes occur, such as inner food cooking, protein denaturation, and starch gelatinization. The formation of a crusty surface continues and contributes to a decreasing in heat transfer. Finally, the bubble end point appears for longer frying periods. There is not presence of bubbles, meaning an absence of moisture loss, and definitive crust formation. This last stage of the frying operation, it is not commonly reached by most of the fried foods, due to the development of a hard texture and presence of fatty flavor, that are undesirable attributes for consumers. Furthermore, the four periods may be divided in only two parts, a non boiling phase that includes initial heating and bubble end point, and a boiling phase, covering the surface boiling and decreasing rate (Budzaki and Seruga, 2007)

2.2. Biochemical and Physico-Chemical Changes

During frying process, several biochemical and physicochemical modifications occur, such as carbohydrates caramelization, oil uptake, moisture loss, protein denaturation, starch gelatinization (Fan *et al.*, 1997), other modifications related with major food components, and some degree of damage in minor components like vitamins. Additionally, there are changes in aroma, flavor, texture, and microstructural reconfigurations.

The crust is the place of cellular changes generating softening of the middle lamella, starch gelatinization, dehydration, cells separation, elongation, volume augment or contraction (Pedreschi *et al.*, 2001; Hickey *et al.*, 2006). The extension of these changes depends of the food item and process parameters.

Furthermore, frying is a complex physical process in which heat and mass transfer phenomena happen simultaneously. Heat is transferred from the oil as the heating medium to food surface by convection, and then heat flows to the center of the food item by conduction. Whereas the mass transfer include a counter flow of moisture by evaporation to the external medium, and penetration of oil to the food.

3. Deep-Fat Frying Studies

Important advances have been completed in relation to deep-fat frying. They cover different aspects, such as chemical deterioration of frying oil, influence of physical properties of vegetable oil on the food products, effect of process variables on properties of the final product, evolution of physical and chemical characteristics of foodstuff, mathematical modeling of mass and heat transfer during frying, among others (Vélez and Xique, 1999). Additionally, the application of frying has been widely extended to contribute for improving of natural and regional raw foods and for developing of new food products.

3.1. New and Regional Products

Some studies related with these aspects are briefly cited next: Chen *et al.* (1998) studied the frying process of gluten balls, a traditional food in Taiwan. The frying process involved three consecutive frying pans, in which different properties were measured. Based on the properties and response surface methodology, the authors concluded that the best temperature ranged from 130-143°C for the first stage, 155-161°C for the second pan, and 195 $\pm$ 3°C for the final frying pan, in order to get very good sensory, texture and color attributes.

Goel *et al.* (1999) modified some blends of corn starch, casein, soy protein, and protein hydrolyzates to study the effect on noodle-like products. In general, as the protein/protein hydrolizates augmented, the oil content increased for all blends, suggesting possible differences in surface tension and hydrophobicity of the analyzed blends.

A first order exponential model for moisture transfer of papad (an oriental snack made from black gram flour, salt and spices) was determined by Math *et al.* (2004). They also determined the apparent diffusivity that was related to the inverse of temperature, computing an activation energy of 45 kJ/mol.

A study of deep-fat frying of gulabjamun balls (an Indian product based on khoa dough, sweet milk, wheat flour and syrup) in vegetal oil at 120, 130 and 140°C was developed by Math *et al.* (2004) for evaluation of color and textural changes. Some parameters, such as luminosity, firmness, and hardness followed a zero-order; other such as yellow and red indexes, and stiffness rise followed a first order kinetics.

Alvis Bermudez (2008) applied the frying process to slices of yam (*Dioscorea alata*) to develop an attractive product. He used soybean oil at 140, 160 and 180°C in the frying of yam, and analyzed the transfer phenomena, as well as the texture (fracture and penetration) and sensory (color, hardness and oilyness) through the frying process to get an acceptable fried item.

3.2. Heat and Mass Transfer

Other set of investigations, in which heat and mass transfer phenomena were importantly analized, are presente at once. There are a good number of works on these aspects of frying

In a very interesting and pioneering study, Gamble *et al.* (1987) established practical relationships for moisture loss and oil uptake during frying of potato slices. Both mass transfer phenomena were correlated as linear functions of the square root of frying time.

Ateba and Mittal (1994a) proposed a mathematical model for frying of beef meatballs considered as a fatty product. They subdivided the spherical geometry into concentric shells, and solved three differential equations to represent moisture, fat and heat transfer at the first stage or fat absorption period, and similarly the analytical approach was applied at the fat desorption or second period, only changing the conditions and adding some assumptions. The model was validated with good agreement between the observed and simulated mass loss data.

Chen and Moreira (1998) proposed a predictive model based on the fundamentals of heat and mass transfer for tortilla chips frying. Temperatures of chips and oil were measured and predicted at different operational conditions, in which the model may be used for a single or batch of chips, getting good agreement.

Krokida *et al.* (2000) obtained an empirical first order kinetic model to describe the mass transfer phenomena, moisture loss and oil absorption of French fries during deep fat frying. Also, their results showed that both mass transfers were significantly affected by oil temperature and thickness of potato samples.

A mathematical model to fit the transport phenomena during frying of tortilla chips was developed by Yamsaengsung and Moreira (2002). The model was a complex set of non-linear partial differential equations that allowed a good agreement between experimental and predicted data.

Oroszváry *et al.* (2005a, 2005b, 2006) conducted some experiments to study the heat and mass transfer phenomena through the frying of beefburgers, as infuenced by composition, comminuted degree, different meat material, pan temperature (100, 150 and 175°C), patty diameter (3-12.7 cm), and meat porosity. They observed a water loss that increased linearly as a function of temperature, suggesting that the available water was not a restricting factor. Additionally, moisture and fat losses influenced heat transfer of the crust and core respectively, in a separated way.

Yildiz *et al.* (2007) carried out an experiment in sunflower oil at 150, 170 and 190°C for immersion frying of potato slices (8.5 x 8.5 x 70 mm). In contrast to other studies, a decreasing heat coefficient was determined as a function of frying temperature (287 at 150, 227 at 170, and 181 W/m^{2}K at 190°C), while the mass transfer coefficient and moisture diffusivity augmented.

The determination of convective heat transfer coefficient (h) has been assessed by several authors including a diversity of food items and process conditions, for instance potato slices fried in sunflower oil at five temperatures contributed to coefficient values of 70-480 W/m^{2}K (Sahin *et al.*, 1999a); potato frying in sunflower oil at two temperatures resulted in h values of 443-750 W/m^{2}K (Costa *et al.*, 1999); mashed potato was fried in sunflower oil at five temperatures with h of 90-200 W/m^{2}K (Sahin *et al.*, 1999b); chicken fillets fried in soybean oil at 185°C allowed the evaluation of h, between 345 and 550 (Pannu and Chinnan, 2003); potato dough balls fried in soybean oil at three temperatures generated h values of 95 to 775 W/m^{2}K (Budzaki and Seruga, 2005); potato slices were fried in sunflower oil at three temperatures finding out an h of 180-290 W/m^{2}K (Yildiz *et al.*, 2007), just to mention a few.

3.3. Physical and Quality Properties

Among the studied properties, some times recognized as physical and other times as quality parameters, Hindra and Baik (2006) included color, nutraceuticals, texture, viscoelastic and volume/density as main properties. Additionally, other properties such as dimensions (diameter, length, surface, thickness), mass diffusivity and transfer coefficient, specific heat, thermal diffusivity and conductivity, porosity (Vélez-Ruiz, 2008) may be characterized through the frying process of the food items.

Miller *et al.* (1994) followed the physical properties of different vegetable oils through the frying process and oil repeatedly used, finding cinematic viscosities in the range of 2.60-3.45 for canola, 2.50-3.95 for palm, and 2.61-3.80 cm^2/s for soybean oil. Oil viscosity obviously influenced the convection mechanism of heat transfer, that were evaluated between 249.5 to 276.2 W/m°C for the tested experimental conditions.

An analytical model to determine thermal and mass diffusivities in cylindrical foods subjected to frying was provided by Dincer and Yildiz (1996). They evaluated a thermal diffusivity of 3.85 x 10^{-7} m^2/s, and a mass diffusivity of 1.31 x 10^{-7} m^2/s for sausages.

Some studies have been conducted on potato products. Sahin *et al.* (1999b) studied the influence of temperature on the frying of potato slices, expressing the color (hue), heat capacity, and thermal conductivity of the samples as a function of time at five oil temperatures (150, 160, 170, 180 and 190°C). Krokida *et al.* (2001b) focused their analysis on color changes of French fries as a function of main frying variables: oil temperature and oil type, and sample thickness. They found a kinetic approach of first order for the color parameters L, a, and b; in it the constant rate was correlated to the main process variables, such oil temperature and sample size. Pedreschi *et al.* (2001) followed the texture kinetics of potato strips through of frying. The determination of texture in situ and normalization of parameters, was used to differentiate the kinetics of core cooking and the kinetics of tissue softening.

Also, maize fried products are important in the market. Kawas and Moreira (2001b) utilized different samples to study the influence of the degree of starch gelatinization on density, porosity, texture and shrinkage of tortilla chips during frying at 190°C. The gelatinization degree affected the oil distribution and final oil content of chips.

Some additional details and other studies on frying of foods are included in table 1.

Table 1. Selected studies on food frying frocesses

Food item	Focus of the study	Studied parameters/Frying temperature	Reference
Chicken patties	Quality parameters	Color, microbial contents, moisture/ 168, 179, 190°C	Yi & Chen, 1987
Potato slice	Mass transfer	Moisture diffusion coefficient/ 145, 165, 185°C	Rice & Gamble, 1989
Tortilla chips	Microstructure	Baked, undercooked, cooked and overcooked chips/ 190°C	McDonough *et al.*, 1993
Meatballs	Quality properties	Crust formation, color and firmness/ 159 ± 1°C	Ateba & Mittal, 1994b
Tortilla chips	Physical properties	Density, specific heat, thermal conductivity, texture/ 190°C	Moreira *et al.*, 1995c
Potato chips	Oil physical properties	Viscosity and surface tension of corn oil/ 100-200°C	Gomes & Singh, 1995
Tapioca chips	Physical changes	Moisture loss, oil absorption and linear expansion/ 110-200°C	Vasanti *et al.*, 1996
Potato samples	Mass transfer	Water loss, two parallel compartments model/ 140 and 180°C	Costa & Oliveira, 1999
French fries	Mass transfer	Moisture and oil contents, mathematical modelling/ 150-190°C	Krokida *et al.*, 2000
French fries	Viscoelastic behavior	Elasticity, maximum strain and stress, viscoleasticity parameters/ 150-190°C	Krokida *et al.*, 2001a

Table 1. Selected studies on food frying processes

Food item	Focus of the study	Studied parameters/Frying temperature	Reference
French fries	Color changes	Lightness, redness, and yellowness parameters/ 150-190°C	Krokida *et al.*, 2001b
Tortilla chips	Quality attributes	Isotherms, glass transition temperature, bulk and solid density, porosity, structure and texture/ 190°C	Kawas & Moreira, 2001a, 2001b
Potato strips	Textural parameters	Puncture tests, force-distance curves/ 160-195°C	Pedreschi *et al.*, 2001
Tortilla chips	Heat and mass transfer	Moisture and temperature profiles, structural changes/ 190°C	Yamsaengsung & Moreira, 2002
Papad	Heat and mass transfer	Water diffusivity and oil uptake, apparent density/ 165-185°C	Math *et al.*, 2004
Beefburguer	Heat and mass transfer	Water and fat fluxes, and other physical parameters/ 100-175°C	Oroszvári *et al.*, 2005a, 2005b
Gulabjamun balls	Color and texture	CIELAB parameters, total color change, and textural attributes (firmness, hardness, and stiffness)/ 120-140°C	Kumar *et al.*, 2006
Potato dough balls	Heat transfer	Convective heat transfer (94-775 W/m^2K) This coefficient was the highest at the beginning of frying/ 160-190°C	Budzaki and Seruga, 2007
Potato slices	Heat and mass transfer	Heat and mass transfer coefficients and diffusivity/ 150-190°C	Yildiz *et al.*, 2007
Yam slices	Product development	Heat and mass transfer, texture and sensory attributes/ 150-180°C	Alvis Bermudez, 2008

4. OIL UPTAKE

One of the most important characteristics of fried foods is the oil absorption, but unfortunately fat has been associated with a diversity of diseases, contributing to a world health concern. Cardiovascular, diabetes type II, obesity, and some cancers have been related with high fat consumption. It is also important that the frying medium is low in saturated and trans fatty acids due to blood lipids implications (Mehta and Swinburn, 2001). And even though, the oil uptake is a generalized phenomenon in fried products, the mechanism of oil gaining is not completed understood. According to Mellema (2003), fat uptake is basically determined by two mechanisms: the condensation effect and capillary effect.

4.1. General Aspects of Oil Absorption

Two mass transfer phenomena are developed during frying, one is the water loss due to evaporation and the other, is the oil uptake. For the second one, a wide spectrum of factors have been recognized: food composition; prefrying treatments and surface treatments; oil quality and composition; initial interfacial tension, porosity and gel strength; frying temperature, food shape, crust formation, and process duration as the main factors in oil gaining (Saguy and Pinthus, 1995).

Other experiments have been done to explain the oil gain. Ufheil and Escher (1996), followed oil uptake during frying of potato slices incorporating a dye to monitor the dynamics of oil absorption. They concluded the oil gain is mainly a surface phenomenon. Annapure *et al.* (1998) modified the composition of the flour used to prepare an Indian snack, they tried to correlate the oil content, oil-holding capacity, and water-holding capacity with the composition of flour blends, without successful results. Moreira and Barrufet (1998) developed a new approach by using capillary pressure theory to describe oil absorption of fried foods, finding good agreement between experimental and predicted results. They concluded that initial moisture and cooling air temperature were more important than surface tension and pore radius in this mechanistic model. Mehta and Swinburn (2001) cited several factors affecting the final oil content of potato chips: freezing treatments, frying fat, moisture content, thickness, volume ratio of potato to oil, and oil temperature among others.

4.2. Fat Uptake Determination

The fat content determination has contributed to the application and development of instrumental techniques. Among the utilized techniques of fat uptake determination and/or observation, the next may be mentioned: Soxhlet extraction method, chromatographic analysis, differential scanning calorimetry, confocal scanning laser and scanning electron microscope observations, magnetic resonance and infrared imaging (Mellema, 2003).

Marikkar *et al.* (2003) utilized gas liquid chromatography, high performance liquid chromatography, and differential scanning calorimetry to analyze the presence of fat, lard or some derivates of them. Hickey *et al.* (2006) applied a magnetic resonance technique to measure the presence of oil in specific sites of potato chips, proving to be as an accurate, fast, non-invasive, and reproducible technique.

4.3. Useful Life of Oils

It is desirable to know how the magnitude of oil abuse affects the frying process and fried food characteristics; therefore, several studies have been focused on this aspect.

Handel and Guerrieri (1990) conducted a study to evaluate the quality of oil heated in frying. The rate at which a frying oil decreased in quality as consequence of heating, measured by acid value and concentration of polar compounds, was roughly correlated to the saturation degree of the oils.

Paul and Mittal (1996a) carried out a research to correlate the fat degradation during frying. The evolution of fat degradation was studied by measuring the concentration of decomposition components and physical properties. Fat degradation affected specific gravity and specific heat, as well as viscosity. Also, Paul and Mittal (1996b) investigated the fat/oil degradation during frying. The evolution of a canola based shortening degradation was studied by measuring the concentration of decomposition components, free fatty acids, polar materials, and optical properties. High significant effects of degradation products on lightness, photometric color index, and turbidity were observed; and no significant change was detected in the refractive index of the samples.

A research to assess corn oil deterioration was carried out by Sebben *et al.* (1998). They applied UV (ultraviolet) and NIR (near-infrared) techniques to measure oil changes developed during frying. Autoxidation was caused by effect of air and frying temperature; suggesting that spectral changes in absorption at certain wavelengths could be used as evidence of oil decomposition through frying processes. On the other hand, Flores *et al.* (2001) concluded that during the frying process, the smoke point is the fastest, most economical, and simplest way for determining oil quality.

Thus, by knowledge of oil uptake and related aspects, it is possible to control the amount of oil or fat present in fried foods. There is a significant effort to decrease the fat content of fried food items, therefore a varied number of processing alternatives have been studied and proposed to get this purpose. But the factors influencing fat absorption during frying are complex and interrelated.

5. OTHER TREATMENTS FOR FRIED PRODUCTS

5.1. General Aspects

Consumer trends are moving towards healthier foods. Therefore, it is essential at this time, to reduce the fat content of fried foods, as well as to maintain their unique organoleptic characteristics. This trend has contributed to look for new alternatives, apply new technologies, and develop combined processing that allows oil reduction of these foods.

Reduction in oil content for fried foods is very outstanding due to the significant role that fried food items play in dietary fat intake. Although the fat reduction in fried foods has been the main objective several researches; process aspects or food properties related with combinations of treatments, including deep fat frying, also have been covered. Between the alternatives for reduction of fat uptake, Mellema *et al.* (2003) cited three: 1) application of coatings and batters, 2) modification of the frying medium, and 3) modification of the frying processes.

Among the studied alternatives, some of them may be mentioned: coating with edible films (Rasco *et al.*, 1987; Balasubramaniam *et al.*, 1995; Balasubramaniam *et al.*, 1997; Williams and Mittal, 1999; Mittal and Zhang, 2000; García-Joffre *et al.*, 2007); substitution of beef tallow (Sinram and Hartman, 1989); restructuring of foods, such as potatoes (Pinthus *et al.*, 1992); modification of food formulations by adding some additives, corn syrup, and salts (Pinthus *et al.*, 1993; Esturk *et al.*, 2000; Moyano *et al.*, 2002); adding of honey and microwave preparing of fat-free potato (Demetriades *et al.*, 1995); using vacuum during the frying process (Garayo and Moreira, 2002); application of oven and dynamic radiant heating (Lloyd *et al.*, 2004); combining blanching and frying (Moyano and Pedreschi, 2006);

combining drying and frying (Moyano and Pedreschi, 2006; Morales-Pérez *et al.*, 2006); combining superheated steam injection and frying (Cenkowski *et al.*, 2008). Food production with low fat level is one of the most important driving forces in the food industry, particularly the snack items production (Esturk *et al.*, 2000). Table 2 summarizes some of the conducted studies in which some treatments are applied in addition or comparison to frying processing.

Table 2. Experimental characteristics of combined treatments

Food item	Dimensions	Treatments	Conclusions/reference
Fish nuggets	34 g weight	Coating with DDGS and frying at 180°C	DDGS showed good attributes. Rasco *et al.*, 1987
Potato cylinders	3.5 cm length	Restructuring of potato mass and controling the gel strength and frying at 170°C	Gel strength affected markedly the mass transfer phenomena. Pinthus *et al.*, 1992
Donuts & Falafel balls	"Normal" donuts Balls 3 cm diameter	Adding of powdered cellulose and methyl cellulose in the formulation of donuts and falafel. Frying at 190 and 170°C	Methocel was significantly more effective than powdered cellulose in reducing oil uptake. Pinthus *et al.*, 1993
Potato chips	1.58 mm thickness	Soaking in a honey solution, and then were heated in a microwave oven	Improving of overall quality (color and texture parameters). Demetriades *et al.*, 1995
Chicken balls	35 and 47 mm diameter 24 and 56 g mass	HPMC films with thickness of 0.05 mm Frying at 175°C	Edible film reduced fat absorption and improved water retention. Balasubramaniam *et al.*, 1995, 1997
Various foods	Discs 1 cm thick, 4.5 cm diameter	Coating with gellan gum, 2-2.5 mm thick to know mass transfer	Fat diffusivity was half that of the non-coated food Williams & Mittal, 1999
Tortilla chips	1.5 mm thickness	Three types of CMC (low, medium and high viscosity) at three levels (0.5, 1.0, 2.0%). Baking at 280°C-180 s, frying at 190°C	Addition of CMC reduced significantly the oil absorption of all samples. CMC of high viscosity was the most effective. Esturk *et al.*, 2000
Coated foods	Infinite slab	Modeling and simulation of coated slabshaped items during deep-fat frying	Temperature at center, and average, fat and moisture contents may be obtained. Mittal & Zhang, 2000
French fries	7 x 7 x 7 mm	Impregnation with corn syrup and NaCl on crust color changes at 160, 170 and 180°C	A first-order rate equation fitted the kinetics of color. Moyano *et al.*, 2002

Table 2. Experimental characteristics of combined treatments

Food item	Dimensions	Treatments	Conclusions/reference
Potato chips	samples with 21.67% of solids content	Vacuum with three levels of pressure 16.7, 9.9 and 3.1 kPa	Vacuum frying is an alternative that could be used to produce a product with lower oil content. Garayo & Moreira, 2002
French fries	0.95 x 0.95 x 10 cm (32-oz packages)	Oven (232°C-8 min) and controlled dynamic radiant heatings (250W-4 min)	Both heating treatments showed promise to produce lower-fat items. Lloyd *et al.*, 2004
Chicken nuggets	1.5 cm thickness 4.0 cm diameter	Addition of soy and rice flours (5%) to the batter formulation	Both flours provided a lower oil absorption as compared to the control. Dogan *et al.*, 2005
Tortilla discs	1.5 mm thickness 15 cm diameter	Drying at temperatures of 35, 48, 60°C and three air rates (2, 4, 6 m/s), followed from frying at 140, 160 & 180°C	Drying contributed to lower fat content that increased with frying temperature. Tortillas dried at 4 m/s resulted in better quality. Morales-Perez *et al.*, 2006
Potato slices	2.5 mm thickness	Blanching with hot water 85°C-3.5 min Drying in a convention oven	Highest oil content at three temperatures Lowest oil uptake. Moyano & Pedreschi, 2006
Chicken slices	1 x 3 x 5 cm	Superheated steam (11.23 kg/h) at 110 and 140°C, 3 and 5 min. Frying at 160°C	Oil absorption was lower for steamed samples, and higher with steam temperature increasing. Cenkowski *et al.*, 2008

DDGS: corn distillers′ dried grains with solubles, D: diameter, HPMC: hydroxyl propyl methyl cellulose, CMC: carboxy methyl cellulose.

5.2. Related Studies on Frying Combined with Other Processes

A group of researchs in which frying is combined with other treatment is summarized. Sinram and Hartman (1989) compared sensory and physicochemical characteristics of French fries that were fried with three different fatty materials (corn oil, flavored oil and tallow), looking for substitution of beef tallow. Their study verified the utility of flavored oil as substitute of beef tallow, based on sensory and physicochemical assessments.

Pinthus *et al.* (1992) carried out an experimental study in which a restructured potato product was fried. Gel strength was controlled and related to moisture loss and oil uptake, that were markedly affected. Oil absorption increased as gel strength augmented. Pinthus *et al.* (1993) added two cellulose derivates (methyl cellulose and powdered) in order to study the effect on two food items during deep fat frying. Both additives were incorporated as part of the formulation of donuts and falafel in order to analyze water loss and oil absorption. These researchers proposed an objective indicator: uptake ratio (U_R) to relate oil gaining and water removal. The uptake ratio was 100% for the control food and decreased in the next order:

84.3 and 92.3 for powdered cellulose; 75.4 and 69.2 for methocel type 1; and 70.9 and 53.8% for methocel type 2, in donuts and falafel balls, respectively.

The application of microwave and addition of honey were combined by Demetriades *et al.* (1995) to improve the quality characteristics of fat-free potato chips. Honey improved the chips quality even though there was not significant difference in the level of utilized honey. There was difference for the oil content quantified in potatoes formulated with honey amounts between 4 and 5 %, but was not different for 5 and 6% of honey.

Balasubramaniam *et al.* (1995 and 1997) carried out experimentalal studies to compare fat uptake and water retention of chicken balls, with edible film and without it. Edible film significantly affected both mass transfer phenomena: moisture loss and oil gaining. The edible film reduced fat absorption (17.9 at surface and 33.7% in the core), also improved moisture retention (8.7 in surface layer and 16.4% in the core) in comparison with the control.

A mathematical model to describe heat and mass transport for the frying process of foods with edible coatings was developed by Williams and Mittal (1999). Experimental data was used to validate the proposed model, finding good agreement ($R^2 = 0.974$), as well as heat transfer ($R^2 = 0.994$). The fat diffusivity of 2 mm gellan gum film was 6.4 x 10^{-10}, whereas of the food was 1.03 x 10^{-9} m^2/s. Moisture content had a lower fitting ($R^2 = 0.918$) between experimental and predicted data.

Carboxymethylcellulose (CMC) was incorporated as part of the tortilla formulation to know the reduction of oil uptake during frying. Baking was also tested as part of the modifications. The reduction of oil absorption ranged from 5 to 40% as a function of CMC type and concentration, and the addition of CMC also favored a decreasing in the initial moisture content of raw tortilla chips (Esturk *et al.*, 2000).

Garayo and Moreira (2002) tested the feasibility of vacuum frying in potato chips. They analyzed the effect of three oil temperatures (118, 132, and 144°C) and three vacuum pressures (16.66, 9.89, and 3.12 kPa), and concluded that a product with lower oil content may be obtained, with vacuum incorporation in the frying process, as well as desirable color and texture. Their results showed that faster moisture loss rate in chips was related to higher oil adhesion, and consequently a higher absorption was observed. Additionally, as the degree of free water was depleted in the item, less oil was gained.

A comparative study for quality measurement of French fry style potatoes was carried out by Lloyd et al (2004). They compared oven heating, immersion frying and controlled dynamic radiant heating and based on quality and sensory determinations; they concluded that both types of heating are promising as an alternative to produce a lower-fat item. The lipid content of potatoes that were oven baked and radiant heated showed lower content (13.8 and 13.0%, respectively), in comparison to immersion frying with 19.2% (wb).

Dogan *et al.* (2005) modified a batter formulation by addition of soy and rice flours (5%) that contributed to a lower oil uptake with respect to the control nuggets. They found that batter viscosity favored the coating formation of the nuggets, forming a resistance to the oil absorption during deep-fat frying at 180°C. When all the quality properties were considered, soy flour added batters increased crispness, improved color and decreased oil content of chicken nuggets.

Moyano and Pedreschi (2006) modeled the kinetics of oil absorption through the frying process of potato slices that were subjected to two different pretreatments: blanching and combined blanching with drying. From their results they observed that an increase in frying temperature contributed to a lower oil uptake, at 120°C the absorbed oil was higher for control and blanched samples than for dried potatoes. At 150 and 180°C, the order of oil gaining was: blanched slices higher than control and the lowest corresponded again to the pre-dried samples, suggesting that the drying generates an external film or crust that lately it represents a barrier to oil gaining in the frying stage.

6. Our Studies with Combined Processes

6.1. Study of a Combined Method (Drying and Frying) for Manufacturing of Toasted Tortillas

6.1.1. Introduction

The world market for snack foods is continuously growing and represents an attractive business; one important snack in Mexico is the toasted tortilla, either in small pieces (chips) or complete tortilla pieces ("tostadas"). This food product includes a frying treatment as part of the manufacturing process, in which oil is incorporated to the product. Taking account of consumer preferences for low and free fat food items as an actual motivator to produce them in the food industry, we carried out a research with the next purposes: to combine advantages of two processes: air drying and frying to produce whole toasted tortillas with lower fat content in comparison with those commercial items existing in the market; to observe the evolution of physical properties, and to model the simultaneous heat and mass transfer phenomena.

6.1.2. Methodology

Commercial Sampling

Four commercial brands of toasted tortillas were characterized in order to establish a reference frame

Drying

Fresh tortillas (initial moisture of 40-42%, w.b.) were dried in a tunnel drier using three air temperatures (35, 48 and 68°C) and three air velocities (2, 4, 6 m/s)

Frying

Dried tortilla samples with 10% of moisture were employed to produce the so named "tostadas" or toasted tortillas. Three temperatures of sunflower oil (140, 160 an 180 °C) and four frying times (30, 60, 90 and 120 s) were analyzed. A commercial batch fryer (Rival^{MR}, Korea) with 4 L of capacity, with electrical resistors as heater was filled with fresh sunflower oil for the deep-fat frying process. When the desired temperature was reached, samples were introduced to the fryer and taken periodically (each 30 s).

Properties Measurement

Moisture content was determined in an oven at 105°C (HDP 334 model, MAPSA, Monterrey, Mexico) following an official method with 5 g of sample (AOAC, 1995).

Oil content was determined using the Goldfisch method (3500100 model, Labconco, Co., Kansas City, MO) by extraction with 30-40 mL of petroleum ether during six hours.

Texture was carried out in a Texture Analyzer instrument TA.XT2 (Texture Technologies, Corp. Scardale, NY). Force needed to rupture the whole toasted tortilla was determined with a spherical probe of 1.25 cm at a velocity of 2.0 mm/s and recording the applied force (F_R).

Color of the tortilla was measured by reflectance on a Hunter tristimulus colorimeter (Colorgard System, BYK Gardner, Inc., Silver Spring, Maryland), and expressed with the Hunter parameters (L, a and b). The net change of color (ΔE) was calculated by:

$$\Delta E = \sqrt{\left[(\Delta L)^2 + (\Delta a)^2 + (\Delta b)^2\right]} \qquad (3)$$

A white tile calibrator was used in the colorimeter, with values L = 92.89, a = -1.05 and b = 0.82. And the difference was calculated between two set of values, one at a given process time and the other at a reference or zero time

Sensory Evaluation

A panel of 20 non trained judges evaluated the acceptability degree (color, taste and texture) of the tostadas, with a nine-points hedonic scale in which a value of 1 corresponded to "dislike extremely", and 9 corresponded to "like extremely" (Larmond, 1970). An ANOVA was conducted in order to evaluate the results.

Experimental Design and Data Analysis

Experimental designs were carried out to study the effects of time and frying temperatures, respectively. A factorial design of 3 x 3 x 3 x 4 levels of air temperature, air velocity, oil temperature and frying time was followed to study the effect of independent variables. Multifactor analysis of variance was performed by using the Minitab 13.20 software. On the other hand, the optimization and fitting of the experimental data, to get the predictive equations and their corresponding correlation coefficients, were obtained from the software Design Expert 6.0.6 by application of a Box-Benken model

6.1.3. Results and Discussion

Product characterization of four commercial brands was carried out; their characteristics, such as moisture (1.8-6.7%) and fat contents (22-28%), color parameters (L: 23-25, a: 0.04-0.18, b: 1.3-2.5) and texture (5.8-7.6 N) were obtained.

For tostadas manufacturing, particularly in the drying process, as expected, lower drying times corresponded at higher temperatures and velocities to get 10% of moisture content; nevertheless at 48 and 68°C a contrary trend was observed for 4 and 6 m/s, requiring a lower drying time (42 and 45 min) at 4 m/s than at 6 m/s (71 and 55 min) to get the same moisture (table 3). Most likely a case hardening phenomena was developed during the drying of tortilla at 6 m/s, avoiding higher water evaporation. Moisture and oil contents, as well as texture, were significantly ($p < 0.05$) affected by the drying temperature, whereas the net color difference (ΔE) was not affected.

For the deep-frying stage a three oil temperatures (140, 160 and 180°C), a dried tortilla with 10% moisture was employed to produce the "tostadas" or toasted tortillas. When the fat content of the fried product was related to the air velocity, it was a significant effect ($p < 0.05$) of this variable. Drying at 2 and 6 m/s produced fried products or tostadas with similar oil contents (8-11 % w.b.), whereas drying with 4 m/s generated tostadas with 16-18%; both processes resulted different to that process with 4 m/s for the air in the drying stage. If these oil contents for tostadas are compared with those shown by commercial products, there is an important difference; drying with 2 or 6 m/s and lately fried are around 50% lower, and even the tostadas obtained with 4 m/s, are lower in this component, indicating that the combination of processes in tostadas manufacture favors the oil decreasing

Table 3. Drying times required to reduce the moisture of tortilla from 40-42% (w.b.) to 10% under different drying conditions

Drying temperature (°C)	Air velocity (m/s)	Drying time (min)
35	2	172
	4	174
	6	200
48	2	95
	4	42
	6	71
68	2	30
	4	45
	6	55

The main data (moisture, oil, net color and texture) obtained for tostadas manufacturing, including the results for the three air velocities used in the drying stage, are included in table 4. Moisture content and texture were significantly affected ($p < 0.05$) by the frying temperature, whereas it did not have effect on the fat content and color (ΔE) of tostadas.

Table 4. Properties range for toasted tortillas*

Frying Temperature (°C)	Moisture content (% w.b.)	Oil content (% w.b.)	ΔE	Texture Force (N)
140-120 s	4-8	8-10	8-12	6.6-12.3
160-120 s	2-4	9-17	6-14	4.2-12.9
180-120 s	1-3	10-17	9-17	6.0-14.1

* The three air velocities (2, 4 and 6 m/s) were included, as previous treatment to frying.

The mass transfer through the drying and frying processes was followed by separated. For drying, a constant drying period of 40 minutes was observed at the three temperatures for the highest air velocity, and also for 4 m/s at 48 and 68°C. While for 35°C and 48°C at the lowest air velocity, the constant drying time augmented up to 80 minutes. Moisture loss ranged from 0.10 to 1.3 x 10^{-2} $kg_{water}/min \cdot kg_{d.s.}$ during the drying operation, being noticeably higher than those obtained for the frying process (0.03 to 0.24 x 10^{-2} $kg_{water}/min \cdot kg_{d.s}$). An inverse relationship was observed among operations, that is to say, faster water evaporation during drying generated slower moisture loss during frying.

For the unstable mass transfer period, in which the solution to the second Fick´s law was utilized, a contrary pattern was shown, lower moisture diffusivity coefficients (D) were computed for drying (4-9 x 10^{-11} m^2/s) than for frying (1.8-4.4 x 10^{-9} m^2/s). The calculated D for the drying operation is comparable with those reported by Moreira *et al.* (1995b) of 9.3 x 10^{-8} m^2/s for tortilla chips frying, and Sosa-Morales *et al.* (2006) of 1.6-30.2 x 10^{-9} m^2/s for pork meat frying.

From the response surface, by using the Design Expert package and following the Box Behnken model, a set of experimental conditions was obtained from that software; from the results the next equations were generated ($R^2 \geq 0.90$):

$$\text{Moisture}(\%) = 4.78 - 0.63A + 0.78B - 0.74C - 0.87D + 1.35A^2 - 0.033B^2 + 0.65C^2 + 4.561AB + 0.03AC + 0.092AD + 0.25BC + 0.20BD - 0.12CD \quad (4)$$

$$\text{Fat}\cdot(\%) = 14.42 + 1.3A - 0.96B + 1.12C - 0.36D + 0.25A^2 - 2.90B^2 - 2.17C^2 - 0.77AB + 1.18AC + 1.10AD - 0.67BC + 1.03BD + 0.033CD \quad (5)$$

$$\text{Texture}\cdot\text{force}(N). = 10.80 - 2.68A + 0.32B + 2.18C - 1.56D + 0.82AB - 4.52AC + 3.15AD - 0.33BC - 0.13BD + 0.53CD \quad (6)$$

$$\Delta E = 36.79 - 1.94A + 4.24B + 3.08C + 5.60D \quad (7)$$

where: A is the air temperature (35-68°C), B is the air velocity (2-6 m/s), C is the frying temperature (140-180°C), and D is the frying time (30-120s).

Then, in order to corroborate the generated information, those aforementioned variables were controlled and used for an experimental run for tostadas manufacturing, in order to check out the predicted parameters. Excellent agreement was found between experimental and predicted values.

6.1.4. Conclusions about this study

Product characterization of four commercial brands was carried out; they were used as a reference frame for the lab manufactured tostadas. Drying with hot air (35, 48 and 68°C) at 2, 4 and 6 m/s were applied to dry tortillas, in which the final moisture contents range of 6 to 12%, taking between 30 and 200 min of process as a function of the studied variables, to reach 10% of final moisture.

Frying was conducted with a controlled initial moisture of 10% at the frying temperatures of 140, 160 and 180°C, only two minutes were needed to get the toasted tortillas. In general terms, at a given frying temperature lower moisture and higher fat contents were recorded for longer frying times; finding that tortillas dried at 4 m/s resulted in better product attributes.

There was significant effect of the drying conditions on the fried product. The obtained toasted product with 2-4% (w.b.) of moisture and 10-12% of oil, resulted lower if fat content that commercial products. Similar quality attributes to those of commercial "tostadas" were developed by the process combination of drying and frying.

Mass transfer parameters were computed for both processes. Finally an optimization was carried out from the software information and experimentally as well.

6.2. Super Heated Steam Treatment and Frying of Chicken Slices

6.2.1. Introduction

To decrease the oil content in fried products, new processes or combination of processes have been developed, but to reach the similar sensory and physical characteristics, with respect to fried products, it is still a challenge. Superheated steam (SHS) impingement has been used to dry and bake foods without oil. Food processing with SHS is a developing technique with numerous advantages, such as reduced environmental impact, increased drying efficiency, and improved product quality (Tang *et al.*, 2000). Application of vapor impingment to foods designated for deep-fat frying such as potato chips has been explored by other researchers, in which the use of SHS has resulted in reduced final oil content and promoted a desirable crunchy texture of fried products (Moreira *et al.*, 1999).

Thus, the objective of this study was to compare the final moisture content, oil uptake and quality attributes (color, density, thermal conductivity, and texture) of deep-fat fried chicken nuggets with those pretreated with SHS prior to frying.

6.2.2. Methodology

Similar methods, to those employed in section 6.1.2, were followed in this research.

Sample Preparation

Fresh Granny`s boneless and skinless chicken breasts were purchased at local grocery stores. The chicken breasts were cut into small slices (5 cm long, 3 cm wide and 1 cm thick). Individual pieces were soaked for 3 min in two-egg batter formulated with 0.5 g of salt and then drained for 5 s. Next, the battered pieces were placed on a layer of bread crumbs for 5 s on each side. Finally, the battered and breaded pieces were placed on a sheet of aluminum foil previously to be deep-fat fried.

Frying Process

An electric kitchen deep-fryer (Model 35015C, Proctor-Silex, Picton, ON) was used in deep-fat frying experiments. The frying pot was filled with 2 L of fresh sunflower oil. The ratio between the chicken nuggets (breaded samples) and oil was maintained at 1/12 allowing 10 nuggets to be fried at the same time. The breaded chicken samples were fried at 160°C for 3 min and served as control. Then, samples were drained in a frying basket from excess oil for 30 s and placed on paper towel for 1 minute to absorb excess oil from the surface, then the pieces were cooled at room temperature on aluminum foil during 20 min. The fried sample nuggets were stored in plastic bags.

Combination of Super Heated Steam Pre-Treatment and Frying

A superheated steam chamber developed at the Biosystems Engineering Department, University of Manitoba, Canada (Pronyk *et al.*, 2008) was employed for the pretreatment of chicken nuggets. One chicken nugget at a time was placed in the SHS chamber in a mesh cell with the No.12 openings. Two temperatures of vapor (110 and 140°C) were used in the pretreatment with exposure time of 3 and 5 min. The distance between the steam injector and the sample was 15 cm. The SHS gauge pressure (5 kPa) and flow rate (11.2 ± 0.2 kg/h) were kept constant in all pretreatments. After the steam treatment, the nuggets were cooled at room temperature for 6 min and stored in plastic bags till frying. The pre-treated nuggets were deep-fat fried in sunflower oil at 160°C for a period of 50-100 s, where the frying time was a function of the pretreatment conditions (table 5). After frying, samples were drained, cooled and stored under the same conditions as non-treated samples.

Table 5. Conditions used for combined SHS and deep-fat frying of chicken nuggets

Variable	Conditions				
	Control	1	2	3	4
Steam temperature (°C)	-	110	110	140	140
SHS treatment time (min)	-	3	5	3	5
Frying time at 160°C (s)	180	100	60	70	50

Temperature Measurements

Temperature history of the nuggets during SHS treatment and deep-fat frying was recorded with type J thermocouples. A calibrated thermocouple was inserted in the geometric centre of a sample. The temperature was recorded every 5 s by a data acquisition system (Agilent Model 34970A, Agilent Technologies, Loveland, CO)

Moisture and Oil Contents Analysis

Crust and core of the chicken nuggets were analyzed separately. Therefore, crust was removed from the individual nuggets. Moisture content was determined by method 7.003 (AOAC, 1995), whereas oil content was quantified by 78 min extraction with petroleum ether using a Soxtec system (Model 2050, Foss Analytical, Hilleroed, Denmark).

Quality Attributes Determination

Texture of fried nuggets was measured by a puncture test using the Texture Analyzer (Model TA-XT2, Texture Technologies, Corp., Scardale, NY). The 2 mm diameter probe was allowed to penetrate 15% of the sample thickness at a velocity of 5 mm/s.

Crust Color was measured at two spots on each nugget using a colorimeter (Model CR-400, Minolta Corp., Ramsey, NJ). The color values were expressed in the L, a, b Hunter scale.

Physical Properties Determination

Apparent Density of nuggets was determined based on mass of a single nugget and its volume. The volume of nuggets was measured with the displaced method using canola seeds as the measuring medium. The volume displacement apparatus was calibrated with metal spheres of known volume.

Thermal Conductivity of nuggets was measured with a Thermal Properties Analyzer (Model KD-2, Decagon Devices, Inc., Pullman WA) where the fine probe was inserted into three nuggets.

Statistical Analysis

All experiments were done in triplicate. The obtained data were subjected to the analysis of variance, and means were separated using Tukey´s method (Minitab Release 14, Minitab Inc., State Collage, PA) at a significance level of 0.05.

6.2.3. Results and Discussion

Moisture Loss and Oil Uptake during Frying

Core moisture content of control nuggets decreased slowly during their deep-fat frying, diminishing from 72.9% to 67.4% (w.b.). Similar trend was reported by Vélez-Ruiz *et al.* (2002) for frying of chicken strips. Moisture content of the crust of the nuggets was reduced from 45.7% to 35.5% (figure 1a). In the first two minutes of frying, the oil content of the nuggets core increased and then it reached a plateau at 2.5% (figure 1b). This characteristic behavior was reported by Balasubramanian (1995) during deep-fat frying of nuggets with edible film, and was attributed to the saturation phenomenon of capillaries. Oil content of nuggets crust increased quickly during the first minute of frying, and at the third minute it reached 24%. Thus, both mass transfer phenomena (moisture loss and oil gain) occurred mainly in the crust, through the deep-fat frying of nuggets due to ingredients type used in batter (fresh eggs containing water), and bread crumbs (with the capacity to absorb oil).

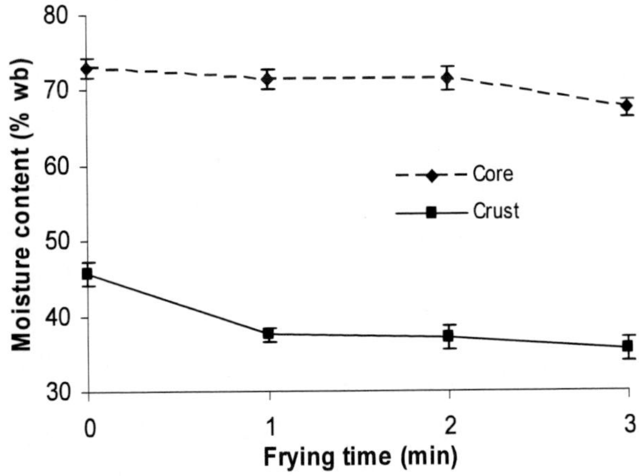

a

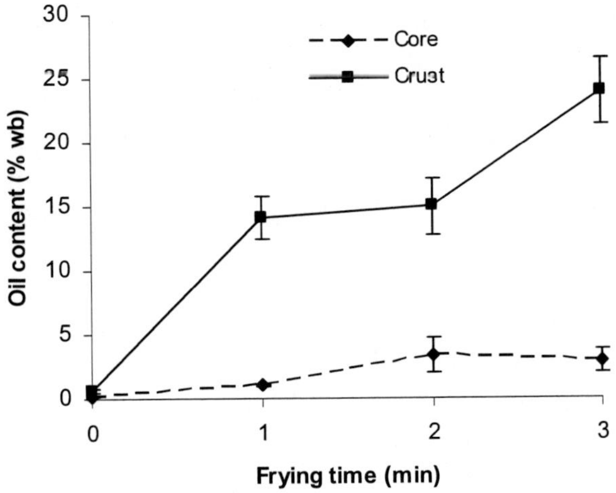

b

Figure 1. (a) Moisture loss and (b) oil gaining during deep-fat frying of chicken cuggets at 160°C.

Crust of control nuggets resulted in 24 % (w.b.) oil content, which was greater than the oil absorbed by crust in the pretreated with SHS ($p < 0.05$). Higher oil uptake was measured when higher steam temperature was applied. It was probably related to a formation of a larger pore size in the SHS treated nuggets. The lowest oil content in crust (7.3%, w.b.) was obtained for nuggets pretreated with SHS at 110°C for 5 min. Crust texture was not affected by the SHS pretreatment. However, color of pretreated samples was lighter in appearance than control.

Moisture and oil contents in the fried nuggets were affected by the SHS treatment ($p < 0.05$). For all the SHS conditions, higher moisture content for both, the core and the crust, were measured with respect to non-treated nuggets (table 6). Increasing in core moisture was in the range of 1.7 to 3.7% w.b. for 110 °C SHS, and 0.9 to 3.5% w.b. for 140 °C. For the same two vapor temperatures, the moisture increase in the crust was in the range of 13.5-18.8% w.b. and 7.0-13.5% w.b., respectively. Higher final moisture levels were attributed to the initial condensation of steam on a sample surface after placing them in the SHS chamber.

Table 6. Moisture and oil contents in non-treated and SSH Pre-treated nuggets after deep-fat frying

Conditions[+]	Final moisture content (% w.b.)[*]		Final oil content (% w.b.)[*]	
	Core	Crust	Core	Crust
Control	67.4 ± 1.2^{a}	37.4 ± 2.6^{a}	2.9 ± 0.9^{a}	24.0 ± 2.6^{a}
1	69.1 ± 0.3^{a}	50.9 ± 0.9^{b}	0.3 ± 0.1^{b}	11.0 ± 0.0^{b}
2	71.1 ± 0.0^{b}	56.2 ± 1.2^{c}	0.4 ± 0.2^{b}	7.3 ± 1.6^{b}
3	70.9 ± 0.1^{ab}	44.4 ± 0.5^{d}	1.2 ± 0.3^{ab}	18.1 ± 0.2^{c}
4	68.3 ± 0.4^{a}	50.9 ± 1.4^{b}	3.2 ± 1.6^{a}	16.1 ± 0.6^{c}

[+] Specific conditions are in Table 5 [*] Same letter indicates no significant difference ($p > 0.05$).

On the other hand, and as a positive result, lower oil contents were measured in all pre-treated nuggets. The steam temperature affected the oil content in fried nuggets, giving a higher oil content in nuggets pre-treated at 140°C (table 6, conditions 3 and 4) in comparison to those pre-treated at 110°C (conditions 1 and 2) giving a decrease in oil uptake. The temperature of 140°C probably caused the expansion of capillaries favoring oil absorption. The lowest crust oil content was measured when nuggets were pretreated with SSH at 110°C for 5 min, reaching only 7.3% w.b. in comparison to 24% w.b. for non-treated nuggets, representing a reduction of 16.7% w.b. in oil absorption or 70 percentage points.

Temperature Profiles

The temperature in the geometric centre (core temperature) of the control samples reached almost 70°C during the deep-fat frying (figure 2a). In all samples that were treated with vapor, the core temperature exceeded 70°C at the end of the pre-treatment (figure 2b). That means the nuggets were cooked after the SHS treatment.

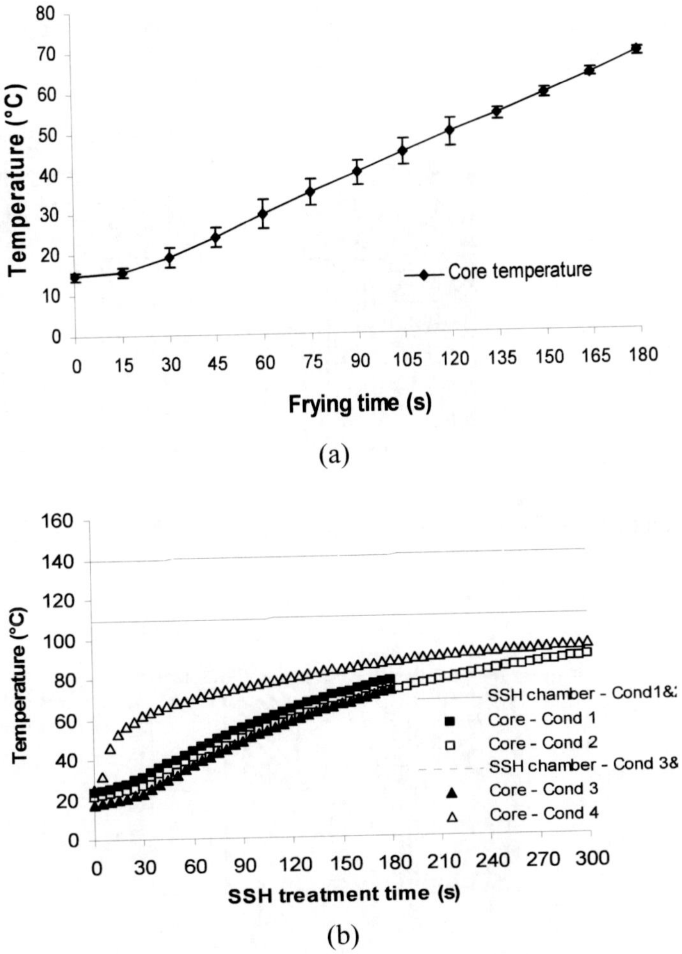

Figure 2. Temperature profiles. (a) Deep-fat frying of chicken nuggets (control), and (b) SHS treatments with conditions (Cond) 1 to 4.

Chicken Nuggets Properties

The SHS treatment time affected the texture, in which a lower penetration force was measured for 3 min-pretreated nuggets in comparison to control ($p < 0.05$). The 5 min SHS treatment of chicken nuggets developed a similar texture to that of the non-treated nuggets (figure 3). Thus, 5 min SHS treatment (for both temperatures; 110 and 140°C) can be used as good pretreatment to get similar texture in the crust of chicken nuggets.

Also, the SHS treatment affected the crust color (table 7) of fried nuggets ($p < 0.05$). No treated nuggets showed lower "L" values and higher "a" values with respect to control. It means lighter appearance and less reddish color. The SHS pretreatments at 110°C for 5 min and pretreatments at 140°C resulted in "b" values similar to non-treated with SHS samples.

As it was expected, density and thermal conductivity of chicken nuggets were not affected by the SHS treatments ($p > 0.05$). Apparent density ranged between 886 and 980 kg/m^3 (figure 4). These values were lower than those reported by Rao and Delaney (1995), who estimated a density of 1040 kg/m3 for breaded chicken pieces (including bone and peel).

Thermal conductivity ranged from 0.46 to 0.49 W/m°C, being similar to those reported by Moreira *et al.* (1999), who reported a value of 0.41 W/m°C.

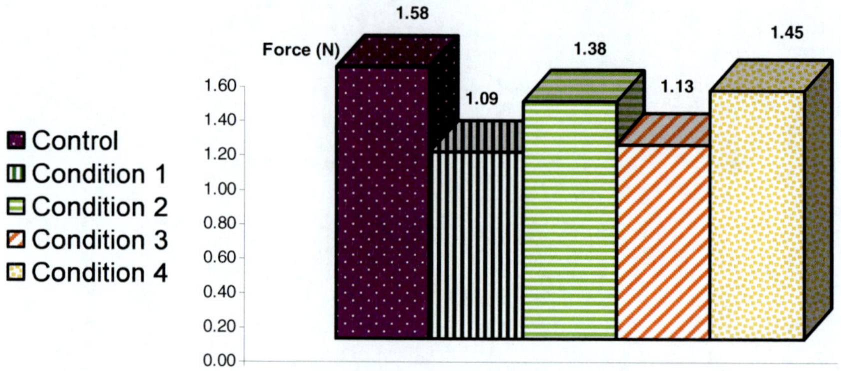

Figure 3. Penetration Force of Chicken Nuggets Pretreated with SHS and Fried (conditions 1 to 4) in Comparison to Control.

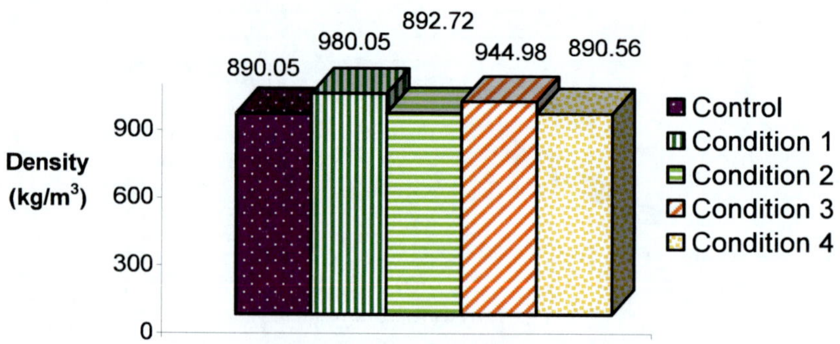

Figure 4. Density of fried nuggets, control and SHS pre-treated samples.

Table 7. Comparison of color parameters for control and SHS treated-fried nuggets*

Conditions	L	a	b
Control	45.2 ± 0.2^{a}	11.1 ± 0.1^{a}	19.4 ± 0.1^{ac}
1	52.6 ± 0.1^{b}	8.1 ± 0.1^{bc}	20.6 ± 0.1^{b}
2	49.2 ± 0.8^{cd}	7.8 ± 0.2^{c}	19.1 ± 0.2^{c}
3	48.0 ± 0.5^{c}	9.1 ± 0.0^{d}	19.7 ± 0.3^{a}
4	50.1 ± 0.4^{d}	8.4 ± 0.1^{b}	19.8 ± 0.1^{a}

*Means with the same letter indicate no significant difference ($p > 0.05$).

6.2.4. Conclusions about this study

Moisture loss and oil gain during the deep fat frying of chicken nuggets occurred mainly in the crust of the chicken nuggets. In comparison to 3-min deep-fat frying at 160°C, the SHS pre-treatment at 110°C-5 min followed by deep-fat frying at 160°C-60 s, reduced an average oil uptake of 2.5% and 16.7% (w.b.) for the core and crust, respectively.

At the same time the SHS treatment of nuggets at 110°C resulted in a higher moisture content of 1.7-3.7% and 13.5-18.8% (w.b.) for the core and the crust, respectively. The crust showed similar texture but lighter and more yellowish color, and due to a higher moisture uptake, the SHS treatment gave a softer product as compared to deep-fat frying only.

SHS did not affect the density and thermal conductivity of the nuggets. Due to its reduction in oil up-take in comparison to deep-fat frying alone, the SHS pre-treatment has potential for processing of chicken nuggets.

FINAL REMARKS

Frying as unique and complicated food process operation has been widely studied during 3 decades (1975-2005). Nowadays there is very good information about process conditions and their effects on fried products, transport phenomena characteristics, mainly moisture loss and oil uptake and heat transfer, and evolution of the food properties. Thus, processing design, product safety and product quality have been improved and it will continue its improvement through of the frying processing.

Precisely, one of the most important aspects of frying is the search for a lower fat content of food products. There are and will have more research on fat decreasing of fried items, consequently the modification or combination of processing methods that allow a lower oil content of fried products, it will be significantly augmented.

The rate of fat decreasing in fried foods, without loss of sensorial attributes, will determine the goodness of the process combinations and the successful of the obtained food products. In this moment, the most promising frying process modifications are: i) modification of original food formulations, ii) covering of foods with edible films, iii) incorporation of vacuum during frying, and iv) combination of food process operations, such as drying and frying, and application of superheated steam and frying.

To discover and implement innovative alternatives or solutions, it is necessary to shift our knowledge about deep fat frying. The fundamentals and process characteristics of this food process operation must be reshaped.

REFERENCES

Alvis Bermudez A. 2008. Heat and mass transfer during the dehydration by immersion of ñame pieces (*Dioscorea alata*) and process effects on quality product. Ph. D. Thesis. Universidad del Valle. Cali, Colombia. (In Spanish).

Annapure R.S., Singhal R.S. and Kulkarni P.R. 1998. Studies on deep-fat fried snacks from some cereals and legumes. *Journal of the Science of Food and Agriculture,* 76: 377-382.

AOAC. 1995. Official Methods of Analysis. *Association of Official Analytical Chemists.* Washington, DC.

Ateba P. and Mittal G.S. 1994a. Modelling the deep-fat frying of beef meatballs. *International Journal of Food Science and Technology,* 29: 429-440.

Ateba P. and Mittal G.S. 1994b. Dynamics of crust formation and kinetics of quality changes during frying of meatballs. *Journal of Food Science,* 59: 1275-1290.

Balasubramaniam V.M., Mallikarjunan P. and Chinnan M.S. 1995. Heat and mass transfer during deep-fat frying of chicken nuggets coated with edible film: influence of initial fat content. Co FE 1995: 103-106.

Balasubramaniam V.M., Chinnan M.S., Mallikarjunan P. and Phillips R.D. 1997. The effect of edible film on oil uptake and moisture retention of a deep-fat fried poultry product. *Journal of Food Process Engineering,* 20: 17-29.

Budzaki S. and Seruga B. 2007. Determination of convective heat transfer coefficient during frying of potato dough. *Journal of Food Engineering,* 66: 307-314.

Cenkowski S., Rosete Hidalgo J., Sosa Morales M.E. and Vélez-Ruiz J.F. 2008. Superheated steam as pretreatment to reduce oil uptake during deep-fat frying of chicken nuggets. CSBE/SCGAB 2008, Annual Conference. July 13-16. Vancouver, BC, Canada.

Chen V. and Moreira R.G. 1998. Modelling of a batch deep-fat frying process for tortilla chips. *Transactions of the Institute of Chemical Engineering,* 75 (C): 181-190.

Chen C.S., Chen J.J., Wu T.P. and Chan C.Y. 1998. Optimizing the frying temperature of gluten balls using response surface methodology. *Journal of the Science of Food and Agriculture,* 77: 64-70.

Costa R.M. and Oliveira F.A.R. 1999. Modelling the kinetics of water loss during potato frying with a compartmental dynamic model. *Journal of Food Engineering,* 41: 177-185.

Costa R.M., Oliveira F.A.R., Delaney O. and Gekas V. 1999. Analysis of the heat transfer coefficient during potato frying. *Journal of Food Engineering,* 39: 293-299.

Demetriades K., Guffey C and Khalil M.H. 1995. Evaluating the role of honey in fat-free potato chips. *Food Technology,* October: 66-67.

Dincer I. and Yildiz M. 1996. Modelling of thermal and moisture diffusion in cylindrically shaped sausages during frying. *Journal of Food Engineering,* 28: 35-43.

Dogan S.F., Sahin S. and Sumnu G. 2005. Effects of soy and rice flour addition on batter rheology and quality of deep-fat fried chicken nuggets. *Journal of Food Engineering,* 71: 127-132.

Esturk O., Kayacier A. and Singh R.K. 2000. Reduction of oil uptake in deep fried tortilla chips. *Food Science and Technology International,* 6 (5): 425-431.

Fan J., Singh R.P. and Pinthus E.J. 1997. Physicochemical changes in starch during deep-fat frying of a molded corn starch patty. *Journal of Food Processing and Preservation,* 21: 443-460.

Farkas B.E., Singh R.P. and Rumsey T.R. 1996a. Modeling heat and mass transfer in immersion frying I, Model development. *Journal of Food Engineering,* 29: 211-226.

Farkas B.E., Singh R.P. and Rumsey T.R. 1996b. Modeling heat and mass transfer in immersion frying II, Model solution and verification. *Journal of Food Engineering,* 29: 227-248.

Flores M.E., Rubio D. and Arellano J.A. 2001. Shelf life evaluation for an oil for industrial frying . *Tecnología de Alimentos,* July-Aug.: 16-22. (In Spanish).

Gamble M.H., Rice P. and Selman, J.D. 1987. Relationship between oil uptake and moisture loss during frying of potato slices from c.v. Record UK tubers. *International Journal of Food Science and Technology,* 22: 233-241.

Garayo J. and Moreira R.M. 2002. Vacuum frying of potato chips. *Journal of Food Engineering,* 55: 181-191.

García-Joffre J., Argaíz-Jamet A., Vélez-Ruiz J.F. and Sosa-Morales M.E. 2007. Doughnuts with reduced fat content based on edible coatings application. Poster: 011-30. IFT Annual Meeting, July 28 – August 1, Chicago, IL, USA.

Goel P.K., Singhal R.S. and Kulkarni P.R. 1999. Deep-fat fried noodle-like products from model individual blends of corn starch with casein, soy protein or their hydrolysates. *Journal of the Science of Food and Agriculture*, 79: 1577-1582.

Gomes M. and Singh R.P. 1995. Viscosity and surface tension of corn oil, at frying temperatures. *Journal of Food Processing and Preservation,* 19: 259-270.

Handel A.P. and Guerrieri S.A. 1990. Evaluation of heated frying oils containing added fatty acids. *Journal of Food Science*, 55 (5): 1417-1420.

Hickey H., MacMillan B., Newling B., Ramesh M., Van Eijck P. and Balcom B. 2006. Magnetic resonance relaxation measurements to determine oil and water content in fried foods. *Food Research International,* 39: 612-618

Hindra F. and Baik O.D. 2006. Kinetics of quality changes during food frying. *Critical Reviews in Food Science and Nutrition,* 41: 133-154.

Kawas M.L. and Moreira R. 2001a. Characterization of product quality attributes of tortilla chips during the frying process. *Journal of Food Engineering,* 47: 97-107.

Kawas M.L. and Moreira R. 2001b. Effect of degree of starch gelatinization on quality attributes of fried tortilla chips. *Journal of Food Science,* 66 (2): 300-306.

Krokida M.K., Oreopoulou V. and Maroulis Z.B. 2000. Water loss and oil uptake as a function of frying time. *Journal of Food Engineering.* 42: 39-46.

Krokida M.K., Oreopoulou V., Maroulis Z.B. and Marinos-Kouris D. 2001a. Viscoelastic behaviour of potato strips during deep fat frying. *Journal of Food Engineering.* 64: 213-218.

Krokida M.K., Oreopoulou V., Maroulis Z.B. and Marinos-Kouris D. 2001b. Colour changes during deep fat frying. *Journal of Food Engineering.* 48: 219-225.

Kumar A.J., Singh, R.R.B., Patel A.A. and Patil G.R. 2006. Kinetics of colour and texture in *Gulabjamun* balls during deep-fat frying. *Food Science and Technology*. LWT, 39:827-833.

Larmond E. 1970. Methods for Sensory Evaluation of Food. Publication 1284. Canada Department of Agriculture.

Lloyd B.J., Farkas B.E. and Keener K.M. 2004. Quality comparison of French fry style potatoes produced by oven heating, immersion frying and controlled dynamic radiant heating. *Journal of Food Processing and Preservation,* 28: 460-472.

Marikkar J.M.N., Ghazali H.M., Long K. and Lai O.M. 2003. Lard uptake and its detection in selected food products deep-fried in lard. *Food Research International*, 36: 1047-1060.

Math R.G., Veli V., Nagender A. and Rao D.G. 2004. Effect of frying conditions on moisture, fat, and density of papad. *Journal of Food Engineering.* 64: 429-434.

McDonough C., Gomez M.H., Lee J.K., Waniska R.D. and Rooney L.W. 1993. Environmental scanning electron microscopy evaluation of tortilla chip microstructure during deep-fat frying. *Journal of Food Science,* 58 (1): 199-203.

Mehta U. and Swinburn B. 2001. A review of factors affecting fat absorption. *Critical Reviews in Food Science and Nutrition,* 46: 239-258.

Mellema M. 2003. Mechanism and reduction of fat uptake in deep-fat fried foods. *Trends in Food Science and Technology,* 14: 364-373.

Miller K.S, Singh R.P., and Farkas B.E. 1994. Viscosity and heat transfer coefficients for canola, corn, palm, and soybean oil. *Journal of Food Processing and Preservation,* 18: 461-472.

Mittal G.S. and Zhang J. 2000. Use of artificial neural network to predict temperature, moisture, and fat in slab-shaped foods with edible coatings during deep-fat frying. *Journal of Food Science,* 65 (6): 978-983.

Morales-Pérez J., Sosa-Morales M.E. and Vélez-Ruiz J.F. 2006. Study of a combined method (drying and frying) for toasted tortilla manufacturing. 13th World Congress of Food Science and Technology (IUFoST). Nantes, France. September 17-21, 2 pages.

Moreira R.G. and Barrufet M.A. 1998. A new approach to describe oil absorption in fried foods: A simulation study. *Journal of Food Engineering,* 35: 1-22.

Moreira R.G., Palau J.E. and Sun X. 1995a. Deep-fat frying of tortilla chips: An engineering approach. *Food Technology,* 49 (4): 146-150.

Moreira R.G., Palau J. and Sun X. 1995b. Simultaneous heat and mass transfer during deep fat frying of tortilla chips. *Journal of Food Process Engineering,* 18: 307-320.

Moreira R.G., Palau J.E., Sweat V.E. and Sun X. 1995c. Thermal and physical properties of tortilla chips as a function of frying time. *Journal of Food Processing and Preservation,* 19: 175-188.

Moreira R.G., Castell-Perez M.E., and Barrufet M.A. 1999. Deep-fat frying. Fundamentals and Applications. Aspen Publishers, Inc.: Gaithersburg, MD.

Moyano P.C. and Pedreschi F. 2006. Kinetics of oil uptake during frying of potato slices: Effect of pre-treatments. *Food Science and Technology.* LWT, 39:285-291.

Moyano P.C., Ríoseco V. and González P.A. 2002. Kinetics of crust color changes during deep-fat frying of impregnated French fries. *Journal of Food Engineering.* 54: 249-255.

Oroszvári B.K., Bayod E., Sjöholm I. and Tornberg E. 2005a. The mechanisms controlling heat and mass transfer on frying of beefburgers. I: The influence of the composition and comminution of meat raw material. *Journal of Food Engineering.* 67: 499-506.

Oroszvári B.K., Bayod E., Sjöholm I. and Tornberg E. 2005b. The mechanisms controlling heat and mass transfer on frying of beefburgers. Part 2: The influence of the pan temperature and patty diameter. *Journal of Food Engineering.* 71: 18-27.

Oroszvári B.K., Bayod E., Sjöholm I. and Tornberg E. 2006. The mechanisms controlling heat and mass transfer on frying of beefburgers. III: Mass transfer evolution during frying. *Journal of Food Engineering,* 76: 169-178.

Pannu K.S. and Chinnan M.S. 2003. Relating food frying to daily abuse. Part 1. Determination of surface heat transfer coefficients with metal balls. In *Transport Phenomena in Food Processing.* (J Welti-Chanes, J.F. Vélez-Ruiz and G.V. Barbosa-Cánovas; Eds.). CRC Press. Boca Raton, USA. pp. 201-221.

Paul S. and Mittal G.S. 1996a. Dynamics of fat/oil degradation during frying based on physical properties. *Journal of Food Process Engineering,* 19: 389-403.

Paul S. and Mittal G.S. 1996b. Dynamics of fat/oil degradation during frying based on optical properties. *Journal of Food Engineering,* 30: 389-403.

Pedreschi F., Aguilera J.M. and Pyle L. 2001. Textural characterization and kinetics of potato strips during frying. *Journal of Food Science,* 66 (2): 314-318.

Pinthus E. J., Weinberg P. and Saguy I.S. 1992. Gel strength in restructured potato products affects oil uptake during deep-fat frying. *Journal of Food Science*, 57 (6): 1359-1360.

Pinthus E. J., Weinberg P. and Saguy I.S. 1993. Criterion for oil uptake during deep-fat frying. *Journal of Food Science*, 58 (1): 204-205, 222.

Pronyk C., Cenkowski S., Muir W.E. and Lukow O.M. 2008. Effect of superheated steam processing on the textural and physical properties of Asian noodles. *Drying Technology*, 26: 192-203.

Rao V.N.M. and Delaney R.A.M. 1995. An engineering perspective on deep-fat frying of breaded chicken pieces. *Food Technology*. 49(4), 138-141.

Rasco B.A., Downey S.E.. Dong F.M. and Ostrander J. 1987. Consumer acceptability and color of deep-fat fried fish coated with wheat or corn distillers′ dried grains with solubles (DDGS). *Journal of Food Science*, 52 (6): 1506-1508.

Rice P. and Gamble M.H. 1989. Technical note: Modelling moisture loss during potato slice frying. *Journal of Food Science and Technology*, 24: 183-187.

Saguy I.S. and Pinthus I.J. 1995. Oil uptake during deep-fat frying: factors and mechanism. *Food Technology*, 49 (4): 142-145, 152.

Sahin S., Sastry S.K. and Bayindiri L. 1999a. The determination of convective heat transfer coefficient during frying. *Journal of Food Engineering*, 39: 307-311.

Sahin S., Sastry S.K. and Bayindiri L. 1999b. Heat transfer during frying of potato slices. *LWT Food Science and Technolog*, 32:19-24.

Sebben E., Slaughter D.C. and Singh R.P. 1998. Optical assessment of corn oil deterioration during frying. *Journal of Food Processing and Preservation*, 22: 265-282.

Singh, R.P. 1995. Heat and mass transfer in foods during deep-fat frying. *Food Technology*, 49 (4): 134-137.

Sinram R.G. and Hartman G.J. 1989. Flavored vegetable oils as a substitute for frying tallow in deep-frying applications. *Food Technology*, June: 90-94.

Sosa-Morales M.E., Orzuna-Espíritu R. and Vélez-Ruiz J.F. 2006. Mass, thermal and quality aspects of deep fat frying of pork meat. *Journal of Food Engineering*, 77(3): 731-738.

Sosa-Morales M.E. and Vélez-Ruiz J.F. 2008. Heat and mass transfer modeling during the deep-fat frying of selected foods. In *Progress in Food Engineering Research and Development*. Nova Science Publishers Inc. New York, NY, USA. pp. 197-222.

Tang, Z., Cenkowski, S. and Muir, W.E. (2000). Dehydration of sugar-beet pulp in superheated steam and hot air. *Transactions of the ASAE*, 43 (3): 685-689.

Utheil L.G. and Escher F. 1996. Dynamics of oil uptake during deep-fat frying of potato slices. *Lebensmittel-Wissenschaft und-Technologie*, 29: 640-644.

Vasanti C.K., Seow C.C. and Sulebele G.A. 1996. Effects of frying parameters on physical changes of tapioca chips during deep-fat frying. *International Journal of Food Science and Technology*, 31: 249-256.

Vélez-Ruiz, J.F. 2008. Notes of Physical Properties of Foods. Unpublished Material. Universidad de las Américas, Puebla. Puebla, Mexico.

Vélez, J.F., and Xique, J. 1999. Review: Frying Process of Foods. *Revista Internacional Información Tecnológica*, 10 (2): 127-134. (In Spanish).

Vélez-Ruiz J.F., Vergara-Balderas F.T., Sosa-Morales M.E. and Xique-Hernández J. 2002. Effect of temperature on the physical properties of chicken strips during deep-fat frying. *International Journal of Food Properties*. 5 (1): 127-144.

Weiss, T.J. 1983. Frying shortenings and their utilization. In *Food Oils and Their Uses*. AVI Publishing Co. Westport, CN, USA.

Williams R. and Mittal G.S. 1999. Low-fat fried foods with edible coatings: Modeling and simulation. *Journal of Food Science,* 64 (2): 317-322.

Yamsaengsung R. and Moreira R.G. 2002. Modeling the transport phenomena and structural changes during deep fat frying. Part II: Model solution and validation. *Journal of Food Engineering*. 53: 11-25.

Yi Y.H. and Chen T.C. 1987. Yields, color, moisture and microbial contents of chicken patties as affected by frying and internal temperatures. *Journal of Food Science*, 52: 1183-1185.

Yildiz A., Palazoglu T.K. and Erdogdu F. 2007. Determination of heat and mass transfer parameters during frying of potato slices. *Journal of Food Engineering.* 79: 11-17.

In: Food Processing and Engineering Topics
Editors: M. E. Sosa-Morales and J. F. Vélez-Ruiz
ISBN: 978-1-60741-788-0

Chapter 9

DRYING OF OSMOTICALLY PRECONCENTRATED FOODS

Jorge F. Vélez-Ruiz
Chemical and Food Engineering Department
Universidad de las Americas, Puebla, Mexico

1. INTRODUCTION

Food preservation and supplying are the main goals for the food industry. Food preservation involves a diversity of knowledge oriented to maintain nutritional, biochemical, physical, and other properties of food items. In order to preserve those food properties at the best conditions, the characteristics of the food process operations should be correctly identified to get maximum benefits.

Among the many transformation processes used for making foods, such as refrigeration, freezing, heating, concentration, frying, and dehydration, among others, they are classified in several groups depending of the main preservation agent. Each one of them offers characteristics and advantages that may be combined to get a better food product. Furthermore, the combination of processes has been and should be oriented to influence positively the properties of the final product.

Most of the studies about the combination of methods, also known as hurdle technology, have covered microbiological, chemical, and quality aspects and a minor quantity of works have focused on engineering and physical aspects. Nowadays, foods based on combined preservation methods are prevalent in developing and industrialized countries. These foods, just to mention some, may be grouped as chilled foods, fermented foods, functional foods, heated foods, intermediate and high moisture foods, healthful foods, and packaged foods depending of the main preservation factor or final purpose; all of them are the result of different and combined preservation processes. This technology of hurdle combination has allowed the preservation or developing of new products, such as fruits and vegetables in Latin America, meats in China, probiotic food items in Europe, and dairy products in India.

Although the combination of food preservation methods is very extensive and covers a diversity of aspects, the aim of this chapter is to cover the most important aspects of two mass transfer process operations, osmotic dehydration and air drying and the combination of both technologies to improve the food properties of the final food products.

2. Mass Transfer Operations

A good number of food process operations are based on the mass transfer phenomenon involving changes in concentration of food components. Depending of the phases and particular components in the food item, considered as a multicomponent mixture, the mass transfer operation must be selected and applied at the best operational conditions. The importance of mass transfer operations, such as Absorption, Adsorption, Dehydration, Distillation, Extraction, Freeze Drying, Humidification, and Lixiviation is profound and determinant for some areas of the food industry in which they are utilized.

The mass transfer is the result of a concentration difference or driven force of a specific component, the component moves out from a portion of the food item or phase of high to one of low concentration. Whatever is the mass transfer operation, there are four major design parameters that are required for engineering calculations of the processes: a number of equilibrium stages for the desired separation, a required process time, a permissible flow rate range, and the correspondent energy needs (Treybal 1980).

Mass transfer is analogous to heat transfer and, depends upon the dynamics of the food systems in which it occurs. There are two mechanisms of mass transfer, the diffusion and convection phenomena; in the first one, the mass may be transferred by a random molecular movement in quiescent food fluids or static solid foods; and in the second one, the mass is transferred from the food surface to a moving fluid, gas or liquid. And such it happens in many food processes, both mechanisms are developed simultaneously. Mass diffusion and convection may be more or less important depending of the specific operation. For example, in distillation and extraction are of less importance, but both are very important in dehydration processes.

2.1. Diffusion

The basic relation for molecular diffusion in a given direction, defines the molar flux related to the component concentration and it is known as the Fick´s first law (Crank, 1975; Welty *et al.*, 1976; Welti-Chanes *et al.*, 2003; Vélez-Ruiz, 2008):

$$J_{iz} = -D_{im}\frac{dC_i}{dz} \tag{1}$$

where: J_{iz} is the molar flux of the i component in the z direction (mol/m^2 s), D_{im} is the mass diffusivity or diffusion constant (m^2/s), being specific for the i component in a given medium, dC_i/dz is the concentration gradient of the i component in the z direction (mol/m^4), dC_i is the concentration difference or driven force (mol/m^3), and dz is the interface separation or separation distance (m), between two points or portions with different concentration of the i component.

The flux of the involved component (i) in equation 1, may be converted to mass units of kilogram by considering the molar weight of it. The diffusion constant has been evaluated at some particular systems, few data are included in table 1 (Welty *et al.*, 1976; Okos *et al.*, 1992, Vélez-Ruiz and Soriano-Morales 2003). As it may be observed, gas diffusion is easier than liquid and solid diffusion, as well as liquid diffusion is easier than solid diffusion.

Table 1. Diffusion coefficients in binary systems and effective diffusion for food items at atmospheric pressure*

System (i component in a medium)	T (°C)	D_{im} or D_{eff} (m^2/s)
Air in water (g/g)	25	2.60×10^{-5}
Carbon dioxide in water (g/g)	25	1.64×10^{-5}
Ammonia (1 mol_{kg}/m^3) in water (l/l)	15	1.77×10^{-9}
Sodium chloride (1 mol_{kg}/m^3) in water (s/l)	18	1.24×10^{-9}
Sodium chloride (5.4 mol_{kg}/m^3) in water (s/l)	18	1.54×10^{-9}
Sodium chloride in raw potato/water (s/s-l)	25	4.00×10^{-10}
NaCl in cucumber/brine (D = 19 mm) (s/s-l)	21	11.9×10^{-10}
NaCl in cucumber/brine (D = 34.9 mm)(s/s-l)	21	9.60×10^{-10}
NaCl in Gouda cheese/water (36.2 % moisture) (s/s-l)	13	1.22×10^{-10}
NaCl in Gouda cheese/water (48.2 % moisture) (s/s-l)	13	2.61×10^{-10}
NaCl in Gouda cheese/water(43.7 % moisture) (s/s-l)	18	2.51×10^{-10}
Moisture from apple	66	6.40×10^{-9}
Moisture from banana chip	25	2.91×10^{-11}
Moisture from carrot cubes	40	6.75×10^{-11}
Moisture from carrot cubes	60	1.21×10^{-10}
Moisture from carrot cubes	80	1.79×10^{-10}
Moisture from garlic slabs	61	1.06×10^{-10}
Moisture from pear slabs	66	9.63×10^{-10}
Moisture from potato	54	2.58×10^{-11}
Moisture from potato	65	4.37×10^{-11}
Moisture from raisin	25	4.17×10^{-11}
Moisture from tapioca root	74	4.80×10^{-11}

* g: gas, l: liquid, s: solid

2.2. Convection

Convective mass transport occurs in fluids as a result from a bulk flow, in which natural and forced motion is involved. It is very similar to heat convection, therefore the properties of the two interacting phases (any of them may be a food material), the supplying medium and the flowing phase, as well as some physical parameters of the system, are involved through of dimensionless groups for the evaluation of the convective mass transfer coefficient.

The molar flux of a given component may be computed from the next equation (Bird *et al.*, 1960; Welty *et al.*, 1976; Welti-Chanes *et al.*, 2003; Vélez-Ruiz, 2008), and as in the case of diffusion, it occurs in the decreasing concentration direction:

$$N_i = -k_m \Delta C_i = -k_m (C_{is} - C_{if}) \qquad (2)$$

where: N_i is the molar flux of the i component in the flow stream direction (mol/m^2 s), k_m is the convective mass transfer coefficient (m/s), ΔC_i is the concentration difference or driving force (mol/m^3), involving a concentration difference between the boundary surface concentration (C_{is}) and the average concentration of the fluid stream (C_{if}).

There are four methods for evaluation of the convective mass transfer: dimensional analysis completed with experimental work; differential or exact analysis of the boundary layer; integrated or approximate analysis of the boundary layer and analogies between transport phenomena (Welty *et al.*, 1976; Geankoplis, 1999; Vélez-Ruiz, 2008). Mass transfer coefficients are expected to vary as a function of the dynamic conditions, geometrical aspects of the involved system, and physical properties of the fluid and solid phases. The main dimensionless groups for the evaluation of the mass transfer coefficient are included in table 2 (Welty *et al.*, 1976; Welti *et al.*, 2003; Vélez-Ruiz, 2008).

Although there are a good number of equations for the evaluation of the convective mass transfer coefficient, food systems and phenomena particularities are demanding for more specific correlations.

Table 2. Dimensionless numbers for convective mass transfer evaluation

Name	Variables	Group	Involved forces
Sherwood (Sh)	k_m, L_c, D	$k_m L_c / D$	Convective force/diffusive force
Reynolds (Re)	v, L_c, ρ, η	$v L_c \rho / \eta$	Inertia force/viscous force
Schmidt (Sc)	ρ, η, D	$\eta / D \rho$	Momentum diffusivity/mass diffusivity
Lewis (Le)	k, ρ, Cp, D	$k / \rho Cp D$	Thermal diffusivity/mass diffusivity
	h, k_m, Cp	$h / k_m Cp$	Thermal convection/mass convection
Fick (Fi)	D, t, L_c	$D t / L_c^2$	Dimensionless time in unsteady state
Biot (Bi)	D, k_m, L_c	$k_m L_c / D$	Mass transfer/solid diffusivity

Cp: fluid specific heat (kJ/kg K), D: diffusion coefficient of the i component (m^2/s), η: fluid viscosity (Pa s), k: fluid thermal conductivity (W/m K), k_m: convective mass transfer (m/s), L_c: characteristic dimension (m), ρ: fluid density (kg/m^3), t: time (s), v: average fluid velocity (m/s).

2.3. Food Dehydration

Food dehydration as a mass transfer phenomenom involves the removal of moisture from the food material. Dehydration may be considered as an intermediate or a final process in which raw agricultural items or intermediate food materials are transformed into retail or final food products with a lower water activity. The dehydration process reduces the moisture content of the food product to different levels, depending of the applied operation and the desired characteristics of the particular item.

Food dehydration is accomplished in the food industry to reach one of the next objectives (Heldman, 1975; Okos *et al.*, 1992; Toledo, 1994; Vélez-Ruiz, 2005): i) to limit microbial growth, ii) to reduce biochemical reactions, iii) to low volume and weight, iv) to preserve quality attributes, v) to facilitate subsequent processing, vi) to produce a convenient food. Dehydration of foods is aimed at producing of high density products, which when are adequately packaged has a long shelf life, after which the food can be rapid and simply rehydrated without substantial loss of sensory and physical properties.

From a general point of view, the moisture removal may be gotten by several food process operations, such as crystallization, evaporation, membrane separation, osmotic concentration, freeze drying, air drying, acoustic and microwave drying, supercritical extraction, etc. Some of them, like crystallization, evaporation, and membrane separation are more used for liquid foods; while others such as osmotic concentration, freeze drying and air drying are meanly applied to solid foods.

Considering the operating characteristics of the drying process, drying of foods is one of the most time and energy consuming processes in the modern food industry. Therefore, there are strong efforts to reduce the processing time, thus facilitating and accelerating the dehydration process, and also lowering energy consumption, thus preserving the environment and saving money. To achieve these two main purposes, a good number of obstacles must be overcome by the food industry.

Nowadays, it is known that the production of high quality products requires the combination of processes to meet the advantages of them. One of the most useful pretreatments for drying of foods, mainly fruits and vegetables, is the osmotic dehydration. With two important advantages, reducing the drying time and contributing to lower energy requirements (Sunjka and Raghavan, 2004).

3. Osmotic Concentration

3.1. General Aspects

Osmotic concentration or osmotic dehydration (OD) process is based on the immersion of foods, whole or in pieces, inside hypertonic solutions of sugars, sodium chloride, sorbitol or glycerol, to reduce water content, while increasing soluble solid content (Matusek and Merész, 2002; Camargo *et al.*, 2004). OD is a preservation method that offers a high quality food product by means of water removal without phase change. Osmotically dehydrated fruits and vegetables have a good retention of nutrients, flavor and aroma (Sunjka y Raghavan, 2004), also OD inhibits polyphenol oxidase, preventing the oxidative browning (Tsamo *et al.*, 2006). The chemical composition and organoleptic characteristics of osmodehydrated fruits and vegetables result in high quality products that can undergo subsequent freezing, dehydrofreezing, air drying, vacuum drying (Matusek and Merész, 2002) or microwave-air/vacuum dehydration (Torringa *et al.*, 2001; Erle and Schubert, 2001). Certainly, the amount of moisture remaining in the osmosed food does not ensure its stability, as water activity is generally higher than 0.9. Thus, when shelf stability is the objective, other complementary process operations, such as convective drying, freezing, freeze drying, refrigeration, or other, should be applied.

OD is a complex process of counter mass transfer phenomena between the plant tissue and hypertonic solution. In a simple manner, in OD a two-way counter-flow of mass exchange takes place. There is water diffusion from the sample to a surrounding hypertonic solution and an opposite flux of osmotic substances, such as sugars, organic acids and mineral salts, plus other nutrients. The flux of osmoactive substances penetrating the food tissue modifies its chemical composition; also, there is a flux of components leaving the food system. The osmotic pressure gradient constitutes the driving force of the mass transfer process, as a function of the concentration and temperature as the main variables to be controlled.

Some fluxes of nutrients, although insignificant as compared to the two main flows of water and solute, can be very important for the sensory and nutritional properties of the final food product (Camargo *et al.*, 2004). The simultaneous material exchange between the food item and the osmotic solution is accompanied by significant shrinkage, deformation, flow interaction, and nutrient loss (Tsamo *et al.*, 2006). Those changes in structure and composition developed during the osmotic operation can influence the food behavior in the next process operation, whatever it is.

OD is effective even at ambient temperature, minimizing thus the heat damage. Although, it has been reported that an elevation in temperature and concentration of the osmotic solution increase the rate of mass transfer up to certain extent, above which undesirable changes in sensory and textural attributes of the product have been observed (Matusek and Merész, 2002).

In a study of OD on pineapple rings at room temperature, that was followed during several hours in order to reach a desired sugar content, by utilizing two different sample thickness: 1 and 2 cm (Vélez-Ruiz *et al.*, 2001). In all experimental runs, the water loss, sugar gain (figure 1) and weight reduction were determined. The evolution of the three process variables during the osmotic treatment (1:4, fruit to syrup ratio), allowed to observe changes in mass transfer rates. Based on water loss and sugar gain parameters, the osmotic process reached constant values after 350 - 400 min. of residence time.

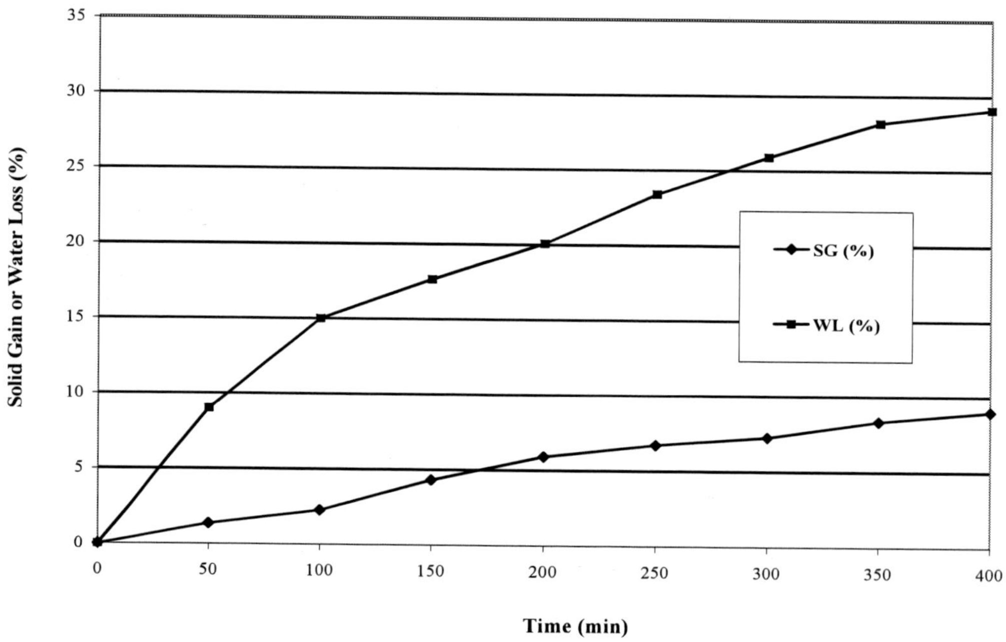

Figure 1. OD on pineapple rings (1 cm), showing the solid gain and water loss kinetics.

At the beginning of the osmotic concentration treatment, both sugar gain and water loss were very quick. Changes in rates suggested a two stage process, a first stage for the initial 150 minutes in which the water loss and sugar gain were higher; a second stage (150 to 400 min) in which the mass fluxes decreased progressively. After 400 min of process, around 29% of water loss, and 9% of sugar gain were observed. Sensory evaluations, quality attributes and process characteristics indicated a better osmotic dehydration for pineapple samples of 1 cm of thickness.

Food dehydration developed during the OD is very sensitive to small differences in concentration of the osmotic agent. In the OD of jicama or yam bean (*Pachyrizus tuberosus*) carried out by Reyes-Herrera *et al.* (2005), with salt and citric acid; different pathways for moisture loss were obtained (figure 2), even with small differences in the percentage of the osmotic agent (3 and 5% for salt, 1 and 2% for citric acid).

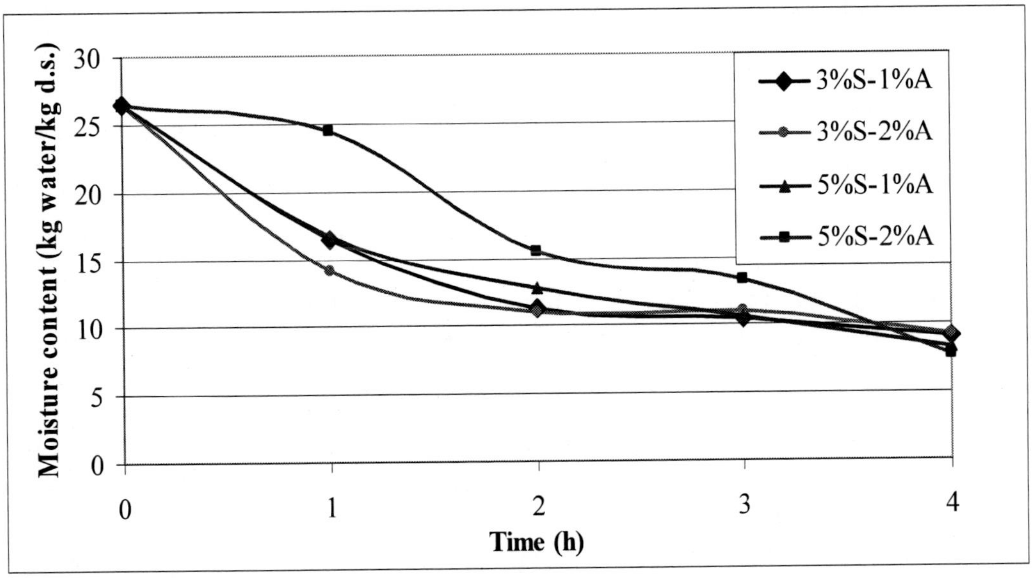

Figure 2. OD of jicama with four osmotic solutions (3% Salt-1% Citric Acid; 3% Salt-2% Citric Acid; 5% Salt-1% Citric Acid and 5% Salt-2% Citric Acid).

Reyes-Herrera *et al.* (2005) reported that the water content of fresh yam bean was 96.4% (w.b.) and after 4 h of osmotic treatment, the moisture content decreased to 88.0-90.0%, the solids content ranged from 7.95 to 9.52%, the weigh loss was 10.97 to 21.19%, while the solid gain was 6.39-11.2% as a function of the solution characteristics. The particular moisture trend of each system at room temperature may be observed in figure 2. A decrease in the moisture content (17.5-18.26 d.b.) of this vegetable was observed after the osmotic dehydration, reaching the highest decrease in the system with 5% NaCl and 2% citric acid. There was not significant difference when that system was compared with the other ones ($p>0.05$), meaning that levels of salt and citric acid did not influence the final moisture content with OD.

The percentage of weight loss was influenced by the salt and acid concentration. The system with 5% salt and 2% acid showed the highest weight loss, while the systems with a lower percentage of solids just reached 12.7 and 11.9%, respectively. And, as a consequence of the osmotic dehydration, in which the water was removed and the solutes were added, the water activity decreased. Thus, with the osmotic dehydration the levels of water activity were not sufficiently reduced to preserve this vegetal product, because this reduction began at

0.970 and decreased to 0.945 and 0.940, being the yam bean susceptible to damage by microorganisms at these activities.

3.2. Process Variables

Among the operational variables involved in the OD process, some of them may be mentioned as the most important: i) type of pretreatment, ii) process time, iii) temperature and mass ratio of the solution, iv) concentration and nature of the solute, v) size and geometry of the solid food, and vi) presence of stirring, among others. In according to Flink (1980), there are some practical observations related to the process variables that should be taken into account:

- The water loss and solids increasing must be controlled by the food material characteristics, and they are strongly influenced by the applied pre-treatments.
- The use of pre-treatments, like sulfites adding, blanching or freezing helps to increase solids content compared to water loss, because the impregnation phenomenon is incited.
- Generally, more than 50% of the initial weight is lost. The greatest water loss takes place during the first two hours of treatment and the maximum gaining of solids occurs during the first 30 minutes.
- The rate of mass transfer increases when the temperature is above 45°C, during which there could be enzymatic browning and flavor changes. With higher temperatures (above 60°C) some characteristics are modified, like flavor impregnation and solids gaining.
- The best temperature for the process depends on the kind of treated food, and it is known that mass transfer is favored by using highly concentrated solutions.
- In fruits, the use of different kinds of sugars as osmotic media strongly determines the kinetics of water loss and increasing of solids.
- The addition of salt to the osmotic solution increases the gradient differential in the dehydration process.
- There are synergic phenomena taking place when sugar and salt are added to the solutions.
- Agitation of the osmotic solution favors the mass transfer phenomenon.

Furthermore, the final results will be a consequence of all the process variables utilized in the specific OD treatment. However, OD is only considered a type of pretreatment, since it does not ensure the osmosed food will possess sufficient shelf and microbiological stability.

3.3. Mathematical Modelling

OD, as a mass transfer process in which important conditions are considered, has a basic modeling supported on material balances. The calculations are based on the general balance of the component concentration, as driving mass transfer force, between the solid food and the liquid solution as the involved phases (Vélez-Ruiz *et al.*, 2001; Li and Ramaswamy, 2005;

Reyes-Herrera *et al.*, 2005). They may be computed with the next resulting expressions (equations 3 to 5) corresponding to water loss (WL), solids gaining (SG) and weight reduction (WR).

$$WL(\%)=\frac{(W_i x_i - W_f x_f)}{W_i}\times 100 \tag{3}$$

$$SG(\%)=\frac{[W_f(1-x_f)-W_i(1-x_i)]}{W_i}\times 100 \tag{4}$$

$$WR(\%)=\frac{W_i - W_f}{W_i}\times 100 \tag{5}$$

where: W_i is the initial weight (kg), x_i is the initial mass fraction, W_f is the final weight (kg), x_f is the final mass fraction, at the end of the OD treatment.

Other two empirical and very useful models have been applied in OD studies to correlate solids content and process time: the Hawkes and Flink (1978) model (Eq. 6) that allows the evaluation of the mass transfer coefficient (k_O), and the Peleg equation (Eqs. 7a and 7b; Peleg, 1988) to evaluate two kinetic parameters.

$$NSC = k_O t^{0.5} \tag{6}$$

where: NSC is the normalized solids content (solids content at time t/initial solids content), k_O represents the overall mass transfer for the osmotic agent ($s^{-0.5}$), and t is the osmotic dehydration time (s).

$$M(t)=M_0 \pm \frac{t}{k_1 + k_2 t} \tag{7a}$$

$$\pm\frac{t}{M(t)-M_0} = k_1 + k_2 t \tag{7b}$$

where: M(t) is the moisture content at time t (min), M_0 is the initial moisture content (g water/100 g sample), k_1 (g water/100 g min)$^{-1}$, and k_2 (g water/100 g sample)$^{-1}$ are kinetic constants.

Both equations have been utilized by Palou *et al.* (1994) for modeling of the OD of papaya, and by Reyes-Herrera *et al.* (2005) for fitting of air drying experiments in pre-concentrated yam plants by OD. There are several mathematical models that have been developed only for OD, as well as experimental conditions normally utilized for OD, they may be consulted in specialized documents related to this process food operation,. Some process characteristics of OD that have been applied as a pre-concentration treatment of the air drying operation are presented in table 3.

Table 3. Studies of drying of osmotically preconcentrated foods at room temperature*

Food item	Conditions in OD	Condition in AD	Observations	Reference
Pineapple rectangles 5x2x0.5 or 1 or 2 cm	Sugar solution 40, 50, 60%, 24 h, 25°C, 1:5 with agitation	Air at 50, 60, 70 and 80°C 6 h, heated oven	D: 0.4-10.5 x 10^{-10} m^2/s shrinkage related to M	Uddin *et al.*, 1990
Apple cubes 1 cm	Sucrose solution 70°Bx, 3 h, 1:6 with agitation	Hot air 30-90°C, 11 h	D: 2.0-9.8 x 10^{-10} m^2/s only falling rate	Simal *et al.*, 1995
Strawberry ellipsoids with three diameters	Cerelose solution 35 and 51%, 25°C, 48 h	15 m/s, 55°C, 6.3 h	D: 11.1-11.7 x 10^{-10} m^2/s calculated with initial R_e	Alvarez *et al.*, 1995
Apple cylinders, 5 cm long, 1 cm diameter	Glucose or sucrose slns. 15, 30 and 45%, 1: 50, 12 h	2 m/s, 55°C, 6 h infinite cylinder	D: 2.5-12.5 x 10^{-10} m^2/s D decreased with OD	Karathanos *et al.*, 1995
Peas of 9 $\pm$ 0.3 mm diameter	Solution: sucrose 60%, 20% citrate + 40% sucrose, 30°C, 24 h, agitated	Air at 65°C-10%RH, 12 h laboratory tray dryer	OD gave a higher drying rate, D: 2.6-4.1x$10^{-10}$$m^2/s$	Ertekin & Cakaloz, 1996
Apple slices 4 x 4 x 0.4 cm	Glucose solution 22, 31.9, 39.5%, 25°C, forced convection, 3 h	16 m/s, 60°C, 1.6 h	D: 1.4-10 x 10^{-10} m^2/s D decrease with glucose	Nieto *et al.*, 1998
Potato cubes 1 cm	10% salt, 40-50% sugar solution (4 kg solution/kg solid), 30°C, 2-6 h	3.6, 5.3 or 7.6 m/s, 120, 140 or 160°C, during 5-15 min fluidized bed dryer	Optimum OD: 10% salt, 50% sugar at 47°C, AD: 140°C-10 min, 5.3 m/s	Ravindra & Chattopa-dhyay, 2000
Apple cubes 1 cm	Sucrose solution 61.5°Bx, 30°C, 1:4, 3 h with mixing (50 rpm)	1.5 m/s, 70°C, 6 h	OD followed by AD affected apple rheology	Lewicki & Lukaszuk, 2000
Papaya slices Infinite slabs 1 cm	Sucrose syrup 32.8 and 42.8°Bx, 120 h, 1:2, not agitated	9 m/s, 65, 75 and 85°C, 3-7.5 h	D: 1.1-2.3 x 10^{-9} m^2/s better quality properties	Castañón *et al.*, 2001a
Mango slices 1 cm thickness	Sucrose syrup, a_w: 0.97, 0.95, and 0.93, 1:2, not agitated	5 m/s, 75°C, 3.3 h lab-scale tunnel dryer	D: 1.3-2.2 x 10^{-9} m^2/s small ΔE values	Castañón *et al.*, 2001b
Pineapple slices 10 cm D, 1 cm thickness	Sucrose solution 60°Bx, 22°C, a_w 0.99, 0.95,0.93, 1:4, 24 h	0.55, 1.10, 1.60 m/s at 65, 75 and 85°C, lab tunnel dryer	D: 4.6-14.8 x 10^{-10} m^2/s effect of OD on color	Vélez-Ruiz *et al.*, 2001
Mango plates 4 x 4 x 0.4 cm	Glucose solution 22.1% (a_w 0.97)	16 m/s, 60°C, 13-14 h	D: 3.1-7.5 x 10^{-10} m^2/s OD decreased drying rate	Nieto *et al.*, 2001
Tomato in quarters	Sucrose soln 61.5°Bx, 30°C, 3 h, with and without $CaCl_2$	2 m/s, 60°C, 6.7-11 h laboratory dryer	AD 20% shorter, $CaCl_2$ increased water removal	Lewicki *et al.*, 2002
Apple, mango and strawberry	Glucose solution 59%, 25°C, 3 h, and 20 h, respectively	15 m/s, 60°C (apple and mango) 55°C (strawberry)	Apple: 5.2, mango 5.4, strawberry 5.0 x $10^{-9}$$m^2/s$	Alzamora *et al.*, 2003
Cherry tomato	Salt solution: 10 and 25% (w/w), 25°C, 6 h, 1:10, with agitation	0.75 and 2.60 m/s, 50, 60 and 70°C in a cabinet dryer	1.2-1.8 x $10^{-10}$$m^2/s$ at 0.75 1.8-2.2 x 10^{-10} at 2.6 m/s	Azoubel & Murr, 2003

Table 3. (Continued)

Food item	Conditions in OD	Condition in AD	Observations	Reference
Papaya cubes 2 cm length	Sucrose solution 70°Bx, 30°C 4 h, 1:10, agitated (110 rpm)	1.25-3.25 m/s, 40 and 60°C, 5 h vertical air flow	D: 1.03-1.78 x 10^{-9} m^2/s D fresh > D osmosed	Abbas El- Aouar, 2003
Jicama slabs of 1 x 1 x 10 cm	Osmotic solution 5-7%, 22°C 4 h, 1:2, not agitated	0.35-0.45 m/s, 45 and 50°C, 6 h laboratory tunnel dryer	k_O by Hawkes-Flink eq. k_1, k_2 by Peleg model	Reyes-Herrera *et al.*, 2005
Acerola 2.1-2.4 cm diameter	Sucrose solution 40°Bx, 26-29°C 4 h, agitated (0-780 rpm)	1.3 m/s, 60-70°C, 6 h fixed bed dryer	D: 8.3-28 x 10^{-7} m^2/s considering shrinkage	Silva *et al.*, 2005
Pear slices, 25 g, 1 cm thickness	Sucrose solution 55°Bx, 1:20 (w/w), agitated 0.5, 3, 48 h	2.2 m/s, 45, 55, 65°C one fruit slice by experiment	Modeled by Fick eq. D: 6.5 x 10^{-12}- 6 x 10^{-10} m^2/s	González-Martínez *et al.* 2006
Onion 0.5 cm thickness	Solution: sucrose (600 g/L), NaCl (300 g/L), mixture (45 g/15g/L), 120 min; 1:10 (w/w), agitated	0.13 m/s and 60°C (dry bulb) cross flow cabinet dryer trays prepared by triplicate	OD 15 min, was enough M_e reduced in drying as a function of OD	Tsamo *et al.*, 2006
Pumpkin slices, 4 mm ($A=20\text{-}25cm^2$)	Sucrose solution: 40, 50 and 60°Bx, 27°C, 1:15, 0.5-3 h, agitated	2 m/s, 50 and 70°C, 1h, laboratory scale dryer	D: 1.34-4.16 x 10^{-10} m^2/s D water > D sucrose	Garcia *et al.*, 2007
Carambola (*Averroha carambola L.*)	Sucrose solution: 40-60% (w/w), 4 h at 32°C	4.2 m/s, 65°C	$K_{solids} > K_{water}$ in OD OD favored AD	Castillo & Cornejo, 2007
Mango plates 4 x 4 x 0.4 cm	Sucrose (60% w/w) and glucose (40% w/w) solutions, 25°C, 1:30	3.5 m/s, 60°C, forced convection, tray dryer	D: 1.90 and 2.30 x 10^{-11} m^2/s, for sucrose and glucose	Maldonado *et al.*, 2007
Chestnuts prisms 1 x 1 x 1.5 cm	Sucrose solution 60%, 25°C 1-24 h; 1:10 (w/w)	2.4 m/s, 65°C, 0.5-6.0 h, drying chamber pilot-plant	Drying was practically independent of OD	Chenlo *et al.*, 2007
Potato and apple cubes of 1 cm^3	Sucrose and salt solutions	35-45°C, 4.1-6.7 h lab vacuum contact dryer,	OD samples in sugar showed higher drying rates	Rahman & Mujumdar, 2007
Tomato slices 1 cm thickness	Corn syrup 60% with and without $CaCl_2$, 25°C, 120 min, 1:3 (w/w), recirculated	1.5 m/s and 80°C, weight reduction: 40, 60, 75 and 85%, pilot dryer	OD increased drying rate of the first falling period, better color	Pani *et al.*, 2008

* OD: osmotic dehydration; AD: air drying; M: Moisture content; D: diffusion coefficient; ΔE: net color change; R_e: equivalent spherical radius; M_e: moisture content at equilibrium.

4. Air Drying

In the course of the last thirty years a good number of activities and studies have been undertaken in order to apply drying for food preservation, particularly for solid foods, as widely as possible. Drying is one of the most used preservation process that extends the food shelf life.

Conventional air-drying (AD) as the most important drying process, is a simultaneous heat and mass transfer process accompanied by phase change. It is also a high cost process due to its energy requirement.

A pre-treatment such as osmotic dehydration can be employed to reduce the initial water content of foods and as consequence requiring less air-drying time (Souza *et al.*, 2007). The organoleptic characteristics of the final product can be improved with this dehydration, because some acids are removed from the food during the osmotic treatment, and also sweeter products can be obtained when the food item has been previously dehydrated by osmotic treatments (Suazo-Abarca *et al.*, 2000).

4.1. General Aspects

A food is considered dried when its moisture content is lower than 30%, and its water activity is lower than 0.6. At this a_w value, there is a very low possibility of microbial growth, only osmophylic yeasts could survive and grow in such media (Badui-Dergal, 1999; Vélez-Ruiz, 2005). Dried foods, especially fruits and vegetables, can be stored and transported at a relatively low cost, although water removal contributes to quality decreasing, nutritive and sensory properties may be importantly affected. Drying as a simultaneous heat and mass transfer process, involves two phenomena: evaporation and diffusion. Water evaporation is associated with a convective mechanism, whereas diffusion is related with the molecular flux of water, resulting in a drying curve with a constant rate period and a falling rate stage, correspondently.

In a study conducted by Vélez-Ruiz *et al.* (2001) to know the influence of process variables on the drying of osmosed pineapples of 1 cm thickness. Three drying temperatures (65, 75 and 85°C), three water activities (0.97, 0.98 and 0.99) and three air velocities (0.55, 1.10, and 1.60 m/s) were analyzed during the drying of pretreated pineapple slices. Figure 3 shows the drying curves with different trends for the osmosed samples with a water activity of 0.98, showing the effect of the different drying conditions.

The drying experiments for osmosed pineapple showed a drying curve with only a falling rate period, in contrast the fresh pineapple exhibited two process rates: a brief constant rate period and a falling rate period, in which the first one was neglected for modeling purposes. From the drying, the mass transfer and effective diffusion coefficients were evaluated as a function of the process conditions. As expected, higher water activities, air temperatures, and air velocities favored the drying process of osmotic dehydrated samples.

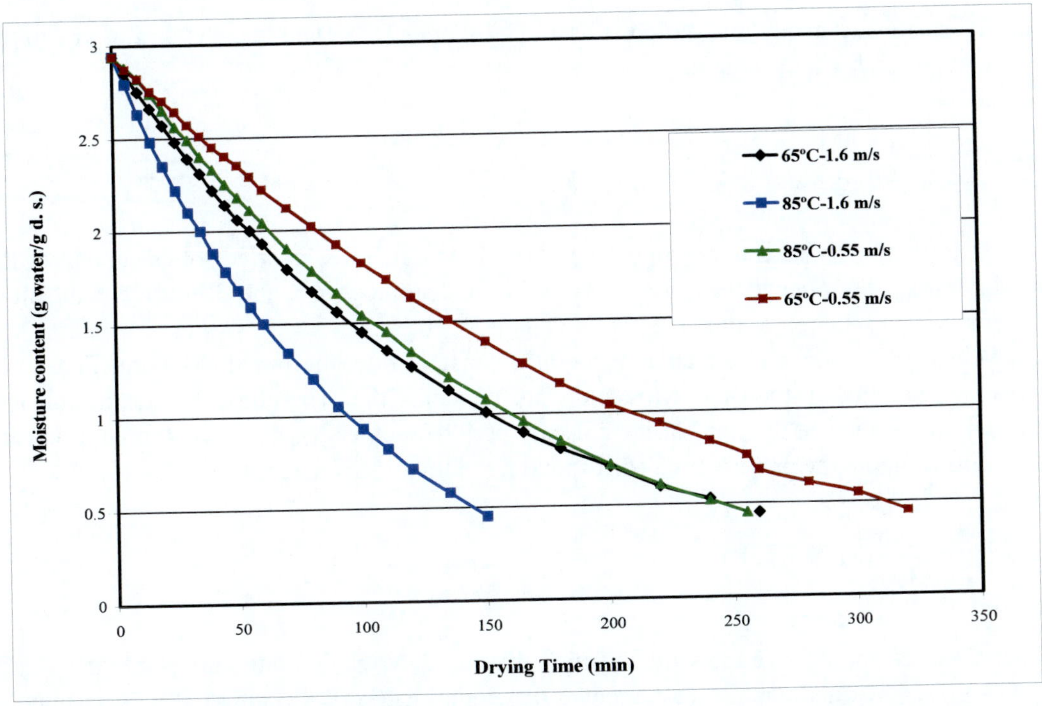

Figure 3. AD Curves for osmosed pineapple (Aw = 0.98) at four experimental conditions.

4.2. Process Variables

There are several factors governing the transfer mechanisms of heat and mass that determine the drying rate and food properties of the final product. Among the operational variables involved in the AD process, some of them may be mentioned as the most important: i) vapor pressure of food material and drying air, ii) process time, iii) air velocity and temperature, iv) water diffusivity in the food material, and v) thickness and surface exposed to drying, among others; being the prediction of the drying time an important and basic value for sizing and optimization purposes of an industrial drying plant.

4.3. Mathematical Modelling

Modeling for AD processes follows the traditional aforementioned equations (1 and 2) for dehydration, but with consideration of the constant-rate period and the falling-rate period and involving the particularities of each drying equipment and food properties.

4.3.1. Constant Rate Period

Heat transfer occurs during evaporation of water coming from the food material that demands energy supply from the air. Thus, mass transfer happens during the moisture removal from food surface and flows up with relative facility, constituting the constant drying rate. A film of water is freely available at the surface and evaporates into the drying medium. It is a mechanism analog to the heat transfer by convection, and due to this, a mass transfer coefficient for the steady rate period is used. Geankoplis (1999) defines the drying rate in the constant period by next equation.

$$R_C = -\frac{Ls}{A}\frac{dM}{dt} \tag{8}$$

where: R_C is the drying rate or flux of water (kg H_2O/h m^2), Ls: is dried solids content of the food item (kg d.s.), A is the exposed area at the food surface (m^2), and dM/dt is the moisture difference as a function of time ((kg H_2O/h kg d.s.)

When the drying process takes place at a constant rate, the initial (M_1) and final (M_2) moistures are bigger than the critic moisture content (M_c). Therefore, by integration and rearranging of the last relation for the constant period, another equation is obtained. It is very useful to compute the process time corresponding to this steady stage (t_c).

$$t_c = \frac{Ls}{AR_c}(M_1 - M_2) \tag{9}$$

where: Rc, Ls and A have the same meaning that Eq. 8, M_1 is the initial moisture content (kg H_2O/kg d.s.), M_2: is the moisture content at different time (kg H_2O/kg d.s), if it correspond to the the final of this constant period then is known as the critical moisture content (M_c).

When the evaporation process, developed during the constant rate period, takes account of the heat convection; the amount of evaporated water may be quantified by the following equation.

$$\frac{dM}{dt} = \frac{R_C A}{L_S} = \frac{h\,A(Ta - Ts)}{L_S h_{fg}} \tag{10}$$

where: dM/dt is the evaporated water or moisture change as a function of time (kg H_2O/ h kg d.s.), h is the convective heat transfer coefficient (W/m^2K), A is the exposed surface (m^2), Ta is the dry air bulb temperature (°C), Ts is the surface temperature for the food (°C), and h_{fg} is the latent heat of vaporization for water (kJ/kg)

It can be noticed that the drying rate, in this stage, is directly proportional to the heat transfer coefficient, h. This heat convective coefficient may be computed by knowledge of the air flow or mass velocity (Charm 1971; Toledo, 1994; Geankoplis, 1999; Vélez-Ruiz, 2005). Therefore, it is possible to know the quantity of evaporated water during this period from the h value, and also the mass transfer coefficient may be evaluated with this information. Additionally, it has been found that the mass transfer coefficient can be empirically obtained by Lewis relationship (equation 11a and b) for experimental conditions existing during the constant rate drying (Heldman, 1975).

$$Le = \frac{h_c}{k_g C_p} \tag{11a}$$

$$4h = k_m \tag{11b}$$

where: Le is the Lewis number (dimensionless), h_c or h is the heat transfer coefficient (W/m^2K), k is the mass transfer coefficient (kg/m^2 s) for moisture as a vapor phase (k_g) or in general for mass (k_m), and C_P is the specific heat for wet air (kJ/kg K).

4.3.2. Falling Rate Period

In the second or falling rate period, the water to be evaporated comes from the internal structure and must be transported to the food surface. For this decreasing drying period, the molecular phenomenon prevails, thus this stage is known as diffusional and is governed by the Fick´s equation. Frequently, the falling rate period is the most important in food dehydration processes for a good quantity of foods, and particularly in OD combined with AD for some foods.

This unsteady state falling period may be modeled by the very known Fick's second law or diffusional equation in which a constant diffusion coefficient is normally considered (Crank, 1975; Welty *et al.*, 1976; Okos *et al.*, 1992; Singh and Heldman, 2001; Vélez-Ruiz, 2008):

$$\frac{\partial C_i}{\partial t} = \frac{\partial}{\partial x}\left(D\frac{\partial C_i}{\partial x}\right) \tag{12}$$

where: $\partial C_i/\partial t$ is the concentration difference (mol or kg/m^3) of the i component as a function of time (s, h), D is the mass diffusivity or diffusion constant for the transferred component (m^2/s), $\partial C_i/\partial x$ is the concentration gradient of the i component in the x (m) direction (mol or kg/m^4)

The analytical solution of this differential equation, for an infinite food slab (Crank, 1975; Okos *et al.*, 1992; Vélez 2005; Tsamo *et al.*, 2006) as one of the most common geometries, gives the next relationship.

$$\frac{M - M_e}{M_0 - M_e} = \frac{8}{\pi^2}\left[\sum_{n=0}^{n=\infty}\frac{1}{(2n+1)^2}\right]\exp-(2n+1)^2\left(\frac{\pi^2 Dt}{4L^2}\right) \tag{13}$$

where: M is the average moisture content of the food sample (kg water/kg d.s.) at any time (h), M_e is the equilibrium moisture content (kg water/kg d.s.) at the end of the process, Mo is the initial moisture content (kg water/kg d.s.), n is the number of series terms, D is the effective or apparent diffusional coefficient (m^2/s), t is the drying time (s, min, h), and L is the half thickness of the slab (m).

The equation (13) has their correspondent expressions for cylinders and spheres (equations 14 and 15), they may be used also for drying processes (Crank, 1975; Okos *et al.*, 1992; Ertekin and Cakaloz, 1996; Alzamora *et al.*, 2003; Vélez-Ruiz, 2005; Silva et al, 2005).

$$\frac{M - M_e}{M_0 - M_e} = \frac{6}{\pi^2} \sum_{n=1}^{n=\infty} \frac{1}{n^2} \exp\left(-n^2 \pi^2 \frac{Dt}{R_s^2} \right) \tag{14}$$

$$\frac{M - M_e}{M_0 - M_e} = \frac{4}{R_{cy}{}^2} \sum_{n=1}^{n=\infty} \frac{1}{\beta_n^2} \exp\left(-D\beta_n^2 t \right) \tag{15}$$

where. M, M_e, Mo, n and D have the same meaning that Eq. 13, R_s is the sphere radius (m), R_{cy} is the cylinder radius (m) and β_n^2 is the Bessel function roots of the first kind and zero order.

For long drying times and for moisture ratios $[(M - M_e)/(M_0 - M_e)]$ less than 0.6, generally one term of the series is used, allowing a simple solution of equations 13 to 15 that facilitates the evaluation of the so called effective diffusion coefficient. Many food drying processes consider two conditions where L is small and t is very large, thus the solution for an infinite slab with n = 0 (the values in the summation for $n \geq 1$ are very small and frequently insignificant) may be presented in the next way for a slab geometry:

$$\ln\left[\frac{M - M_e}{M_0 - M_e} \right] = \ln\left(\frac{8}{\pi^2} \right) - \frac{\pi^2 D_{ef} t}{4L^2} \tag{16}$$

And although the previous simplified equation is normally utilized for the diffusion coefficient evaluation, for the case in which D_{ef} is known, the extension of the falling film period (t_f) may be obtained by the equation (17)

$$t_f = \frac{4L^2}{\pi^2 D_{ef}} \ln(0.8106)\left(\frac{M_0 - M_e}{M - M_e} \right) \tag{17}$$

Abbas El-Aouar *et al.* (2003) utilized an analytical solution for water transport in the AD of a solid cube, corresponding to the product of three perpendicular infinite slabs solution (equation 13). Therefore the moisture content during air drying treatment may be fitted by this equation.

$$\overline{M} = \frac{M - M_e}{M_0 - M_e} = \frac{8}{\pi^2} \left(\left[\sum_{n=0}^{n=\infty} \frac{1}{(2n+1)^2} \right] \exp - (2n+1)^2 \left(\frac{\pi^2 D_{ef} t}{4L^2} \right) \right)^3 \tag{18}$$

where: $\overline{M}$ is the dimensionless moisture ratio in the food sample, D_{ef} is the effective diffusional coefficient of water (m^2/s), and the rest of symbols have the same meaning as in previous equations.

There is another model that has been applied in many dehydration processes of biological materials, and it is cited as Page´s equation (Abbas El-Aouar, 2003; Vélez-Ruiz *et al.*, 2004;

Sogi *et al.*, 2005; Singh and Gupta, 2007), being a simplified solution of the Fick´s law as a modified diffusion approach:

$$\frac{M - M_e}{M_0 - M_e} = \exp(-kt) \tag{19}$$

where: M, M_0 and M_e have the same meaning, k is the drying rate constant (h^{-1}), and t is the process time (h).

Dandamrongrak *et al.* (2002), Singh and Gupta (2007), among others have reported and utilized other empirical drying models. A two term equation is shown below

$$\overline{M} = A\exp^{-k_1 t} + B\exp^{k_2 t} \tag{20}$$

where: $\overline{M}$ is the dimensionless moisture ratio in the food sample, and A, B, (dimensionless), k_1 and k_2 (1/t) are constants of the model.

This exponential two term model was developed by Glenn in 1978 (Dandamrongrak *et al.*, 2002) to improve the accuracy of the Page's model, specifically in the last drying period. This model derives from the general diffusion equation, and is applied in few geometries and limited layer conditions, where constant diffusivity is assumed, and constant temperature is required during the whole process.

Furthermore, it is a common practice to relate the temperature effect with the diffusion coefficient by means of the Arrhenius relationship that allows the evaluation of the activation energy.

$$\ln(D_{ef}) = \ln(D_0) - \frac{E_A}{RT} \tag{21}$$

where: D_{ef} is the effective diffusional coefficient of water (m^2/s), D_0 is the reference coefficient, E_A is the activation energy (kJ/mol), R is the universal gas constant, and T is the absolute temperature.

Tables 3, 4 and 5 show data of experimental conditions, some observations, modeling and other data related with AD applied to OD of food plants.

5. COMBINED METHODOLOGY

Osmotic dehydration of fruits and vegetables by immersion in solutions with lower water activity, has received considerable attention in the last twenty years as an air drying pretreatment (Vélez-Ruiz, 2005, 2008). Osmotic dehydration offers several advantages, avoiding structural collapse, getting better textural attributes, preserving nutritional value, decreasing color and flavor damages, shortening the air drying time up to 60% and saving energy with a minimum of 20% (González-Martínez *et al.*, 2006; Tsamo *et al.*, 2006) as a previous process operation followed by a drying process.

The combination of OD and AD or Osmoconvection, as it has been identified by several researchers (Marcotte *et al.*, 2004; Silva *et al.*, 2005) has broad applications on fruit and vegetables processing and provides some unique benefits. OD is a more economical process operation than AD and the quality of the partly dewatered food item is better. The loss of heat labile nutrients and undesired changes in appearance and texture are prevented as well. On the other side, traditional AD process consumes intensively high energy and cause significant loss of flavor and nutrients due to high temperature exposures. Thus the combination of both dehydration processes has been widely explored, from the nineties to the date.

Partial dehydration and solute intake can be achieved by OD, thus by modification and controling of the partial dewatering and syrup composition, the final product not only can be diversified, also its chemical, functional and physical properties may be improved.

5.1 Proposed Models

In according to some researchers, the combination of OD and AD may be modeled in order to optimize the process conditions to obtain a better food product. Teles *et al.* (2006), Fernandes *et al.* (2006), and Oliveira *et al.* (2006) developed a model divided in three parts in which the optimization was completed by using the method of Levenger-Marquardt for solution of the simultaneous equations and a FORTRAN computer program to minimize the total process time (OD time + AD time).

The equations for the OD, considered the mass transfer between the fruit and the osmotic solution (sucrose in water), loss of water and gain of sugar:

$$\frac{dM_{FR}^{W}}{dt} = -K_{m}^{W} \cdot A_{FR} \cdot \left(C_{FR}^{W} - C_{OS}^{W}\right) \cdot V_{FR} \tag{22}$$

$$\frac{dM_{FR}^{S}}{dt} = -K_{m}^{S} \cdot A_{FR} \cdot \left(C_{FR}^{S} - C_{OS}^{S}\right) \cdot V_{FR} \tag{23}$$

where: $\frac{dM_{FR}^{W}}{dt}$ is the change of the mass of water in the fruit with respect to time (kg/h), K_m is the mass transfer coefficient (1/h m^2) for water (K_{m}^{W}) and for sucrose (K_{m}^{S}), A_{FR} is the surface area of fruit phase (m^2), $\left(C_{FR}^{W} - C_{OS}^{W}\right)$ is the difference of water concentration (kg/m^3) between fruit (FR) and osmotic solution (OS), V_{FR} is the fruit volume (m^3), $\frac{dM_{FR}^{S}}{dt}$ is the change of the mass of sucrose in the fruit with respect to time (kg/h), and $\left(C_{FR}^{S} - C_{OS}^{S}\right)$ is the difference of sucrose concentration (kg/m^3) between fruit (FR) and osmotic solution (OS).

A relationship between the shrinkage and the water mass change in the fruit was included.

$$\frac{dV_{FR}}{dt} = \alpha \frac{dM_{FR}^{W}}{dt} \tag{24}$$

where: $\frac{dV_{FR}}{dt}$ is the change of the fruit volume with respect to time (m^3/h), α is the shrinkage factor of the fruit (m^3/kg), and $\frac{dM_{FR}^{W}}{dt}$ has the same meaning that equation 22.

The loss of water by evaporation was also accounted, with the next equations.

$$\frac{dM_{OS}^{W}}{dt} = -\frac{dM_{FR}^{W}}{dt} - \frac{K_m^{EVAP} A_{ODD}}{\lambda} \tag{25}$$

$$\frac{dM_{OS}^{S}}{dt} = \frac{dM_{FR}^{S}}{dt} \tag{26}$$

$$\frac{dV_{OS}}{dt} = \frac{1}{\rho^{W}} \cdot \frac{dM_{FR}^{W}}{dt} \tag{27}$$

where: $\frac{dM_{OS}^{W}}{dt}$ is the change of water in the osmotic solution with respect to time (kg/h), K_m^{EVAP} is the mass transfer coefficient ($1/h\ m^2$) for evaporated water, A_{ODD} is the surface area of the osmotic dehydration device (m^2), λ is the vaporization heat, $\frac{dM_{OS}^{S}}{dt}$ is the change of the mass of sucrose in the osmotic solution with respect to time (kg/h), $\frac{dV_{OS}}{dt}$ is the change of volume for the osmotic solution with respect to time (m^3/h), and ρ^{w} is the density of water (kg/m^3).

Finally, a couple of traditional relations were considered for the AD process:

$$\frac{dH}{dt} = -\frac{RA_{FR}}{M_s} \tag{28}$$

$$\frac{dH}{dt} = -\frac{2\pi}{\delta^2} D\left(H - H_{eq}\right) \tag{29}$$

where: $\frac{dH}{dt}$ is the change of water on dry basis with respect to time ($kg_{water}/h\ kg_{d.s.}$), R is the drying rate or flux of water at the constant period ($kg_{water}/h\ m^2$), A_{FR} is fruit surface area (m^2), M_S is the mass of dry solids (kg), δ is the fruit thickness (m), D is the diffusivity (m^2/h), H is the humidity of the fruit at any time ($kg_{water}/kg_{d.s.}$), and H is the water content at equilibrium ($kg_{water}/kg_{d.s.}$).

Teles *et al.* (2006) found several process combinations for melon processing to get the same final humidity of 30%: i) 1750 min (just of AD), ii) 900 min (270 min of OD with

76.9°Brix and 630 min of AD at 65°C), iii) 715 min (175 min of OD with 71.3°Brix and 540 min of AD at 65°C), and iv) 670 min (220 min of OD with 73.7°Brix and 450 min of AD at 65°C), being the last three as equivalent process optimization.

5.2 Some Reported Studies

Additional information of OD-AD combinations on vegetal foods is included in Table 3 for room temperature, in Table 4 for higher temperatures, and in Table 5 for activation energies and non-Fickian modeling.

Table 4. Studies of drying on osmotically preconcentrated foods at high temperatures

Food item	Conditions in OD	Condition in AD	Results/Observations	Reference
Potato cubes 1 cm	Salt solution 10%, sugar solution 40-50%, 1:4, 40 and 50°C, 2-6 h	3.6, 5.3 and 7.6 m/s, 120, 140 and 160°C, during 5 - 15 min, fluidized bed dryer	Optimum OD: 10% salt, 50% sugar at 47°C, AD: 140°C-10 min, 5.3 m/s	Ravindra & Chattopadhyay, 2000
D'Anjou pear cubes	Sucrose syrup 55°Bx, 40°C 1:10 recirculation, 6 h	1-2 m/s, 40, 60 and 80°C, 7.5 h tray drier,	D: 1.9-8.1x10^{-10} m^2/s, greater drying rates	Park *et al.*, 2003
Blueberries and cranberries	Sucrose syrup 67.5-70°Bx, 1:5 (w/w), 35-65°C	1.4 m/s, 60-100°C, 15% w.b convective dryer	OD: 1.6-4.0 x 10^{-10} m^2/s 10-60 min of drying	Marcotte *et al.*, 2005
Acerola 2.1-2.4 cm diameter	Sucrose solution 40°Bx, 35-40°C 30-90 min, agitated (780 rpm)	1.3 m/s, 60, 65, 70°C, 6 h fixed bed dryer	D: 1.0-2.8 x 10^{-10} m^2/s considering shrinkage	Silva *et al.*, 2005
Apple slices 2 cm diam, 1.1 cm height	Sucrose and fructose solutions 30 and 45%, 40°C, 1:1.5, 18 h	2 m/s and 55°C, 7.5 h air oven, four samples	OD with higher concentration. had better physical properties	Mandala *et al.*, 2005
Banana cubes 3.0 cm average side	Sucrose solution 45, 55 and 65°Bx, 65°C, 1:2 and 1:4 (w/w), 3 h, agitation	2 m/s and 65°C, 16 h shelf dryer	OD: 34-158 for water, 13-47 1/h m^2 for sucrose AD: D = 8-10 x 10^{-10} m^2/s	Oliveira *et al.*, 2006
Melons cylinders 2.0 cm diam., 1 cm thickness	Sucrose solution 45, 55 and 65°Bx, 65°C, 1:2 and 1:4 (w/w), 5 h agitation	0.12 m^3/s and 65°C, 18 h shelf dryer	OD: 72-121 for water, 21-42 1/h m^2 for sucrose AD: D = 4.6 x 10^{-9} m^2/s	Teles *et al.*, 2006
Tomato 45 g, 5 cm lenght, 3.3 cm.diam.	Sucrose solution (600 g/L), NaCl (300 g/L), mixture (45 g + 15g/L), 100 h	0.13 m/s and 60°C trays were prepared and measured by triplicate	20 h of OD was enough, AD occurred in two falling rate periods	Tsamo *et al.*, 2006
Carrot cubes 1 x 1 x 1 cm	NaCl 10%, sucrose 55°Bx, mixture 10%+50°Bx, 35-55°C, 90-180 min	1-2 m/s, 55, 65, 75°C, 10.3- 15.0 h, convective drying	D: 7.8-13.2 x 10^{-10} m^2/s OD augmented AD time	Singh & Gupta, 2007

AD: Air drying; OD: Osmotic dehydration.

Table 5. Non Fickian models and activation energy applied in dried foods osmotically preconcentrated

Food item	Condition in AD	E_A (kJ/mol)	Observations	Reference
Pineapple rectangles 5 x 2 x 0.5 or 1 or 2 cm	Heated oven, with air at 50, 60, 70 and 80°C, 6 h	35.5	D augmented with thickness and air temperature	Uddin *et al.*, 1990
Apple round slices 2 mm thick	Lab tunnel drier, 1.5 m/s at 50, 60 , 70°C, until 20% moisture (1-2 h)	-----	Two falling rate periods Generalized curve was used	Vergara *et al.*, 1997
Pineapple slices 10 cm D, 1 cm thichness	Lab tunnel drier, 0.55, 1.10, 1.60 m/s at 65, 75 and 85°C	27.2 for OD (a_w 0.98), 33.3 for OD (a_w 0.97)	Final a_w 0.70 or 0.45. One falling drying rate period	Vélez-Ruiz *et al.*, 2000
Pineapple slices 10 cm D, 1 cm.thickness	Lab tunnel drier, 0.55,1.10, 1.60 m/s at 65, 75 and 85°C	-----	Generalized curve model less accurate than Fick's law	Vélez-Ruiz *et al.*, 2001
Cherry tomato (spheres)	Cabinet drier 0.75 and 2.60 m/s, 50, 60 and 70°C	18.3 at 0.75 m/s 10.6 at 2.6 m/s	Samples did not show a constant rate period of drying	Azoubel & Murr, 2003
D'Anjou pear cubes	Convective tray drier, 1-2 m/s, 40, 60 and 80°C, 7.5 h	24.34 at 1 m/s 28.20 at 2 m/s	Drying rates increased with air velocity in OD samples	Park *et al.*, 2003
Apple slices 2 cm D, 1.1 cm height	Air oven, 2 m/s and 55°C, 7.5 h four samples	-----	Modeled by Page eq., k: 0.039-0.075 (1/min) for osmosed-dried	Mandala *et al.*, 2005
Carrot cubes 1 x 1 x 1 cm	Convective drying on a thin layer 1-2 m/s, 55, 65, 75°C, 10.3-15.0 h	10.0 for NaCl 14.8 for sucrose 16.2 for mixture	Def estimated from Fick´s second law	Singh & Gupta, 2007

AD: Air drying; OD: Osmotic dehydration.

6. Examples of Application

6.1. Illustrative Example 1

Sodium chloride is diffusing through two systems under steady state conditions. Consider, an aqueous salt solution as the first system, and a cucumber (of 3.6 cm of diameter) placed inside of a brine as the second. Both systems are exposed to room temperature and atmospheric pressure, and the concentration of the solution (first system) or brine (second system) is 31.59% (w/v). Calculate the rate of diffusion, supposing that water and solution are initially separated by a distance of 1.8 cm.

Solution

i) As consequence of contacting inside a container, a diffusion phenomenon starts. Thus, by using the equation 1, with the next data: NaCl concentration of 5.4 mol_{kg}/m^3, Dim = 1.54 x 10^{-9} m^2/s (Table 1).

$$C_{water} = 0;\text{ and}$$

$$C_{sol} = \left(\frac{31.59\,g}{100\,mL} = \frac{315.9\,g}{L} = \frac{0.3159\,kg}{L} x \frac{1000\,L}{1m^3}\right) = \frac{315.9\,kg}{m^3} x \frac{1mol_{kg}}{58.5\,kg} = 5.4\frac{mol_{kg}}{m^3}$$

$$J_{NaClZ} = 1.54x10^{-9}\frac{m^2}{s}\left(\frac{5.4-0\frac{mol_{kg}}{m^3}}{0.018\,m}\right) = 4.62x10^{-7}\frac{mol_{kg}}{m^2s}\left(x\frac{3600\,s}{1h}\right) = 1.66x10^{-3}\frac{mol_{kg}}{m^2h}$$

$$J_{NaClZ} = 1.66x10^{-3}\frac{mol_{kg}}{m^2h}\left(x\frac{58.5kg}{mol_{kg}}\right) = 9.73x10^{-2}\frac{kg}{m^2h}$$

ii) For the system 2 following a similar reasoning, and considering that in this case the center of the cucumber as the final point. The brine has the same concentration of 5.4 mol_{kg}/m^3, but the effective diffusion is different, Deff = 9.60 x 10^{-10} m^2/s (Table 1).

$$C_{cucumber} = 0;\cdot\text{and} \rightarrow C_{sol} = 5.4\frac{mol_{kg}}{m^3}$$

$$J_{NaClr} = 9.60x10^{-10}\frac{m^2}{s}\left(\frac{5.4-0\frac{mol_{kg}}{m^3}}{0.018m}\right) = 2.88x10^{-7}\frac{mol_{kg}}{m^2s}\left(x\frac{3600s}{1h}\right) = 1.04x10^{-3}\frac{mol_{kg}}{m^2h}$$

$$J_{NaClZ} = 1.04x10^{-3}\frac{mol_{kg}}{m^2h}\left(x\frac{58.5kg}{mol_{kg}}\right) = 6.06x10^{-2}\frac{kg}{m^2h}$$

Conclusion: From the developed calculations and by comparison of the magnitudes, it can be summarized that, the diffusional process in water is higher than in a food item, basically as a consequence of the D value. D may be a reflect of the structure, simple solution against a vegetal item. The flux of salt in the cucumber was 63% of the salt in water.

6.2. Illustrative Example 2

Estimate the rate of water evaporated from a pan (0.9 m of diameter) full of water. Air with 30% RH, 27°C and a velocity of 1.2 m/s is flowing over the top of the reservoir.

Solution

This is a case of convection process in which an empirical correlation must be used to evaluate the mass transfer coefficient and, lately the evaporated water may be computed. Firstable, the Reynolds and the Schmidt numbers (Table 2) need to be quantified, data for air were taken from Appendix I (Welty *et al.*, 1976). At 300 K: $\rho = 1.177$ kg/m^3, $\eta = 1.846 \times 10^{-5}$ Pa s, D = 2.6 x 10^{-5} m^2/s. Thus, the mass transfer coefficient is computed.

$$\mathrm{Re} = \frac{1.5\frac{m}{s}\mathrm{x}0.9\mathrm{mx}1.1769\frac{kg}{m^3}}{1.8464\mathrm{x}10^{-5}\frac{N\cdot s}{m^2}} = 86049;\mathrm{and}\cdot \mathrm{Sc} = \frac{1.8464\mathrm{x}10^{-5}\frac{N\cdot s}{m^2}}{2.6\mathrm{x}10^{-5}\frac{m^2}{s}\mathrm{x}1.1769\frac{kg}{m^3}} = 0.603$$

$$\rightarrow \mathrm{Sh} = 0.664\,\mathrm{Re}^{0.5}\,\mathrm{Sc}^{0.333} \rightarrow \mathrm{Sh} = 0.664(86049)^{0.5}(0.60341)^{0.333} = 164.62 = \frac{k_m L_c}{D}$$

$$\therefore k_m = \frac{164.62\mathrm{x}2.6\mathrm{x}10^{-5}\frac{m^2}{s}}{0.9\mathrm{m}} = \mathbf{4.76x10^{-3}\frac{m}{s}}$$

Finally, the rate of evaporated water is estimated by mean of equation 2, by taking C_{is} for the water as the concentration under saturated conditions, 0.0255 kg/m^3 for saturated vapor at 27°C in table A-4 (Cengel and Boles, 2006), and C_{if} for the moisture in the air stream, 0.0077 kg/m^3 (= C_{is}xRH/100)..

$$N_{water} = -4.76\mathrm{x}10^{-3}\frac{m}{s}(0.0255-0.0077)\frac{kg}{m^3} = 8.47\mathrm{x}10^{-5}\frac{kg}{m^2 s}\left(\mathrm{x}\frac{3600s}{1h}\right) = 0.305\frac{kg}{m^2 h}$$

And by considering the pan area (A= π/4 x d^2 = 0.7854 x (0.9m)^2 = 0.64 m^2)

$$N_{evaporated\cdot water} = 0.305\frac{kg}{m^2 h}\mathrm{x}0.64\mathrm{m}^2 = \mathbf{0.194\frac{kg}{h}}$$

Conclusion: From the developed calculations and by analysis of the driving force, it can be observed that: if the temperature of the air is increased, the ΔC_{water} will augment (because the C_{is} is higher); whereas if the RH is increased, then the ΔC_{water} will decreased (because the C_{if} is higher).

6.3. Illustrative Example 3

In order to apply Peleg's model, moisture content data as a function of time for OD of papaya in a solution 50% (w/w) were taken from the figure 1a (Palou *et al.*, 1994). Determine the Peleg´s parameters (equation 7)

Solution

Taking data and applying the equation 7b, the next information was obtained.

Table 6. Data and information for example 6.3

t (min)	M (g ag/100g)	t/(M(t) - Mo)	
0	90.0	0	
148	83.0	-21.1429	
241	81.0	-26.7778	corr. -0.9331
400	80.0	-40.0000	interc= k1 -0.0245
750	72.5	-42.8571	
1450	61.0	-50.0000	slope = k2
2833	56.5	-84.5672	-17.5374

Conclusion: The used values for time and moisture content are approximate, because they were taken from the graphic data. Even though, the correlation coefficient, k_1 and k_2 constants are very close to those reported by the authors (0.982, 0.022, and 17.563).

FINAL REMARKS

The combination of osmotic dehydration (OD) and air drying (AD), from the technical and practical point of view, is completely proved and accepted, there are a good number of advantages and benefit of this couple of food unit operations. One of the most important is that drying accounts for up to 15-20% of all industrial energy usage of a country, besides of the low thermal efficiency, in a range of 30-60%.

The literature results conclude that influence of OD on AD has showed differences that may be attributed to the tissue characteristics of each foodstuff. The increasing need for high quality products will guarantee the incorporation of osmo-convected food items into bakery products, mixed with ready to eat cereals, their use as powder food ingredient, or well, as shelf stable foods.

Even though, there are a good number of investigations applying OD and AD models, there is not a proper modeling of the combined processes. A couple of the consulted papers included OD and AD equations to optimize the total process time, but it is not a general model that may be considered for combination of both unit operations. Optimization of the whole process aiming the time reduction has been estimated from experimental runs at specific operating conditions, for banana, melon, and potato.

The application of vacuum has been reported as convenient for osmotic process acceleration, as well as microwave has been studied for time reduction of air dehydration. Both incorporations have contributed with very good improvements of the OD/AD processes, but normally not economical concerns have been involved in the analysis of the whole process to establish the degree of convenience for use it.

ACKNOWLEDGMENTS

To Alicia Reyes-Herrera, M.Sc., for her contribution to this chapter. Initially we planned to be coauthors, but unfortunately she did not have time to participate. Anyway I appreciate her ideas and effort.

REFERENCES

Abbas El-Aouar A., Moreira Azoubel P. and Xidieh Murr E. 2003. Drying kinetics of fresh and osmotically pre-treated papaya (*Carica papaya* L.). *Journal of Food Engineering,* 59: 85-91.

Alvarez C.A., Aguerre R., Gómez R., Vidales S., Alzamora S.M. and Gerschenson L.N. 1995. Air dehydration of strawberries: Effects of blanching and osmotic pretreatments on the kinetics of moisture transport. *Journal of Food Engineering,* 25: 167-178.

Alzamora S.M., Nieto A. and Castro M.A. 2003. Structural of blanching and osmotic dehydration pretreatments on air drying kinetics of fruit tissues. In *Transport Phenomena in Food Processing.* (J Welti, J. Vélez and G. Barbosa; Eds.). CRC Press. Boca Raton, FL, USA. Pp. 69-94.

Azoubel P.M. and Murr F.E.X. 2003. Effect of pre-treatment on the drying kinetics of cherry tomato (*Lycospersicon esculentum* var. Cerasiforme). In *Transport Phenomena in Food Processing.* (J Welti, J. Vélez and G. Barbosa; Eds.). CRC Press. Boca Raton, *USA*. Pp. 137-151.

Badui-Dergal, S. 1999. Food Chemistry. Pearson Educación, Mexico D.F., Mexico. (In Spanish).

Bird R.B., Stewart W.E. and Lightfoot E.N. 1960. Transport Phenomena. John Wiley and Sons, Inc. New York, NY, USA

Camargo, G., Moretti, R. and Ledo, C. 2004. Quality of dried tomato pre-treated by osmotic dehydration, antioxidant application and addition of tomato concentrate. Drying 2004. *Proceedings of the 14th International Drying Symposium*, Sao Paulo, Brazil, August 22-25, Vol. C: 2207-2215.

Castañón X., Ibarz A., Welti-Chanes J., Palou E. and López-Malo A. 2001a. Air drying behavior of osmotically dehydrated papaya slices. ICEF8. *Proceedings of the Eighth International Congress of Engineering and Food.* Technomic Pub. Co. Inc. Vol. II: 1061-1065.

Castañón X., Ibarz A., Welti-Chanes J., Palou E. and López-Malo A. 2001b. Water activity effects on air drying behavior and quality of mango. ICEF8. *Proceedings of the Eighth International Congress of Engineering and Food.* Technomic Pub. Co. Inc. Vol. II: 1066-1070.

Castillo M.E. and Cornejo F. 2007. Study of the effect of osmotic dehydration in the production of dried pieces of carambola (*Averroha carambola L.*). CIBIA VI. *Proceedings of the 6th Ibero American Congress of Food Engineering.* Universidad Técnica de Ambato. Vol. 16 (3): 190-193. (In Spanish).

Charm S.E. 1971. Fundamentals of Food Engineering. AVI Pub. Co., Inc. Westport, CN, US. Pp. 294-348.

Crank, J. 1975. The Mathematics of Difussion. Oxford University Press, Wiltshire, UK.

Chenlo F., Moreira R. and Torres M.D. 2007. Rheological properties of chesnuts processed by osmotic dehydration and convective drying. *Food Science and Technology International,* 13 (5): 369-374.

Dandamrongrak R., Young G. and Mason R., 2002 Evaluation of various pre-treatments for dehydration of banana and selections of suitable drying models. *Journal of Food Engineering.* 55:139-146.

Erle U. and Schubert H. 2001. Combined osmotic microwave-vacuum dehydration of apples and strawberries. *Journal of Food Engineering.* 49:193-199.

Ertekin F.K. and Cakaloz T. 1996. Osmotic dehydration of peas. II. Influence of osmosis on drying behavior and product quality. *Journal of Food Processing and Preservation,* 20: 105-119.

Fernandes F.A.N., Rodrigues S., Gaspareto O.C.P. Oliveira E.L. 2006. Optimization of osmotic dehydration of bananas followed by air-drying. *Journal of Food Engineering.* 77:188-193.

Flink, J.M. 1980. Dehydrated carrot slices: Influence of osmotic concentration on drying behavior and product quality. In *Food Process Engineering.* (P. Linko, Y. Malkki, J. Olkku and J. Larinkari, Eds.). Academic Press Inc., London, England. Vol. I: 412-418.

García C.C., Mauro M.A. and Kimura M. 2007. Kinetics of dehydration and air-drying of pumpkins (*Cucurbita moschata*). *Journal of Food Engineering,* 55:139-146.

Geankoplis C.J. 1999. Transport Processes and Unit Operations. CECSA Mexico, D.F. (In Spanish).

González-Martínez C., Cháfer M., Xue K. and Chiralt A. 2006. Effect of the osmotic pre-treatment on the convective air drying kinetics of pear var. blanquilla. *International Journal of Food Properties,* 9 (3): 541-549.

Hawkes, J., and Flink, J. 1978. Osmotic concentration of fruit slices prior to freeze dehydration. *Journal of Food Processing and Preservation,* 2: 265-275.

Heldman D.R. 1975. Food Process Engineering. The AVI Publishing Company, Inc. Westport, CN, USA.

Karathanos V.T., Kostaropoulos A.E. and Saravacos G.D. 1995. Air-drying kinetics of osmotically dehydrated fruits. *Drying Technology,* 13 (5-7): 1503-1521.

Lewicki P.P. and Lukaszuk A. 2000. Effect of osmotic dewatering on rheological properties of apple subjected to convective drying. *Journal of Food Engineering,* 45: 119-126.

Lewicki P.P., Le H.V. and Pomaranska-Lazuca W. 2002. Effect of pre-treatment on convective drying of tomatoes. *Journal of Food Engineering,* 54: 141-146.

Li H. and Ramaswamy H.S. 2005. Study of deviation from diffusion controlled solid penetration in osmotic drying of apple slices. IADC 2005. *Proceedings of the 3th Inter American Drying Conference*, Montreal, Québec, Canada. August 21-23,. Paper IV-3, 16 pages.

Maldonado S., Bertuzzi M.A. and Arnau E. 2007. Influence of the osmotic dehydration previous to dehydration on the rehydration process of mango. CIBIA VI. *Proceedings of the 6th Ibero American Congress of Food Engineering.* Universidad Técnica de Ambato. Vol. 16 (1): 193-195. (In Spanish)

Mandala I.G., Anagnostaras E.F. and Oikonomou C.K. 2005. Influence of osmotic dehydration conditions on apple air-drying kinetics and their quality characteristics. *Journal of Food Engineering,* 69: 307-316.

Marcotte M., Grabowski S., Quan D., Taherian A.R., Zareifard M.R., Pourier M. and Kudra T. 2004. Quality aspects of Canadian blueberries and cranberries dried by osmo-convective method. IADC 2005. *Proceedings of the 3th Inter American Drying Conference,* Montreal, Québec, Canada, August 21-23. Paper IV-2, 16 pages.

Matusek, A. and Merész, P. 2002. Modelling of sugar transfer during osmotic dehydration of carrots. *Periodica Polytechnica Series of Chemical Engineering.* 46 (1-2): 83-92.

Nieto A., Salvadori D., Castro M.A. and Alzamora S.M. 1998. Air drying behaviour of apples as affected by blanching and glucose impregnation. *Journal of Food Engineering,* 36 (1): 63-79.

Nieto A., Castro M.A. and Alzamora S.M. 2001. Kinetics of moisture transfer during air drying of blanched and/or osmotically dehydrated mango. *Journal of Food Engineering,* 50: 175-185.

Okos M.R., Narsimhan G., Singh R.K. and Weitnauer A.C. 1992. Food dehydration. In *Handbook of Food Engineering* (D.R Heldman. and D.B. Lund; Eds.). Marcel Dekker, Inc. NY, NY, USA. Pp. 437- 562.

Oladele, A. and Odedeji 2008. Osmotic dehydration of catfish (*Hemisynodonis membranaceus*): effect of temperature and time. *Pakistan Journal of Nutrition.* 7(1): 57-61.

Oliveira I.M., Fernandes F.A.N., Rodrigues S., Sousa P.H.M., Maia G.A. and Figueiredo R.W. 2006. Modeling and optimization of osmotic dehydration of banana followed by air-drying. *Journal of Food Process Engineering,* 29: 400-413.

Palou E., López-Malo A., Argaiz A. and Welti J. 1994. The use of Peleg´s equation to model osmotic concentration of papaya. *Drying Technology,* 12(4): 965-978.

Pani P., Leva A.A., Riva M., Maestrelli A. and Torreggiani D. 2008. Influence of an osmotic pretreatment on structure-property relationships of air-dehydrated tomato slices. *Journal of Food Engineering,* 86: 105-112.

Park K.J. Bin A. and Brod F.P.R. 2003. Drying of pear d'Anjou with and without osmotic dehydration. *Journal of Food Engineering,* 56: 97-103.

Peleg, M. 1988. An empirical model for the description of moisture sorption curves. *Journal of Food Science,* 53:1216-1219.

Rahman S.M.A. and Mujumdar A.S. 2007. Effect of osmotic treatment with concentrated sugar and salt solutions on kinetics and color in vacuum contact drying. *Journal of Food Processing and Preservation,* 31: 671-687.

Ravindra M.R. and Chattopadhyay P.K. 2000. Optimisation of osmotic preconcentration and fluidised bed drying to produce dehydrated quick-cooking potato cubes. *Journal of Food Engineering,* 44: 5-11.

Reyes-Herrera A., Sosa-Morales M. E. and Vélez-Ruiz J. F. 2005. Osmotic dehydration and drying of jicama (*Pachyrizus tuberosus*) impregnated with citric acid. IADC 2005. *Proceedings of the 3th Inter-American Drying Conference,* Montreal, Québec, Canada. August 21-23. Paper A-4, 8 pages.

Silva V.A., Alsina O.L.S. and Pereira J.A. 2005. Osmo-convective drying of acerola: Modeling of the convective kinetics. IADC 2005. *Proceedings of the 3th Inter American Drying Conference,* Montreal, Québec, Canada. August 21-23. Paper IV-4, 10 pages.

Simal S., Deyá E., Frau M. and Roselló C. 1995. Simple modeling of air drying curves of fresh and osmotically pre-dehydrated apple cubes. *Journal of Food Engineering,* 24:139-150.

Singh B. and Gupta A.K. 2007. Mass transfer and determination of effective diffusivity during convective dehydration of pre-osmosed carrot cubes. *Journal of Food Engineering,* 79: 459-470.

Singh R.P. and Heldman D.R. 2001. Introduction to Food Engineering. Academic Press. Glasgow, Great Britain, UK. 625-670.

Sogi D.S. Wani A.A. and Shivhare U.S. 2005. Effect of tray load on the drying kinetics of watermelon seeds. IADC 2005. *Proceedings of the 3th Inter American Drying Conference,* Montreal, Québec, Canada. August 21-23. Paper A-8, 9 pages.

Souza J., Medeiros F., Margarida M., Magalhaes M., Rodrigues S., Fabiano A. and Fernandes A. 2007. Optimization of osmotic dehydration of tomatoes in a ternary system followed by air drying. *Journal of Food Engineering,* 83:501-509.

Suazo-Abarca, S., Sánchez-Pineda H., Pineda-Ramos R., Ramírez-Torres V. and Robledo-Bautista I. 2000 Comparison of two drying processes for pretreated vegetal products. *XXXIII Mexican Congress of Food Science and Technology.* September 25-27, Uruapan, Mich. Mexico. (In Spanish).

Sunjka, P. and Raghavan, G. 2004 Assessment of pretreatment methods and osmotic dehydration for cranberries. *Canadian Biosystems Engineering.* 46: 3.35-3.40

Teles U.M., Fernandes F.A.N., Rodrigues S., Lima A.S., Maia G.A. and Figueiredo R.W. 2006. Optimization of osmotic dehydration of melons followed by air-drying. Food *International Journal of Food Science and Technology,* 41: 674-680.

Toledo R.T. 1994. Fundamentals of Food Process Engineering. Chapman and Hall. New York, NY, USA. 470-493.

Torringa E., Esveld E., Scheewe I., van der Berg R. and Bartels P. 2001. Osmotic dehydration as a pre-treatment before combined microwave-hot-air drying of mushrooms. *Journal of Food Engineering.* 49:185-191.

Traybal R.E. 1980. Mass Transfer Operations. McGraw-Hill Book Company. York, NY, US.

Tsamo C.V.P., Bilame A.F. and Ndjouenken R. 2006. Air drying behaviour of fresh and osmotically dehydrated onion slices (*Allium cepa*) and tomato fruits (*Lycopersicon esculentum*). *International Journal of Food Properties,* 9 (4): 877-888.

Uddin M.S, Hawlader M.N.A. and Rahman S. 1990. Evaluation of drying characteristics of pineapple in the production of pineapple powder. *Journal of Food Processing and Preservation,* 14: 375-391.

Vélez-Ruiz J.F. 2005. Notes of Food Engineering III. Universidad de las Américas, Puebla. Puebla, Mexico. Unpublished document.

Vélez Ruiz J.F. 2008. Notes of Transport Phenomena in Food Engineering. Universidad de las Américas, Puebla. Puebla, Mexico. Unpublished document.

Vélez-Ruiz J.F., Kosegarten-Conde C. and Lopez-Malo A. 2000. Drying of pineapple slices previously treated by osmotic dehydration (A-373). IDS 2000. *Proceedings del 12th International Drying Symposium.* August 28-31 Noordwijkerhout, Netherlands, 4 pages.

Vélez-Ruiz J.F., Kosegarten-Conde C. and Lopez-Malo A. 2001. Effect of process variables on the drying of osmotic dehydrated pineapple slices. ICEF8. *Proceedings of the Eighth International Congress of Engineering and Food.* Technomic Pub. Co. Inc. Vol. II: 1106-1110.

Vélez-Ruiz J.F. and Soriano-Morales A.L. 2003. Evaluation of physical properties of foods by using a computer program. *Revista Internacional Información Tecnológica,* 14 (4): 23-28. (In Spanish).

Vélez-Ruiz J.F., Lima Carrera M.E. and Macedo y Ramírez R.C. 2004. Air drying of three aroma herbs (basil, dill and tea herb). IDS 2004..*Drying 2004 – Proceedings of the 14th International Drying Symposium.* São Paulo, Brazil, August 22-2, Vol. C: 1914-1921.

Vergara F., Amézaga M., Bárcenas M.E. and Welti J. 1997. Analysis of the drying processes of osmotically dehydrated apple using the characteristic curve model. *Drying Technology,* 15 (3-4): 949-963.

Welti-Chanes J., Mújica Paz H., Valdez Fragoso A. and León Cruz R. 2003. Fundamentals of Mass Transport. In *Transport Phenomena in Food Processing.* (J Welti, J. Vélez and G. Barbosa; Eds.). CRC Press. Boca Raton, USA. Pp. 3-23.

Welty J.R., Wicks C.E. and Wilson R.E. 1976. Fundamentals of Momentum Heat and Mass Transfer. John Wiley and Sons, New York, NY, USA.

PART IV: OTHER IMPORTANT TOPICS IN FOOD PROCESSING

In: Food Processing and Engineering Topics
Editors: M. E. Sosa-Morales and J. F. Vélez-Ruiz
ISBN: 978-1-60741-788-0

Chapter 10

ULTRASOUND AS AN EMERGING TECHNOLOGY FOR FOOD PROCESSING

María Teresa Jiménez-Munguía, Aurelio López-Malo and Enrique Palou
Chemical and Food Engineering Department
Universidad de las Américas, Puebla, Mexico

1. INTRODUCTION

Among the emerging technologies used as alternatives to traditional thermal processing, ultrasound has demonstrated to be a good option to provide safe food with high quality characteristics.

Applications of ultrasound in food processing can be classified in regards of the applied intensity. Low-intensity ultrasound (high frequency, 2 - 10 MHz) does not cause chemical or physical changes in the material where the waves pass through, being a non-destructive technique which is commonly used for diagnostic tests. In contrast, high-intensity ultrasound (low frequency, 20-100 kHz) can cause physical damage and promote certain chemical reactions. The advantages provided by high-intensity ultrasound had led to investigate numerous applications in food and chemical engineering: emulsion homogenization, cell rupture, microbial destruction, enzyme inactivation, enhanced filtration and extraction, among others.

This chapter will present high-intensity ultrasound research/ advances in food processing for microbial destruction (bacteria, yeasts and molds) non-thermal juice processing and homogenization and pasteurization of milk for yoghurt processing. Microbial destruction studies include mold inhibition in model systems subjected to different stress factors (a_w, pH, synthetic or natural antimicrobials), and prediction of microbial survival. Predictive modeling in fruit juices (orange and grapefruit) was utilized to validate ultrasound treatments. For dairy processing, studies concerning the application of ultrasound as a pre-treatment of pasteurization will be discussed, as well as milk homogenization by means of ultrasound for yoghurt preparation.

2. Ultrasound: Definition and Applications in the Food Industry

Ultrasound is defined as sound waves with frequencies above the threshold for human hearing (>16 kHz). Depending on the energy amount generated by ultrasound, characterized by the sound power (W), sound intensity (W/m^2) or sound energy density (Ws/m^3), applications of ultrasound have been classified basically into two groups: low intensity and high intensity ultrasound (McClements, 1995; Knorr *et al.*, 2004). Low intensity ultrasound (LI-ULTS) involves frequencies higher than 100 kHz at intensities below 1 W/cm^2. This type of ultrasound does not cause physical or chemical alterations in the properties of the material through which the ultrasound waves pass, therefore is considered as non-destructive. The main application of LI-ULTS in the food industry is as an analytical technique for providing information about the physicochemical properties of foods such as composition, structure and physical state. Advantages of ultrasound as analytical technique over conventional techniques are that measurements can be carried out faster, are more precise, non-destructive and can be performed on-line (McClements, 1995). On the other hand, high intensity ultrasound (HI-ULTS) involves intensities higher than 1 W/cm^2 at frequencies between 20 and 100 kHz. In contrast to LI-ULTS, HI-ULTS can cause physical, chemical and biological alterations in the material where it is applied, changing several properties (Mason, 1990). HI-ULTS has been used for many years to disrupt cells, homogenize emulsions, clean material and disperse aggregated materials. More recently, various areas have been identified with great potential for future development, for example, modification and control of crystallization processes, degassing of liquid foods, enzyme inactivation, induction of oxidation reactions, and food freezing (McClements, 1995; Knorr *et al.*, 2004, Zheng and Sun, 2006).

The effects caused by HI-ULTS in liquid media are attributed to the cavitation phenomena. When ultrasound waves travel through a liquid, regions of alternating compression and expansion are created. These regions of different pressure cause cavitation to occur and consequently, gas bubbles are formed in the medium. These bubbles have a larger surface area during the expansion cycle, which increases the diffusion of gas, causing the bubble to expand. A point is reached where the provided ultrasonic energy is not sufficient to retain the vapor phase in the bubble; therefore, rapid condensation occurs. The condensed molecules collide violently, creating shock waves. These shock waves create regions of very high temperature and pressure, reaching up to 5500°C and 50MPa (Suslick, 1988; Sala *et al.*, 1995). Ability of ultrasound to cause cavitation depends on ultrasound characteristics (frequency, intensity), product properties (viscosity, surface tension) and ambient conditions (temperature, pressure) (Suslick, 1988; Dolatowski *et al.*, 2007).

One of the constant challenges of food technologists is the development of new food processing technologies and new food products. Development of non-thermal processing methods has been an area of study in the last decades, applied to the food industry, which includes high pressure processing, pulsed electric and magnetic fields, and ultrasound among others (Povey and Mason, 1998; Butz and Tauscher, 2002; Mason, 2003; Mañas and Pagán, 2005).

2.1. Food Safety

Food preservation methods currently used by the industry rely either on the inhibition of microbial growth or on microbial inactivation. The most common techniques used to inactivate microorganisms in food products are conventional thermal pasteurization and sterilization. However, the severity of the treatments carries out nutrient loss, functional properties deterioration and degradation of organoleptic characteristics (Pisayena *et al.*, 2003; Zenker *et al.*, 2003).

It is widely recognized that the bactericidal effect of HI-ULTS is attributed to the cavitation phenomena, which causes cell disruption (López-Malo *et al.*, 2005). The high temperatures produced during cavitation may also have some effect, but since temperature changes occur momentarily, only the liquid in the immediate surroundings is heated and therefore only a small number of cells are affected (Sala *et al.*, 1995). Resistance of different microorganisms species varies widely as a result from specific effect of ultrasound on the cell wall and differences in the cell wall structures among the microorganisms (Earnshaw, 1998). Bacterial spores are much more resistant than vegetative bacteria (Alliger, 1975).

Nevertheless, application of HI-ULTS, in order to provide significant microbial reduction in food systems like pasteurization, would require high intensities. Therefore, HI-ULTS has been used in combination with other processing techniques like temperature up to 60°C (thermosonication), using hydrostatic pressure up to 600 kPa (manosonication), or both, temperature and pressure (mano-thermosonication), providing important microbial destruction (Raso *et al.*, 1998). Effectiveness of an ultrasound treatment depends on the type of microorganism, culture or food composition, and HI-ULTS processing conditions (volume processed, time, temperature and amplitude of the ultrasonic waves) (Piyasena *et al.*, 2003; Mañas and Pagán, 2005). Several studies provide information about the effectiveness of the HI-ULTS for microbial destruction in combination with other food preservation techniques, applying the hurdle technology approach (table 1).

Furthermore, the hurdle technology approach may be enhanced by modifying some other environmental factors of the media, like reducing pH or water activity (a_w), and the addition of antimicrobials to improve microbial destruction or inhibition. Some studies carried out by López Malo *et al* (2005) demonstrated mold (*Aspergillus flavus* and *Penicillum digitatum*) inactivation in model systems, combining HI-ULTS with potassium sorbate or vanillin, under different pH (5.5 or 3.0) and a_w (0.99 or 0.95) conditions. Survival curves were obtained, and from them, D and z values were calculated. Results from this study encouraged research on HI-ULTS treatments in combination with some other factors, in order to generate the basis of feasible ultrasonic pasteurization treatments.

Since many factors contribute to microbial inactivation or inhibition, surface response methodology is a good alternative to predict microbial behavior among the levels of the evaluated factors. We describe how this predictive model can be obtained; in order to study a wide number of process parameters (temperature, wave amplitude, time) and media conditions (antimicrobial concentration). Results from a study of Jiménez-Munguía (2000) are taken as an example where two different species of *Penicillum* were submitted to thermoultrasonic treatments, evaluating their inhibition at different incubation time (2, 4 or 6 days), in model systems formulated with selected antimicrobial agents (sodium benzoate or vanillin).

Table 1. Some applications of ultrasound combined with other process treatments, used for the inactivation of different microorganisms

Treatment / studied factor	Microorganism/ medium of treatment	Reference
Ultrasound Resistance to thermal treatment	Bacillus cereus, Bacillus licheniformis / culture suspensions	Burgos *et al.*, 1972
Resistance of cellular wall	Alternaria alternata, Aspergillus flavus, Arthrinium francii, Penicillum frequentans, Rhizopus arrhizus, Candida albicans, Rhodotorula glutinis / culture suspensions	Idrissi *et al.*, 1996
Ultrasound + Thermal treatment (Thermosonication) Simultaneous application and resistance to thermal treatment	Streptococcus faecium, Streptococcus durans / saline solution	Ordóñez *et al.*, 1984
	Bacillus subtilis var . niger, Bacillus subtilis ATCC 6051 / water, glycerol and whole milk	García *et al.*, 1989
Treatment temperatures and simultaneous application	*Salmonella typhimurium* / Brain heart infusion, skim milk and whole egg	Wrigley & Llorca, 1992
Temperature and ultrasound's intensity	Saccharomyces cerevisiae / water	Ciccolini *et al.*, 1997
Temperature, amplitude and pH Cellular structure	Saccharomyces cerevisiae / Sabouraud broth	Guerrero *et al.*, 2001
Temperature, and ultrasound's intensity (inactivation kinetics)	Saccharomyces cerevisiae / water	López-Malo *et al.*, 1999
Ultrasound + High pressure (Manosonication) Ultrasound amplitudes and hydrostatic pressure Lethal prediction	*Listeria monocytogenes* / citrate-phosphate buffer	Mañas *et al.*, 2000a
Ultrasound + temperature + high pressure (manothermosonication) Temperature, ultrasound's intensity and hydrostatic pressure	*Yersenia enterocolitica* / citrate-phosphate buffer	Raso *et al.*, 1998
	Streptococcus faecium, Listeria monocytogenes, Salmonella enteritidis, Aeromonas hydrophila / citrate-phosphate buffer	Pagán *et al.*, 1999
Ultrasound + Bactericidal agents		

Table 1. (Continued).

Treatment / studied factor	Microorganism/ medium of treatment	Reference
Bactericidal effect of hydrogen peroxide	Salmonella enteritidis, Salmonella typhimurium, Salmonella senftenberg / citrate-phosphate buffer and whole egg	Mañas *et al.*, 2000b
	Fusobacterium nucleatum, Veillonella parvula, Streptococcus sanguis, Staphylococcus aureus, Clostridium sporogenes, Bacillus cereus, Candida albicans / water	Ahmed & Russell, 1975
Bactericidal effect of chlorine	*Salmonella typhimurium* / saline solutions for chicken's skin disinfection	Lillard, 1993
	Escherichia coli / water	Phull *et al.*, 1997

Adapted from Jiménez-Munguía, 2000.

Response surface model was applied in combination with logistic regression for modeling of the responses. Logistic regression is a probabilistic model which allows the analysis of the responses as binary or dicotomic variables. Therefore, the response of "growth" or "not growth" was registered for every experiment. Logistic regression relates the probability of an event to occur (Y) which is function of conditional or independent variables (x_i). The term $\pi(x)$ represents the conditional mean of Y for a given x_i for a logistic distribution, given by:

$$\pi(x) = \frac{\exp(\sum \beta_i x_i)}{1 + \exp(\sum \beta_i x_i)} \qquad (1)$$

where β_i are the parameters for each variable x_i.

A transformation for this expression is needed in order to obtain a model with linear regression characteristics for the included variables, being continuous and depending on the x magnitude. The transformation allowing this to happen, is the *logit* transformation (Hosmer and Lemeshow, 1989), leading to a model like:

$$logit(p) = g(x) = \ln\left[\frac{\pi(x)}{1 - \pi(x)}\right] = \sum \beta_i x_i \qquad (2)$$

Applying this model, a value of "1" was given for the responses of "growth" and for the "not growth" responses a value of "0" was given. The independent variables in this case were: the ultrasound amplitude (A; 0, 50 or 75% of the maximum of 120 μm), treatment temperature (T; 45, 50 or 55°C), treatment time (t; 0, 10 or 15 min), type of microorganism (M; 0 or 1), concentration of the antimicrobial (C; 3 levels depending on the antimicrobial type), and incubation time (I; 2, 4 or 6 days). The type of microorganism M, is a discrete variable, assigning a different value for each of the two *Penicillum* species, "0" for *Penicillum italicum* or "1" for *Penicillum digitatum*. Evaluated antimicrobial concentrations were below to the minimal inhibitory concentrations reported by López-Malo (2000) for sodium benzoate (30-150 ppm) and vanillin (200-1000 ppm).

The adjusted logistic regression for the growth of both *Penicillum*, was developed (SPSS Inc., Chicago IL), obtaining a polynomial (Eq. 3) which includes each one of the independent variables, and interactions of the studied variables.

$$g(x)=\beta_0 + \beta_1 A + \beta_2 T + \beta_3 t + \beta_4 M + \beta_5 C + \beta_6 I + \beta_7 A T + \beta_8 A t + \beta_9 A C + \beta_{10} A I + \beta_{11} T t + \beta_{12} T C + \beta_{13} T I + \beta_{14} t C + \beta_{15} t I + \beta_{16} C I + \beta_{17} A T t + \beta_{18} A T C + \beta_{19} A T I + \beta_{20} A t C + \beta_{21} A t I + \beta_{22} A C I + \beta_{23} T t C + \beta_{24} T t I + \beta_{25} T C I + \beta_{26} t C I + \beta_{27} A t C + \beta_{28} A t I + \beta_{29} A T t C + \beta_{30} A T t I + \beta_{31} T t C I + \beta_{32} T C I A \quad (3)$$

The reduced polynomial model for each antimicrobial was obtained keeping the independent variables and only those variable interactions showing a significant effect ($p > 0.10$) on the growth response. An example of one logistic regression model for the probability of growth of *Penicillum italicum* or *Penicillum digitatum* in a model system of pH 3.5 and a_w 0.99 with vanillin for 2 to 6 days of incubation is presented in table 2.

Table 2. Reduced polynomial model coefficients used to predict the probability of growth of *Penicillium italicum* and *Penicillium digitatum* subjected to thermoultrasonic treatments, inoculated in potato dextrose agar with different concentrations of vanillin

Variable	Units	Range	Coefficient	Standard error	Significance
Constant			-6.0149	3.8921	0.1222
Temperature (T)	°C	40 – 50	0.1780	0.0766	0.0201
Treatment time (t)	min	0 – 15	0.9722	0.1903	0.0000
Wave amplitude* (A)	%	0 – 75	0.0000	0.0000	0.0000
Concentration (C)	ppm	200-1000	-0.0106	0.0016	0.0000
Incubation time (I)	days	2 - 6	17.9386	2.6243	0.0000
Type of microorganism (M)	Penicillium italicum		1.0029	0.1780	0.0000
					0.0000
	Penicillium digitatum		0.0000		
Interactions					
T.t			-0.0246	0.0042	0.0000
T.A			-0.0004	0.0001	0.0000
T.I			-0.3454	0.0522	0.0014
C. I			-0.0252	0.0037	0.0637
T.A.C			6.69E-07	3.04E-07	0.0279
T.C.I			0.0005	7.19E-05	0.0000

*100 % of wave amplitude corresponds to 120μm. (Adapted from Jiménez-Munguía, 2000).

Examples of the surface responses obtained from the logistic regression, showing the effect of the treatment temperature (T), ultrasonic wave amplitude (A) and treatment time (t) on the probability of growth for *Penicillum italicum* and *Penicillum digitatum* are presented in figures 1-4. The model obtained for sodium benzoate (figures 1, 2) demonstrates that *Penicillum italicum* is less resistant to the thermoultrasonic treatments than *Penicillum digitatum* since a higher number of conditions, when combining T and A, results in no-growth (p = 0) than for *Penicillum italicum*.

An important effect of temperature (T) on the probability of growth (p) was observed, rather than wave amplitude (A), showing that when both parameters increase resulted on a decrease of the probability of growth. Examples of the response surfaces obtained for vanillin are presented in figures 3-4, where the effect of treatment time (t) and wave amplitude (A) at 50°C, for both *Penicillum italicum* and *Penicillum digitatum* can be observed. Notice that increasing the treatment time from 2 to 14 min, the probability of growth (p) decreases from 0.8 to 0.1, for both *Penicillum* species.

Validation of these models was carried out in a model media and a grape juice, demonstrating that combination of the process variables predicted with a $p<0.05$, resulted in the inhibition of the studied molds, after 6 days of incubation at 25±1°C. With this example, we demonstrated that response surface models allow the analysis of the effect of different processing variables on microbial survival.

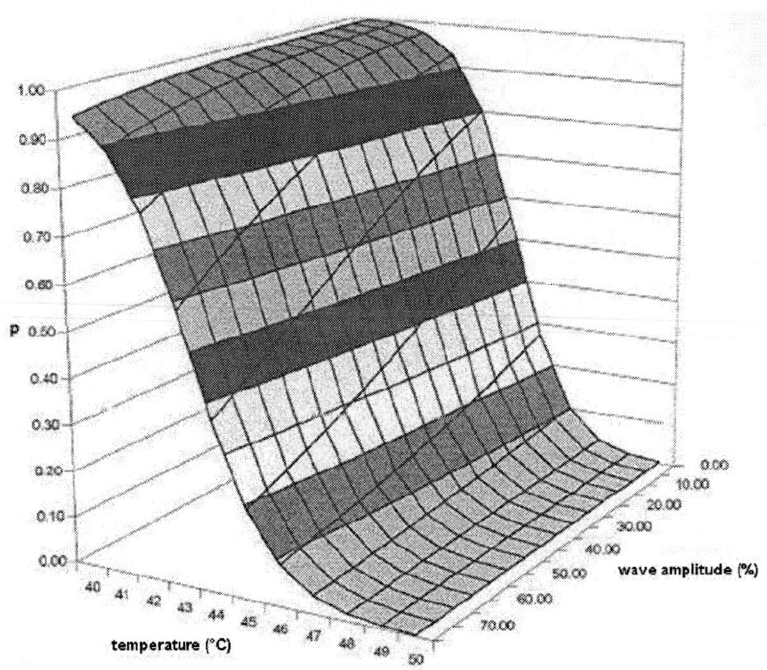

Figure 1. Probability of growth (p), after 4 days of incubation, of *Penicillum italicum*, subjected to selected high ultrasound treatments (100% of wave amplitude corresponds to 120 μm), in media formulated with a_w 0.99, pH 3.5 and 60 ppm of sodium benzoate. (From Jiménez-Munguía, 2000).

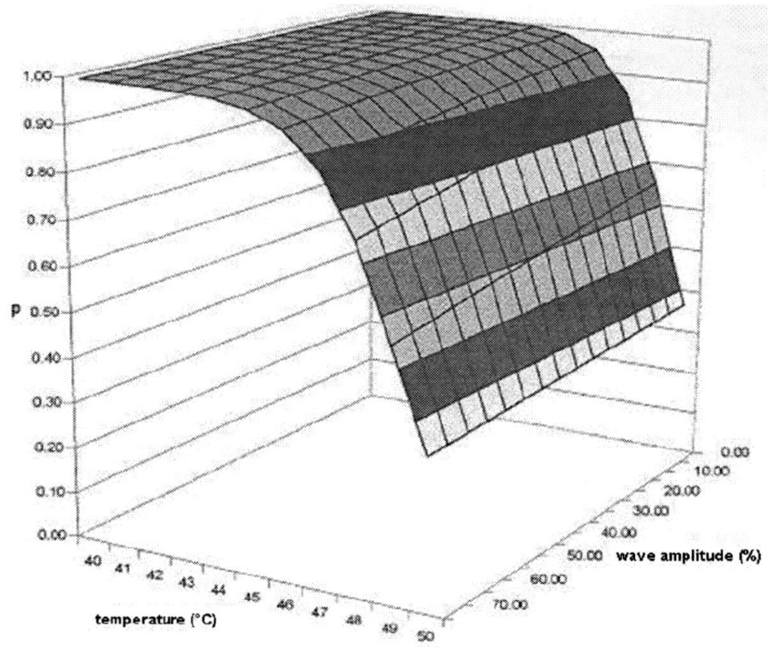

Figure 2. Probability of growth (p), after 4 days of incubation, of *Penicillum digitatum*, subjected to selected high ultrasound treatments (100% of wave amplitude corresponds to 120 μm), in media formulated with a_w 0.99, pH 3.5 and 60 ppm of sodium benzoate. (From Jiménez-Munguía, 2000).

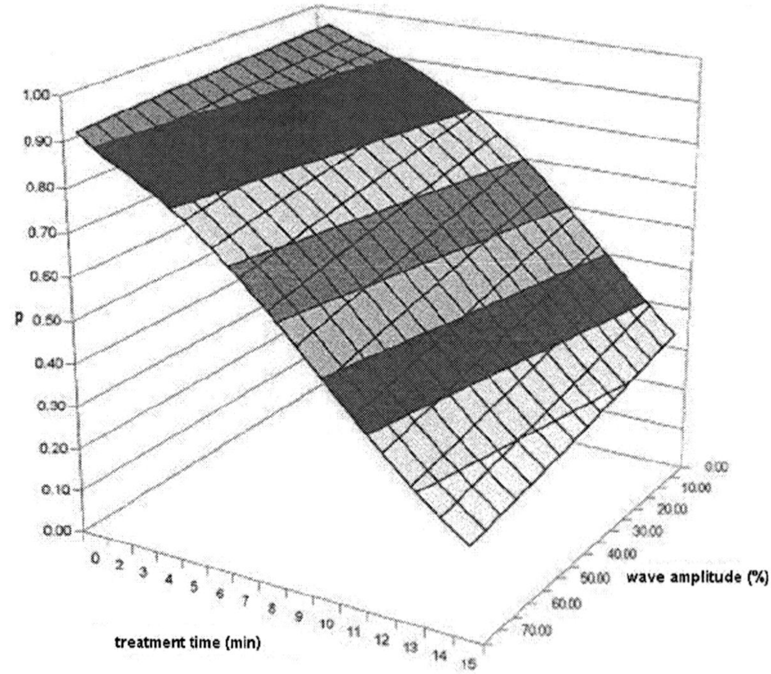

Figure 3. Probability of growth (p), after 4 days of incubation, of *Penicillum italicum*, subjected to selected high ultrasound treatments (100% of wave amplitude corresponds to 120 μm), in media formulated with a_w 0.99, pH 3.5 and 350 ppm of vanillin. (From Jiménez-Munguía, 2000).

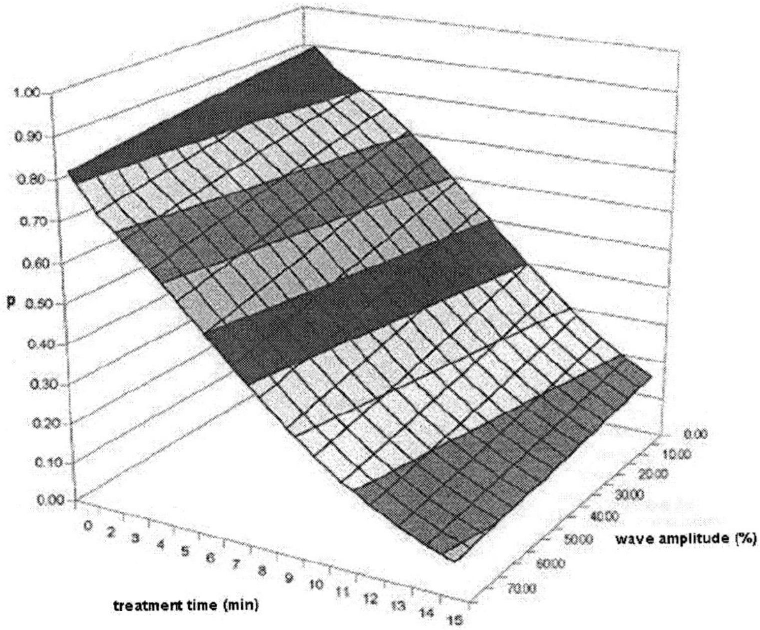

Figure 4. Probability of growth (p), after 4 days of incubation, of *Penicillum digitatum*, subjected to selected high ultrasound treatments (100% of wave amplitude corresponds to 120 μm), in media formulated with a_w 0.99, pH 3.5 and 350 ppm of vanillin. (From Jiménez-Munguía, 2000).

2.2. Juice Processing

As stated before, HI-ULTS treatment applied individually would require high intensities in order to reach significant microbial inactivation, but introducing the hurdle technology approach, many studies have demonstrated that HI-ULTS can be a promising technique for food processing.

Based on several studies related to yeast and mold inactivation by thermoultrasonication (Jiménez-Munguía *et al.*, 2001; Guerrero *et al.*, 2001; Arce-García *et al.*, 2002) other studies on orange juice (Barranco-Romero, 2002) and grapefruit juice (Alvarez-López, 2003) were carried out in order to evaluate microbial inactivation and possible physicochemical changes after HI-ULTS treatments. HI-ULTS treatments were applied with different wave amplitude (60, 72, 90 or 108 μm) at 25°C or 30°C during 15 min. In table 3 physicochemical characteristics of both juices, before and after HI-ULTS treatment applied with the highest levels of the evaluated variables (108 μm, 30°C, 15 min) are presented. No significant changes ($p>0.05$) were obtained after the HI-ULTS treatment on juice characteristics. Even though, a reduction of 55.7 ± 0.4 % on their viscosity was observed for every evaluated HI-ULTS treatment. This change is attributed to the cavitation phenomena, which promotes an important homogeneization or emulsification effect on the liquid (Mason *et al.*, 1996).

Moreover, inactivation curves were obtained for the native flora (standard plate and yeast and molds counts) every 2 min for the 15 min of the HI-ULTS treatments for both juices, at different wave amplitudes (60, 72, 90 or 108 μm), at 25°C or 30°C. From the inactivation kinetics, D values were calculated and are summarized in table 4.

Table 3. Physicochemical characterization of orange and grapefruit juice before and after high intensity ultrasound (HI-ULTS) treatments (108 μm, 30°C, 15 min)

Physicochemical property	Orange juice		Grapefruit juice	
	Before HI-ULTS	After HI-ULTS	Before HI-ULTS	After HI-ULTS
pH	3.63±0.02	3.65±0.02	3.30±0.02	3.50±0.02
a_w	0.99±0.01	0.99±0.01	0.99±0.01	0.99±0.01
Acidity (% citric acid)	0.99±0.01	1.00±0.02	0.86±0.02	0.81±0.03
Ascorbic acid* (mg/mL)	0.32±0.02	0.31±0.02	0.31±0.02	0.29±0.02

* Naturally present, not added. (Adapted from Barranco-Romero, 2002 and Alvarez-López, 2003).

Table 4. D values for different high intensity ultrasound (HI-ULTS) treatments applied on orange or grapefruit juice for standard plate count (STP) and yeast and molds

HI-ULTS treatments	Orange juice D (min)		Grapefruit juice D (min)	
Wave amplitude (μm)	STP	Yeast and molds	STP	Yeast and molds
20°C				
60	58.48	78.74	128.20	111.10
72	51.02	74.63	*	*
90	41.32	57.80	41.66	50.76
108	32.90	33.90	4.54	4.35
30°C				
60	15.24	23.69	*	*
72	14.31	21.14	*	*
90	8.06	17.06	*	*
108	7.87	15.34	*	*

* Not determined. (Adapted from Barranco-Romero, 2002 and Alvarez-López, 2003).

D values for yeast and mold were higher than those for mesophilic bacteria. According to previous studies (Earnshaw, 1998; Alliger, 1975) fungi are more resistant than vegetative bacterial cells. An expected effect was that higher wave amplitude will produce a higher microbial inactivation as reported by previous studies (Scherba *et al.*, 1991; Sala *et al.*, 1995; Pagán *et al.*, 1999; Guerrero *et al.*, 2001), since this parameter is related to ultrasound intensity, enhancing cavitation as wave amplitude is increased. Besides, the D values obtained with these HI-ULTS treatments demonstrate the synergistic effect of temperature and ultrasound applied simultaneously, in order to reduce microbial inactivation. From the results obtained with orange juice we can see that with an increase of 10°C for treatments at 90 μm of wave amplitude, a reduction of 30 min for D value is obtained; while an increase of 30 μm in the wave amplitude (60 to 90 μm) at 30°C results in a reduction of only 7 min for the D value. These results confirm the conclusion obtained in the predictive models, where increasing temperatures showed a higher inactivation effect than ultrasound intensity.

Based on the hurdle technology concept, cell disruption can be deteriorated easily if other factors are applied. Multifactorial microbial inactivation applying thermoultrasonic treatments has been suggested by López-Malo *et al.* (2005) whom obtained favorable results for deteriorative molds in different model media formulated with or without antimicrobial agents.

Therefore, addition of antimicrobial agents in grapefruit juice was evaluated in combination with HI-ULTS treatments to enhance juice preservation. Since the purpose of combining different processing factors is to reduce the level used of each factor, the HI-ULTS treatment chosen was at 90 μm of wave amplitude and 20°C. The antimicrobial agents used were potassium sorbate (90, 120 or 150 ppm) or vanillin (800, 1000 or 1200 ppm). From inactivation kinetics, the D values obtained for these treatments are presented in table 5.

Results demonstrated that antimicrobial inactivation was enhanced when antimicrobial agents were added. D values were reduced ten times in average, for both antimicrobial agents (potassium sorbate or vanillin) and microbial groups (standard plate or yeast and mold). From these studies we can suggest that HI-ULTS treatments can be a safe technology for minimally processed juices.

Table 5. D values for grapefruit juice under high intensity ultrasound (HI-ULTS) treatments at 90 μm of wave amplitude and 20°C formulated with different antimicrobial agents

Antimicrobial Agent	Concentration (ppm)	D (min)	
		SPC*	Yeast and molds
Control	0	41.66	50.76
Potassium sorbate	90	6.20	8.56
	120	4.34	6.35
	150	2.20	4.25
Vanillin	800	6.35	8.70
	1000	4.34	6.08
	1200	2.10	2.14

*Standard count plate. (Adapted from Barranco-Romero, 2002 and Alvarez-López, 2003).

2.3. Dairy Products

Colloidal milk proteins and fat globules can absorb or reflect ultrasound energy, therefore energy attenuation has been attributed to the structure of the casein present in milk (Paci, 1953; Hueter *et al.*, 1953). Thermoultrasonication in dairy products promotes homogenization of milk, improves protein coagulation and fat globule size reduction (Sala *et al.*, 1995; Villamiel and Jong, 2000, 2002; Dolatowski *et al.*, 2007). The emulsions generated with ultrasound are often more stable than those produced conventionally and often require little or no surfactant addition (Mason *et al.*, 1996).

Milk

HI-ULTS promoted homogenization and creates stabilized oil in water emulsion, with fat globule sizes between 0.1 to 10 μm. The cavitation phenomena produced by HI-ULTS, leads to formation of bubbles, which may collapse near the interface of the immiscible liquids causing a mixing effect of both phases (Povey and Mason, 1998).

An important effect of the homogenization of milk when applying thermosonication (T-ULTS) is a larger fat globule size reduction in comparison to the application of heat treatments. Villamiel and Jong (2000) observed a substantial reduction (up to 60%) of fat globules in the median diameter (D[v,0.5]) when applying a continuous flow (11-28 mL/min) of milk subjected to T-ULTS treatments (55°C, 40.2 s of residence time, at a wave amplitude of 120 μm) with respect to heat treatments (HT) under same temperature and flow processing conditions. Similar reduction of fat globule size were obtained by Hernández-Juárez (2003) when analyzing fat globule size in milk after batch heat treatments (63°C, 30 min) and after batch T-ULTS treatments (63°C, 70 μm, 30 min) (see figures 5 and 6). The D[v,0.5] obtained were 13.78 ± 1.8 μm and 7.65 ±0.8 μm, after HT and T-ULTS treatments respectively; which represents a 44.5 % size reduction. In both studies, the synergistic effect of ULTS and temperature on fat globule size reduction was confirmed. High temperature modifies milk viscosity, enhancing penetration of ultrasound waves through the liquid (Earnshaw *et al.*, 1995).

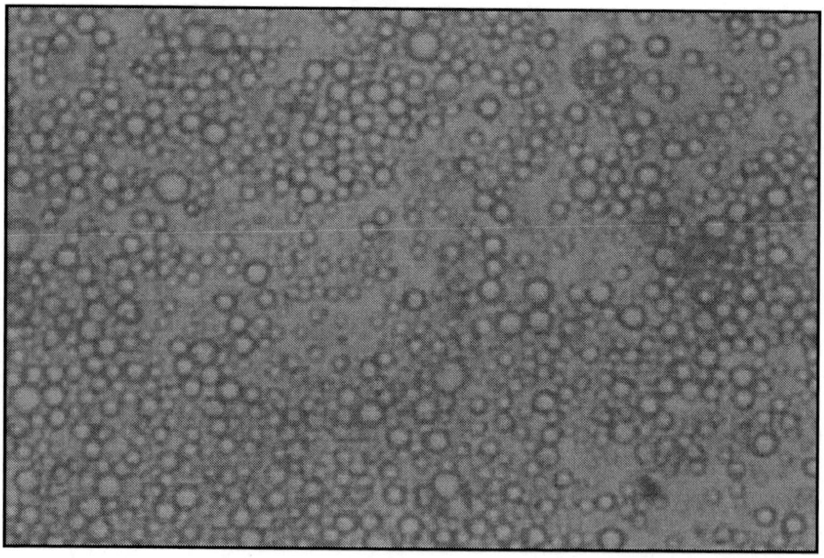

Figure 5. Optical microscope image (63x) of milk after heat treatment (63°C, 30 min), for yoghurt production. (From Hernández-Juárez, 2003).

Figure 6. Optical microscope image (63x) of milk after H-UTLS treatment (63°C, 70 μm, 30 min) for yoghurt production. (From Hernández-Juárez, 2003).

Besides the benefits of milk homogenization by HI-ULTS, the use of thermoultrasonication on milk also reduces microbial populations (Ordoñez *et al.*, 1987; Villamiel and Jong, 2000; Knorr *et al.*, 2004). Results obtained by Hernández-Juárez (2003) demonstrated a reduction up to 62.5% in D values for batch T-ULTS (63°C, 108 μm of wave amplitude, 30 min) compared to batch HT (63°C, 30 min) for native microflora on milk. Table 6 presents lower D values for T-ULTS when increasing wave amplitude. A similar reduction (up to 70%) of D values was obtained by Zenker *et al.* (2003), for *E. coli* with batch H-ULTS treatments at 60°C, 110 μm, in milk when compared to HT.

Table 6. Comparison of D values for native microflora in milk, with heat treatments (HT) and thermoultrasonic treatments (T-ULTS) at different wave amplitudes (μm) at 63°C during 30 min

Treatment	D (min)
HT	29.33
T-ULTS, 60 μm	20.79
T-ULTS, 84 μm	17.36
T-ULTS, 108 μm	10.98

Adapted from Hernández-Juárez, 2003.

Yoghurt

Nowadays there is a wide variety of fermented products from milk which include different types of yoghurt. Physical properties of these dairy products depend on processing conditions, being homogenization of milk one of the most important. Homogenization creates a stabilized emulsion of oil in water, with fat globule size between 0.1 to 10 μm and increases casein micelle adsorption, promotes protein coagulation and diminishing syneresis (Tamime and Robinson, 1991).

Milk homogenization treated by T-ULTS enhanced fat globule disruption, creating an emulsion with low fat globules sizes (1-20 μm diameter), depending on the ULTS intensity (Villamiel and Jong, 2000; Hernández-Juárez, 2003). As consequence of this fine milk emulsion, some physical properties may be modified. Jiménez *et al.*, (2004) generated predictive models for different physical characteristics of yoghurt like texture force (F), luminosity (L) and syneresis (S), when applying T-ULTS treatments on milk; evaluating processing variables such as temperature treatment (T) from 63 to 83°C, wave amplitude (A) from 60 to 109 μm and milk solid content (SC) from 2 to 5%. From this study, it was concluded that increasing SC in milk, yoghurt texture was enhanced while diminishing syneresis (figure 7). Diminish of syneresis in yoghurt is attributed to the water retention effect when increasing solid content in milk and homogenization effect on casein protein coagulation (Tamime and Robinson, 1991). Better yoghurt qualities were obtained when T-ULTS was applied at 60-70 μm and 63°C respectively, stabilizing the yoghurt gel. Luminosity was also directly related with the homogenization effect of ultrasound, resulting in a brighter yoghurt gel (higher L) when increasing wave amplitude. This result is due to the fat globule size reduction during homogenization, which increases light reflection, generating brighter yoghurt.

Furthermore, the effect of T-ULTS treatments on yoghurt processing was evaluated (Hernández-Juárez, 2003), measuring pH decrease (down to 4.6) on milk, during fermentation. T-ULTS treatments applied on milk at 63°C, 60 or 84 μm of wave amplitude, diminished fermentation time in 1 hour, which demonstrates that the application of ultrasound improves yoghurt processing, as stated by Wu *et al.* (2001).

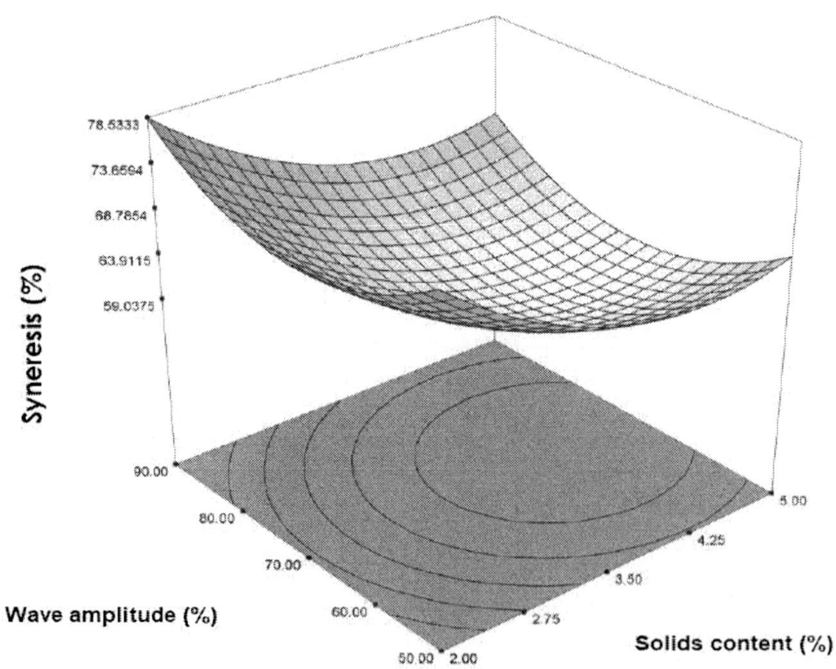

Figure 7. Syneresis in yoghurt when milk was formulated with different solids content and homogenized with thermoultrasonic treatments (T-ULTS) at 63°C during 30 min. 100% wave amplitude corresponds to 120 μm. (From Jiménez *et al.*, 2004).

3. Final Remarks

Studies carried out with selected microorganisms demonstrate that predictive models are useful to predict microbial inactivation, a wide number of different conditions can be evaluated and results can be extended to different food products.

Application of HI-ULTS treatments for juice processing needs further studies to evaluate microbial adaptation or damage regeneration during storage.

Regarding the application of thermoultrasonication treatments on milk to enhance microbial reduction and improve homogenization, results show that is a promising technique for dairy processing.

Further studies are needed, since scarce information on food systems has been evaluated.

References

Ahmed F.K. and Russell C. 1975. Synergism between ultrasonic waves and hydrogen peroxide in the killing of microorganisms. *Journal Applied Bacteriology,* 39: 31-40.

Alliger H. 1975. Ultrasonic disruption. *American Laboratory,* 10: 75-85.

Alvarez-López J.A. 2003. Grapefruit juice preservation by low frequency ultrasound and antimicrobial agents. Undergraduate Thesis. Universidad de las Américas, Puebla. Mexico. (In Spanish).

Arce-García R., Jiménez T., Palou E. and López-Malo A. 2002. Ultrasound treatments and antimicrobial effects on *Zygossacharomyces rouxii.* Institute of Food Technologists Annual Meeting. Anaheim, U.S.A.

Barranco-Romero A.F. 2002. Low frequency ultrasound as a method for orange juice preservation. Undergraduate Thesis. Universidad de las Américas, Puebla. Mexico. (In Spanish).

Burgos J., Ordoñez J.A. and Sala F.J. 1972. Effect of ultrasonics waves on the heat resistance of *Bacillus cereus* and *Bacillus licheniformis* spores. *Applied Microbiology,* 24: 497-498.

Butz P., Tauscher B. 2002. Emerging technologies: chemical aspects. *Food Research International,* 35(2-3): 279-284.

Ciccolini L.P., Taillandier A.M., Wilhem H., Delmas H. and Strehaiano P. 1997. Low frequency thermoultrasonication of *Saccharomyces cerevisiae* suspensions: effect of temperature and of ultrasonic power. *Journal of Chemical Engineering,* 65: 145-149.

Dolatowski Z.J., Standnik J. and Stasiak D. 2007. Applications of ultrasound in food technology. *Acta Scientiarum Polonorium Technologia Alimentaria,* 6(3): 89-99.

Earnshaw R.G. 1998. Ultrasound a new opportunity for food preservation. In *Ultrasound in Food Processing* (M.J. Povey and T.J. Mason, Eds.), Blackie Academic & Professional, London. pp. 183-192.

Earnshaw R.G., Appleyard J. and Hurst R.M. 1995. Understanding physical inactivation process: combined preservation opportunities using heat, ultrasound and pressure. *International Journal of Food Microbiology,* 28: 197-219.

García M.M., Burgos J., Sanz B. and Ordoñez J.A. 1989. Effect of heat and ultrasonic waves on the survival of two strains of *Bacillus subtilis*. *Journal of Applied Bacteriology,* 67: 619-628.

Guerrero S., López-Malo A. and Alzamora S.M. 2001. Effect of ultrasound on the survival of Saccharomyces cerevisiae: influence of temperature, pH and amplitude. *Innovative Food Science and Emerging Technologies,* 2: 31-39.

Hernández-Juárez, J.Y. 2003. Yoghurt manufacture using milk treated by ultrasound. Undergraduate Thesis. Universidad de las Américas, Puebla. Mexico. (In Spanish).

Hosmer D.W. and Lemeshow S. 1989. Applied Logistic Regression. John Wiley and Sons, New York, U.S.A.

Hueter T.F., Morgan H. and Cohen M.S. 1953. Ultrasonic attenuation in biological suspensions. *Journal Acoustics Society of America,* 25: 1200-1201.

Idrissi F.Z., Agut M., Larrondo J. and Calvo M.A. 1996. Effect of ultrasound on fungal cells. *Cytobios,* 88: 119-122.

Jiménez-Munguía T. 2000. Combined methods for the inactivation/inhibition of mold spores using ultrasound of low frequency. MSc Thesis. Universidad de las Américas, Puebla, Mexico. (In Spanish).

Jiménez-Munguía T., Arce-García R., Argaiz A., Palou E. and López-Malo A. 2001. Mold spore inactivation during cavitation due to ultrasound treatments. Institute of Food Technologists Annual Meeting, New Orleans, LO, U.S.A.

Jiménez T., López-Malo A., Argaiz A. and Hernández J. 2004. Optimizing quality characteristics of yoghurt with ultrasound. Proceedings of the 9th International Congress on Engineering and Food (ICEF9), March 7-11, Montpellier, France.

Knorr D., Zenker M., Heinz V. and Lee D.U. 2004. Applications and potential of ultrasonics in food processing. *Trends in Food Science and Technology,* 15: 261-266.

Lillard H.S. 1993. Bactericidal effect of chlorine on attached *Salmonellae* with or without sonification. *Journal of Food Protection,* 56(8): 716-717.

López-Malo A., Guerrero S. and Alzamora S.M. 1999. *Saccharomyces cerevisiae* thermal inactivation kinetics combined with ultrasound. *Journal of Food Protection,* 62(10): 1215-1217.

López-Malo A. 2000. Multitarget preservation of food: effect of traditional and emerging factors in the growth response of *Aspergillus flavus*. Ph.D Thesis. Universidad de Buenos Aires, Argentina. (In Spanish).

López-Malo A., Palou E., Jiménez-Fernández M., Alzamora S.M. and Guerrero S. 2005. Multifactorial fungal inactivation combining thermosonication and antimicrobials. *Journal of Food Engineering,* 67: 87-93.

Mañas P. and Pagán R. 2005. Microbial inactivation by new technologies of food preservation. *Journal Applied of Microbiology,* 98: 1387-1399.

Mañas P., Pagán R. and Raso J. 2000a. Predicting lethal effect of ultrasonic waves under pressure treatments on *Listeria monocytogenes* ATCC 15313 by power measurements. *Journal of Food Science,* 65(4): 663-667.

Mañas P., Pagán R., Raso J., Sala F.J. and Condón S. 2000b. Inactivation of *Salmonella enteritidis, S. typhimurium, S. senftenberg* by ultrasonic waves under pressure. *Journal of Food Protection,* 63(4): 451-456.

Mason T.J. 1990. Chemistry with ultrasound. *Critical Reports on Applied Chemistry,* 28: 1-25.

Mason T.J. 2003. Sonochemistry and sonoprocessing: the link, the trends and probably the future. *Ultrasonics Sonochemistry,* 10: 175-179.

Mason T. J., Paniwnyk L. and Lorimer J.P. 1996. The uses of ultrasound in food technology. *Ultrasonics Sonochemistry,* 3: S253-S260.

McClements J. 1995. Advances in the application of ultrasound in food analysis and processing. *Trends in Food Science and Technology,* 6: 293-297.

Ordoñez J.A., Sanz B., Hernández P.E. and Lopez-Lorenzo P. 1984. A note on the effect of combined ultrasonic and heat treatments on the survival of thermoduric streptococci. *Journal of Applied Bacteriology,* 56: 175-177.

Ordoñez J. A., Aguilera M.A., Garcia M.L. and Sanz B. 1987. Effect of combined ultrasonic and heat treatment (thermoultrasonication) on the survival of a strain of *Staphylococcus aureus*. *Journal of Dairy Research,* 54: 61-67.

Paci C. 1953. Use of ultrasound to sterilize milk. *Lait,* 33:610-615.

Pagán R., Mañas P., Raso J. and Condón S. 1999. Bacterial resistance to ultrasonic waves under pressure at nonlethal (manosonication) and lethal (manothermosonication) temperatures. *Applied and Environmental Microbiology,* 65:297-300.

Phull S.S., Newman A.P., Lorimer J.P., Pollet B. and Mason T.J. 1997. The development and evaluation of ultrasound in the biocidal treatment of water. *Ultrasonic Sonochemistry*, 4: 157-164.

Pisayena P., Mohareb E. and McKellar R.C. 2003. Inactivation of microbes using ultrasound: A review. *International Journal of Food Microbiology,* 87: 207-216.

Povey M. and Mason T.J. 1998. Ultrasound in Food Processing. Ed. Blackie Academic & Professional, London, U.K.

Raso J., Palop A., Pagán R. and Condón S. 1998. Influence of temperature and pressure on the lethality of ultrasound. *Applied and Environmental Microbiology,* 64: 465-471.

Sala F.J., Burgos J., Condón S., López P. and Raso J. 1995. Effect of heat and ultrasound on microorganism and enzymes. In *New Methods of Food Preservation,* G.W. Gould (Ed.), Blackie Academic & Professional, London, U.K. pp 176-204.

Scherba G., Weigel R.M. and O'Brien J.R. 1991. Quantitative assesment of the germicidal efficiency of ultrasonic energy. *Applied and Environmental Microbiology,* 57:2079-2084.

Suslick K.S. 1988. Homogeneous sonochemistry. In *Ultrasound. It's Chemical, Physical and Biological Effects,* K.S. Suslick (Ed.), VCH, New York, N.Y., U.S.A. pp 123-146.

Tamime A. Y. and Robinson R. F. 1991. Yoghurt, Science and Technology. Ed. Acribia. Spain. (In Spanish).

Villamiel M. and Jong de P. 2000. Influence of high-intensity ultrasound and heat treatments in continuous flow on fat, proteins, and native enzymes of milk. *Journal of Agriculture and Food Chemistry,* 48:472-278.

Villamiel M. and Jong de P. 2002. Inactivation of *Pseudomonas fluorescens* and *Strepococcus thermopilus* in trypticase soy broth and total bacteria in milk by continuous-flow ultrasonic treatment and conventional heating. *Journal of Food Engineering,* 45(3):171-179.

Wrigley D.M. and Llorca N.G. 1992. Decrease of *Salmonella typhimurium* in skim milk and egg by heat and ultrasonic wave treatment. *Journal of Food Protection,* 55: 678-680.

Wu H., Hulbert G.J. and Mount J.R. 2001. Effects of ultrasound on milk homogenization and fermentation with yogurt starter. *Innovative Food Science and Emerging Technologies,* 1: 211-218.

Zenker M., Heinz V. and Knorr D. 2003. Application of ultrasound-assisted thermal processing for preservation and quality retention of liquid foods. *Journal of Food Protection,* 66: 1642-1649.

Zheng L. and Sun D.W. 2006. Innovative applications of power ultrasound during food freezing processes- A review. *Trends in Food Science and Technology*, 17: 16-23.

In: Food Processing and Engineering Topics
Editors: M. E. Sosa-Morales and J. F. Vélez-Ruiz
ISBN: 978-1-60741-788-0

Chapter 11

PROCESSING AND ADDITIVES FOR FROZEN DOUGH AND THEIR INFLUENCE ON THE QUALITY OF FINAL BAKED PRODUCTS

María E. Sosa-Morales[1], Hugo González-Loo[2] and Jorge F. Vélez-Ruiz[1]
[1] Chemical and Food Engineering Department
Universidad de las Americas, Puebla, Mexico
[2] Pabisan, S.A. Puebla, Mexico

1. INTRODUCTION

Bakery is one of the oldest food industries. In Mexico, the sales of bakery products in 2000 were estimated over US$1,700 millions, and for 2003 they reached up US$2,300 millions (INEGI, 2008). As it can see, bakery industry keeps a big growing up trend, not only in Mexico, also around the world. Research and development have been conducted to increase the diversity of breads types, formulations and presentations, even preservation technologies for the bakery product . Because bread is a perishable product (mainly conceived as softness loss), some resources have been employed to avoid its deteriorate, such as use of additives, temperature reduction for dough or pre-cooked products and packaging in modified atmospheres.

Breads taste is much better when they are just baked, due to aroma and texture are optimal. Frozen dough may be an alternative to reduce losses and costs; only the required dough pieces are placed into the oven and thus, the consumer can get just baked products at any time. To get this advantage, the design of process for frozen dough and potential use of adequate additives to improve the final bread characteristics must be investigated.

Pabisan, S.A. is a Mexican local industry devoted to the manufacturing of frozen bread, which is commercialized in the rest of the country. Several studies have been conducted to get products of good quality such as biscuits, sponge cake bread and Danish bread. In this Chapter, conditions for processing, storage and tested additives blends for this kind of sweet breads are presented. The effect on density, texture, color and sensory perception of the final product are discussed. Some engineering aspects, such as dough freezing point and description of freezing process are included as well.

2. FROZEN DOUGH

2.1. Freezing of Dough

Freezing is the process through the water in a substance (or food) changes of physical state,from liquid to ice, as energy is removed by cooling below freezing temperature (Rahman and Vélez-Ruiz, 2007). The freezing point is the temperature at which the first ice crystals appear at a point where liquid (water) and solid phase (ice) coexist at equilibrium. The temperature of the food undergoing freezing process remains relatively constant (plateau) until most of the water is frozen; then after, the temperature decreases to reach the storage temperature or the temperature of the freezing medium (Desrosier, 1970). The freezing curve of biscuits is showing in figure 1, where the three typical zones of a freezing process can be distinguished: I, decreasing of temperature till freezing point; II, freezing, and III, decreasing of temperature till temperature of the freezing medium.

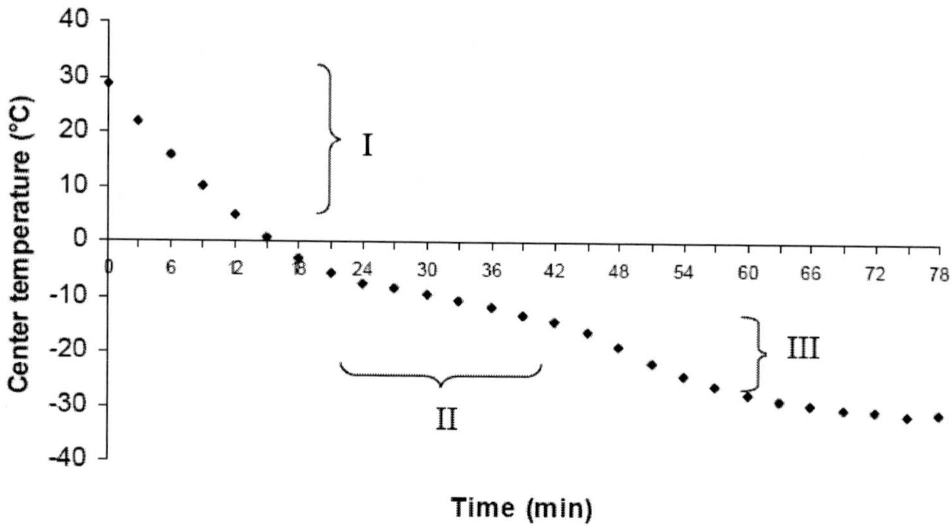

Figure 1. Freezing curve for biscuits in a tunnel freezer at –32°C (from Sosa-Morales *et al.*, 2004).

Freezing point and freezing time are important to establish the conditions and duration of the process during the production of the food. There are different methods to measure the freezing point, such as cooling curve, by using of a cryoscope and by using of a differential scanning calorimeter. Cooling curve is one of the most simple, accurate, and widely used methods (Rahman *et al.*, 2009). Another choice to determine the freezing point is its calculation. By using relations from freezing-point depression that involves Raoult's Law, and considering the food as a saturated aqueous solution, the freezing temperature can be calculated from (Heldman and Singh, 1981):

$$\frac{\lambda}{R}\left(\frac{1}{T_w}-\frac{1}{T_f}\right)=\ln(x_w) \tag{1}$$

where λ (kJ/kg·mol) is the latent heat of fusion for water, R (8.314 kJ/kmol ·K) is the universal gas constant, T_w (K) is the freezing point of pure water, T_f (K) is the freezing point of the food, and x_w is the mole fraction of water in the food.

Knowing the freezing point, the freezing time (t_f) can be predicted with the Plank´s equation (Rahman, 1995):

$$t_f=\frac{\rho\lambda}{T_f-T_m}\left[\frac{Pa}{h}+\frac{Ra^2}{k}\right] \tag{2}$$

where ρ (kg/m^3) is the density of the food sample, T_m (K) is the temperature of the freezing medium, P and R (dimensionless) are Plank´s constants, with different values depending of the geometry of the sample (slab, cylinder or sphere), a (m) is the dimension (1/2 of the thickness for slab, or the radium for cylinder and sphere geometries), h (W/m^2·K) is the convective heat transfer coefficient and k (W/m·K) is the thermal conductivity of the sample. Table 1 shows both freezing point and time measured or calculated for the bread types to review in this chapter.

Table 1. Freezing point and time for breads determined during the freezing in a forced air tunnel at –20°C

Bread/dough type	Freezing point (°C)	Method	Freezing time (min)[a]	Reference
Precooked biscuits	-7.25	Raoult´s Law	40.0	Sosa-Morales *et al.*, 2004
Sponge cake dough	-2.90	Cooling curve	48.1	Morales de la Peña, 2005
Dough for Danish bread	-1.50	Cooling curve	38.3	Encinas-Banda *et al.*, 2005

[a] Calculated with Plank´s equation.

2.2. Changes due to Freezing and Frozen Storage in Dough

Physical properties of the food, such as density and thermal properties, are affected as a consequence of the freezing process. Also, chemical, nutritional, physicochemical, textural and sensorial changes are induced (Rahman *et al.*, 2009).

Freezing process affects the baking performance of frozen dough. Havet *et al.* (2000) cited yeast and gluten network as the two most important factors affected by freezing. Yeast activity was strongly related to the freezing rate, thus, slow freezing rate must be preferred to preserve its activity. Modification in gluten network affects the ability to retain CO_2, minimizing the bread volume.

Major textural changes are related to gelatinization and retrogradation of amylopectin in the dough during the frozen storage (Ribotta *et al.*, 2003). Bárcenas *et al.* (2003) reported that when bread is partially baked, retrogradation of amylopectin was not observed through a DSC analysis, however some structural changes in amylopectin occurred during the frozen storage, crumb hardness in re-baked bread increased with the time of frozen storage. Other changes are related to the rheological properties variation during frozen storage of dough resulting in physical damage of the gluten network, caused by ice crystallization and recrystallization during freeze-thaw cycles (Inoue and Bushuk, 1991; Battacharya *et al.*, 2003; Sharadanant and Khan, 2003). Large variations in the temperature during the frozen storage have been reported as a cause of loss quality in both dough and bread. Le Bail *et al.* (1999) reported that with stable temperature, the shelf life of frozen French bread was 37 days at –22°C, while with large temperature variations, 9 days of storage resulted in poor quality products. Phimolsiripol *et al.* (2008) observed that variations of ±0.1°C and even ±1°C did not have effect on the CO_2 production rate by yeasts in frozen dough stored at –18°C up to 112 days. But, large variations (±5°C) resulted in significant lower quality.

In summary, many changes occur in dough as a result of both freezing process and frozen storage conditions. The goal to get good quality breads has encouraged to search for agents which could improve the characteristics of the final products. Major studies are focused to avoid the lost of yeast activity and the lost of softness and volume. Respect to yeasts, more freeze-tolerant strains are required. Gélinas *et al.* (1994) did not find out differences in the freeze-thaw tolerance of cream yeasts and compressed yeasts. Variations in the resistance of the strain were more dependent of the batch than the way in that yeasts were commercialized. Trehalose has been related with the tolerance of the yeast to frozen storage, due to this dissacharide is considered an energetic reserve and a stress metabolite. Meric *et al.* (1995) observed that 4-5% of trehalose content in yeast was enough to prevent chilling injury. Also, it has been reported that lipid composition of the yeast affect its cryoresistance. Murakami *et al.* (1996) reported that freeze-tolerant strains had lower sterol-phospholipid ratio than those freeze-sensitive strains.

2.3. Additives to Avoid Changes During Frozen Storage

There are different kinds of additives recognized as improvers for dough and consequently, for bread performance. The most employed are gums, emulsifiers, enzymes, oxidant agents, reducer agents and preservatives.

Gums

They are polisaccharydes with the ability to produce gels, solutions with high viscosity and more stable systems. Combination of gums can result in better results, due to a synergic effect (Badui-Dergal, 1999). Some hydrophylic gums are suitable for bakery products, such as arabic gum (to increase the consistency) and guar gum (to stabilize and improve the dough consistency).

Emulsifiers

They emulsify the shortening or butter in the dough, getting better uniformity (Tramstorf, 1991; Matz 1992). Usually, authorized emulsifiers in bread production are separated in two categories: dough conditioners and crumb softeners. Dough conditioners or fortifiers are employed to have stronger dough, making it more resistant to the stress when are subjected to mixing, sheeting, forming or cutting, and all the steps subjected to mechanical forces during the process. Better gas retention is a consequence and the final products have larger volume, with more uniform crumb. DATEM (Diacetil tartaric acid esters of monoglyceride) is one of the most recognized dough fortifier. Crumb softeners are mainly used in dough which will be deposited in molds and will produce breads with large self-life. Between them, lactilates offer good results (Tejero, 1998). Joensson and Toernaes (2004) reported that emulsifiers are the more efficient softeners using 0.3-0.6% (taking 100% as the weight of the flour)

Enzymes

Fungal and malt alpha-amylases have been employed for years in bakery products. They can degrade the starch in small dextrines, developing better color in the crust of breads (due to more available sugars for the Maillard reaction) and increasing the shelf-life of the products (volume increase and softer texture) (Badui-Dergal, 1999). Qi Si (1997) cited a wide variety of enzymes with benefic effects in breads: amylases from malt avoid the aglomeration in the dough, xylanases work on the pentosans and improve the handle of the dough, lipases as dough fortifiers, proteases become softer dough from hard wheat. Enzymes from soybean flour, such as lipoxygenases, which have a fortifier effect in the dough and a blanching effect in the final product, with good color and increasing the shelf-life. Also, synergistic effects of enzymes have been observed, for example, the combination of alpha-amylase and xylanase achieve optimal dough consistency, stability and bread quality. However, an excess of enzymes will result in a very sticky dough, difficult to handle for bakers or baking equipment (Novozymes, 2008).

Oxidant Agents

Probably, oxidant agents are the most effective dough improvers (AIB, 2001). Potassium bromate action has been related to oxidation of the disulfide groups of gluten protein in dough. This change improves gas-cell structure, retention of carbon dioxide and consequently, loaf volume. 5-10 ppm of bromate results in optimal loaf volume (Panozzo *et al.*, 1994). Ascorbic acid (AA) produces a similar effect than bromate, in doses between 20-150 ppm, while the azo-dicarbonamide (ADA) is employed from 10 to 20 ppm (Tramstorf, 1991). Tejero (1998) reported that dough oxygenation increases the effect of the oxidant agents, and better results have been observed when traditional machines are used for breadmaking respect to high power mixer machines.

Reducer Agents

A reducer agent for breadmaking can be defined as that compound able to reduce the time of mixing, getting a handling dough. L-cisteine is a popular additive that reduces the mixing time between 15 and 20% of the total time. Sorbic acid, employed mainly as preservative, is effective as reducer agent used in low levels (15-30 ppm), reducing till 20% the total time for mixing step (AIB, 2001).

Preservatives

Molds are able to grow up in dough and breads. Sodium and calcium propionates are the most employed antimicrobial agents in the bakery products. The FDA allows their use in no more than 0.32% (Matz, 1992). Some studies about the use of additives for improving the quality of frozen dough and/or bread are summarized in table 2.

Table 2. Examples of additives tested to improve the quality of frozen dough and final baked bread

Additive	Storage conditions	Effect on bread characteristics	Reference
Sodium caseinate	------	Decreasing of proof time, increasing of loaf volume and improving of texture. At 4% effects were similar to adding AA or DATEM	Kenny *et al.*, 2001
WPC and		It interferes with the gluten network, negative effects	
WPC submitted to heat treatment		Similar volume than control	
Carboxy methyl cellulose	16 weeks	Better than control	Sharadanant and Khan, 2003
Arabic gum		Better than control	
Locust bean gum		The best, with brighter color and softer loaves	
κ-carrageenan		Darker color, detrimental effect	
DATEM and guar gum	-18°C	Improving volume and texture	Ribotta *et al.*, 2004
Guar gum, xanthan gum and their combination	28 days	Reduction of fusion enthalpy respect to control	Guinoza Matuda *et al.*, 2008

AA: Ascorbic acid, DATEM: Diacetil tartaric acid esters of monoglycerides, WPC: Whey protein concentrate.

From our side, part of this chapter is devoted to present those results when commercial blends of additives were employed to produce frozen dough or partially cooked bread and the quality evaluation of the baked products. This is a collaboration between Pabisan, S.A., a small industry located at Puebla, Mexico, and our University (Universidad de las Américas, Puebla) located in Cholula, Mexico.

3. Formulation and Elaboration Process

Three kinds of products have been studied: biscuits –also called "English muffins", leavened with salts (baking powders), and two yeast-leavened breads: sponge cake breads (known in Mexico with the Spanish word "conchas") and Danish type breads (with forms of "horn" and "braid"). Figure 2 shows the aspect of these kinds of bakery products. There are variations in the name of the breads from one country to other..

Common formulations by Pabisan were employed (table 3). In the case of biscuits, the re-formulation consisted in the addition of improvers of crumb and improvers of volume (soybean flour, DATEM, AA, ADA and calcium carbonate, Magnopro© from Safmex, Lesaffre Group).

Figure 2. Aspect of different sweet breads: (a) biscuits, (b) sponge cake bread ("conchas"), and Danish type bread with form of (c) horn and (d) braid.

Table 3. Original formulations of biscuits, sponge cake and Danish type breads

Ingredients	Biscuits	Sponge cake	Danish type
Wheat flour	100	100	100
Water	55.1	30.9	29.6
Sugar	8	27.2	22.7
Salt	1.7	1.0	1.3
Dried Yeast	0	1.0	1.5
Baker powder	3.7	–	–
Vegetal Shortening	25	4.5	–
Butter (to paste)	–	–	18.20
Milk powder	5.5	3.4	3.2
Liquid whole egg	–	30.9	29.6
Flavoring (vanilla)	–	0.4	0.4

In the case of sponge cake breads, the control formulation involves a blend of emulsifiers (soybean flour and DATEM), fungal alpha-amylase and AA (Magnopro© from Safmex). Three new commercial blends were tested, to replace the control mixture:

1. A blend of alpha-amylase, DATEM and pectin (Improver Power Freeze© from Danisco).
2. A blend of DATEM (Panodan© from Danisco) and malt alpha-amylase (Novamil© from Novo Nordish).
3. A blend of DATEM, alpha-amylase, AA, ADA and gums (Congelix© from Safmex).

For Danish type breads, four different re-formulations were tested:

Formulation A: increasing the yeast level to 4% and addition of a mixture of soybean flour, DATEM, calcium carbonate, AA, ADA and fungal alpha-amylase (Magnopro© from Safmex).
Formulation B: addition of a mixture of gluten, gums, emulsifiers, oxidant agents, enzymes and reducer agents (Congelix© from Safmex).
Formulation C: addition of an emulsifier (only DATEM, Panodan© from Danisco).
Formulation D: adding emulsifiers, proteins, hydrocolloids and an agent for cryo-protection (Texturizante GC© and Eurocongelado© from Estrali).

Formulations B, C and D also included a modified starch maintaining 1.5% of yeast level. Elaboration process of every product is presented in figures 3 to 5.

For the final baking of precooked biscuits, two baking processes were carried out in conventional oven and in microwave oven. Sponge cake method was used for sponge cake bread; it consists in the elaboration of the sponge (the fermented dough got from a part of flour, and a part of water with all the yeast) and then, mixing with the rest of the ingredients (Pyler, 1988).

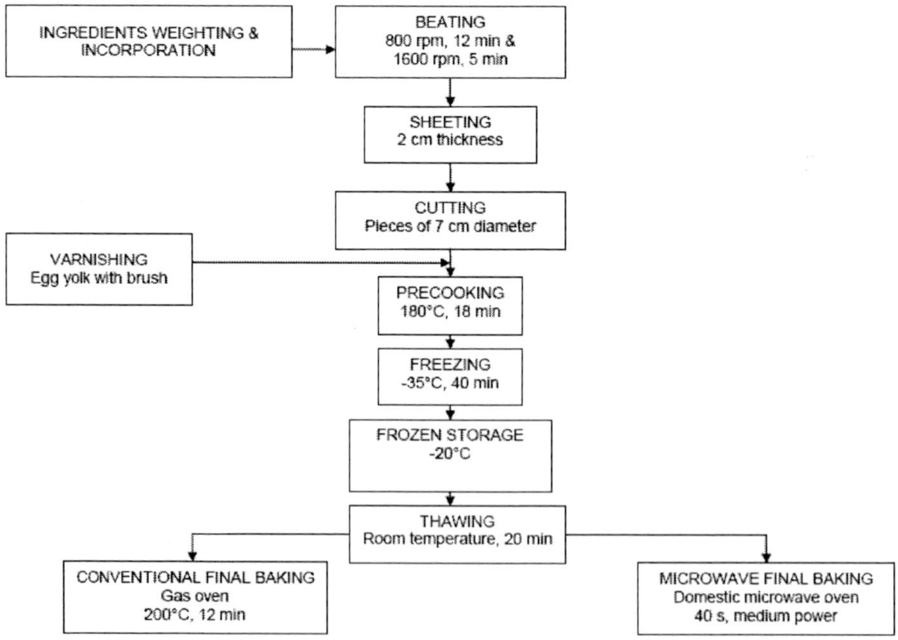

Figure 3. Elaboration process for biscuits.

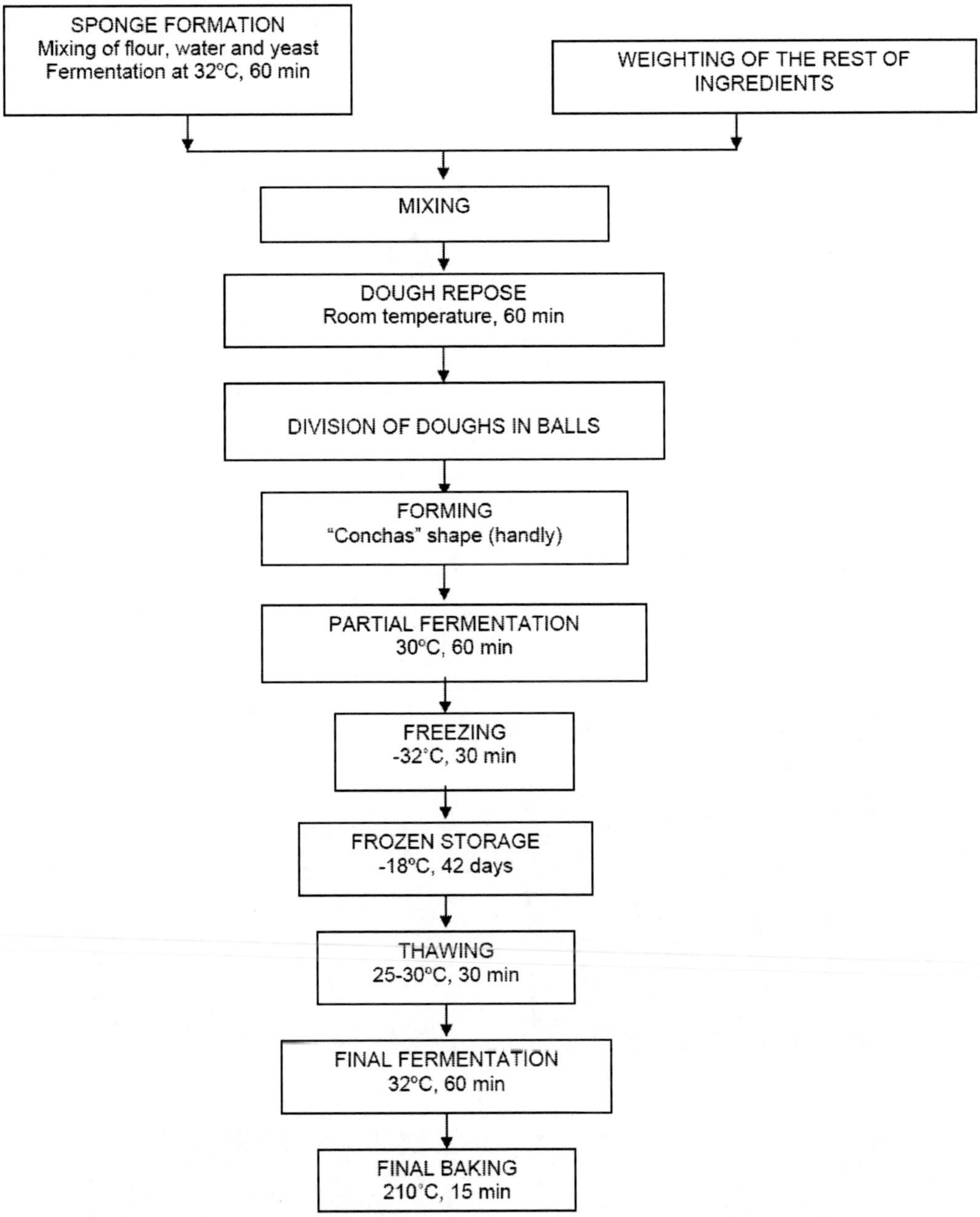

Figure 4. Elaboration process for sponge cake bread ("conchas").

In the elaboration processes, some steps, as weighting, mixing, freezing, are common for the three products. Others are specific of each product, such as the varnishing for biscuits; covering for sponge cake dough, or pasting for the Danish bread. For this last one, the formation of the paste is critical to reach up the desirable texture in the final product. When the paste is ready, dough is sheeted again and cut on triangles to shape the "horn", this operation is done by people. Figure 6 shows the paste and the formation by hand.

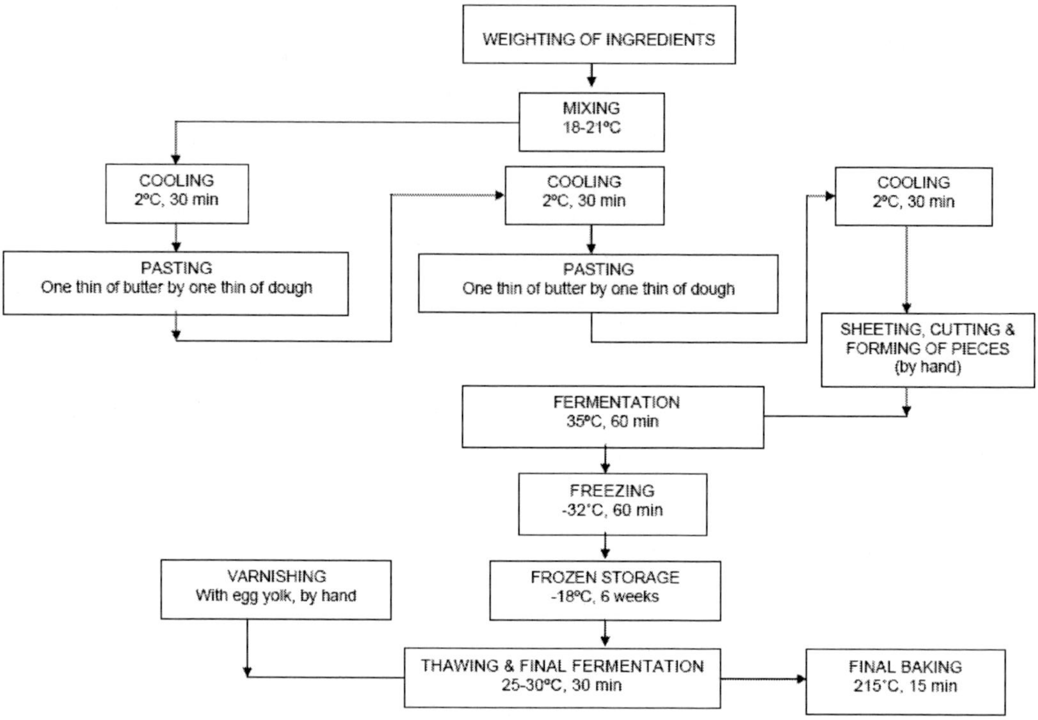

Figure 5. Elaboration process for Danish type breads.

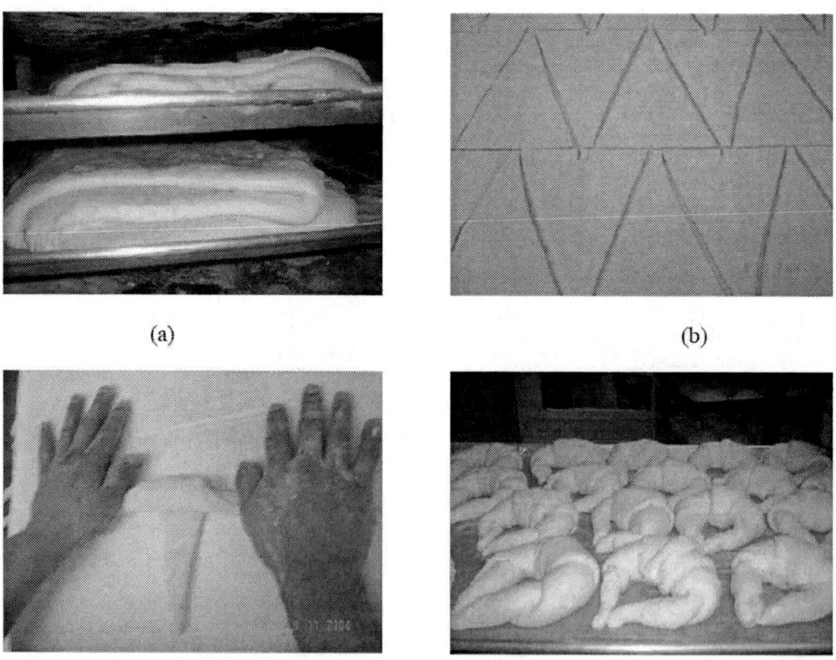

Figure 6. Several steps for making Danish type bread. (a) pasting, (b) cutting, (c) forming, and (d) finished pieces with "horn" shape.

Common equipment, such as mixers, fermentation chamber, and freezer, were employed for the three processes. The tunnel freezer (T25, Gouet Española, Spain) worked with a forced air flow of 25000 m^3/h. There were other kind of equipments used only for a specific breadmaking, such as the machinery to do balls for sponge cake dough, or sheeter, for the biscuits and croissant type breads. The electrical sheeter (Rondostar 3000, Rondo, Spain) worked at a rolls velocity of 85 cm/s. Sponge cake and croissant type breads required a fermentation process before the freezing, which was completed after the freezing storage. The used freezing chamber (Europan, Spain) kept a relative humidity of 70%. Some of the used machines are presented in the figure 7, all of them are property of Pabisan, S.A. de C.V.

Samples were taken to finish the process at different storage times: 7, 14, 21 or 28 days. Then, bread pieces were baked and their physical properties were evaluated.

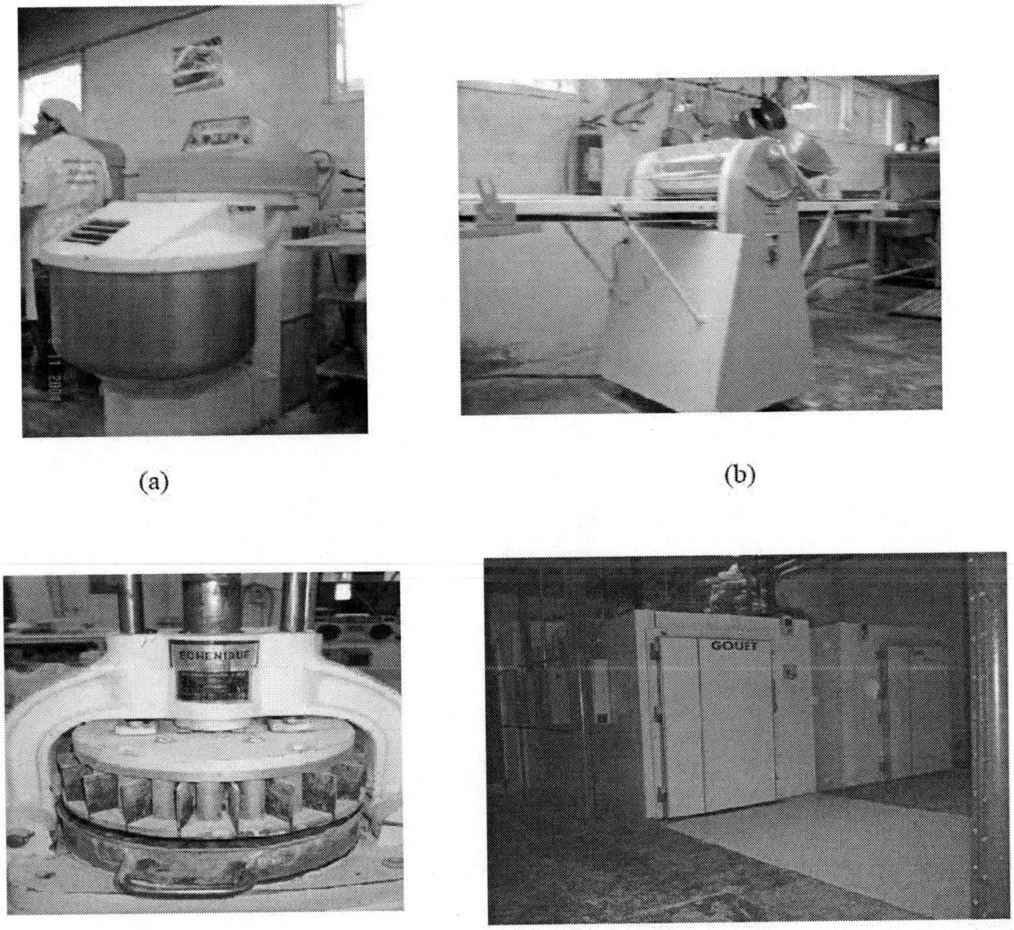

(a) (b)

(c) (d)

Figure 7. Some equipments used for different breadmaking. (a) Dough mixer, (b) sheeter (for biscuits and croissant type processes), (c) equipment to do balls of dough to form the sponge cake bread, and (d) tunnel freezer.

4. METHODOLOGY FOR PHYSICAL AND SENSORY ANALYSIS

Those methods employed for bread characterization, are presented next, in order to know the specifications.

Volume of breads was calculated by the volume displacement method (AACC, 1986).

Density was calculated by the relation of the mass and the volume (kg/m^3). The specific volume is also a common determination in bakery products, and is calculated as the inverse of the density (usually expressed in cm^3/g).

Color of crumb breads was measured with a Color Gard System Colorimeter (Hunter Lab., Reston, VA). Color was measured by reflectance mode and expressed by L (luminosity), a (green-red) and b (blue-yellow) Hunter parameters. The colorimeter was calibrated utilizing the black tile and the white standard (L= 92.89, a=-1.05, b=0.82).

Textural test adapted from the official method 74-09 (AACC, 1986) was conducted on the final product, using a TA.XT2 Texture Analyzer (Texture Technologies Corporation, United Kingdom) and Texture Expert software (1.22 version, 1999). Firmness was measured when breads were compressed at 25% by a plate of 36 mm diameter with a head velocity of 1.7 mm/s.

A sensory analysis using a nine-points hedonic scale, in which an affective sensory test (Larmond, 1976), was conducted with 20 non-trained panelists. Judges evaluated color, texture and general acceptance of the breads from control formulation and breads from the tested formulations, containing the blends of additives.

For data analysis, ANOVA and Tukey's pairwise comparisons were conducted with the Minitab software (10.5 version, 1995) in order to determine significant difference between treatments for frozen storage, final baking and sensory tests, with a significance level of 95%.

5. RESULTS ON THE QUALITY OF BAKED PRODUCTS

5.1. Biscuit Bread

The modification in biscuit formulation was conducted to avoid volume loss. Figure 8 shows the volume of precooked biscuits made with the modified formulation at every week during 8 weeks of frozen storage at –18°C. There was no significant change in the volume through the storage ($p>0.05$). Firmness of baked biscuits in conventional oven (after 8 weeks of frozen storage) was 6.05 N and ranked an overall acceptability qualification of 7.6 by the sensory panel, which was similar than the control formulation and better than the exhibited by biscuits baked in microwave oven (table 4). Even though lower texture was measured in biscuits baked in microwave oven, the sensory acceptance was lower than the other, control and biscuits baked in gas oven. Thus, the selected blend of additives (emulsifiers and oxidant agents) was suitable to get loaf volume in this product, and the best baking was obtained with conventional oven upon the microwave baking.

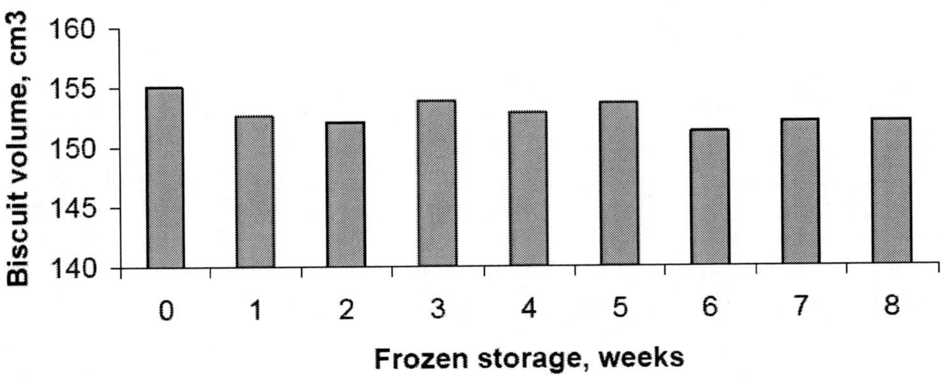

Figure 8. Volume of biscuits (formulated with the blend of additives) during 8-weeks frozen storage at −18°C.

Table 4. Texture and sensory evaluation of biscuits*

Biscuit formulation	Firmness (N)	Overall acceptability
Control	6.41 ± 0.45[a]	7.50 ± 1.55[a]
Modified (baked in gas oven)	6.05 ± 0.36[a]	7.65 ± 1.13[a]
Modified (baked in microwave oven)	5.13 ± 0.16[b]	6.90 ± 1.74[a]

* Means with the same letter are not significantly different (p>0.05).

5.2. Sponge Cake Bread

Table 5 shows some properties of the sponge cake bread, baked after 42 days of frozen storage at −18°C. Density of the sponge cake bread ranged between 158.7 and 167.2 kg/m^3, indicating a low density of the product. These values are similar to those reported for malt bread 221 to 261 kg/m^3 by Magee and Bransburg (1995) and for bread crumb 88-252 kg/m^3 by Zghal *et al.* (1999). There was not effect of the type of additives blend on the density of the final product (p>0.05). Also, no effects were observed in the crumb color; the luminosity parameter was similar between control and formulations, being around of 38 (table 5).

Sponge cake bread texture was affected by the kind of used additives (p<0.05). Lower firmness was observed in control and formulation 2 with values of 4.87 and 4.33 N, in comparison with the obtained for formulations 1 and 3 (5.86 and 6.45 N, respectively). The reason for a softer texture was related to the presence of enzymes (alpha-amylases) in the employed blends for control and formulation 2. More even, softer texture was observed when malt alpha-amylase (formulation 2) was employed respect to fungal alpha-amylase (used in the control set). This observation agreed with the reported by Qi Si (1997), who cited that malt alpha-amylase has an effect against bread hardening (avoiding the amylopectin retrogradation) that improves the elasticity of the bread crust.

As result of the sensory analysis, higher scores were given to the control (7.0) and breads under formulation 2 (7.85). This result was strongly related to the texture of the breads, thus, it was concluded that texture is the most critical parameter for this kind of breads.

Table 5. Several physical properties of sponge cake breads* baked after 42 days of frozen storage at –18°C

Formulation	Density (kg/m^3)	Crumb Luminosity	Firmness (N)	Sensory acceptability
Control	159.50 ± 7.96 ^{a}	38.10 ± 1.78 ^{a}	4.87 ± 0.56 ^{a}	7.00 ± 0.97 ^{ab}
Formulation 1	167.23 ± 7.94 ^{a}	38.90 ± 1.04 ^{a}	5.86 ± 0.61 ^{b}	6.50 ± 1.60 ^{b}
Formulation 2	158.79 ± 11.87 ^{a}	38.56 ± 1.10 ^{a}	4.33 ± 0.29 ^{a}	7.85 ± 0.93 ^{a}
Formulation 3	167.00 ± 3.30 ^{a}	38.82 ± 1.05 ^{a}	6.45 ± 0.28 ^{b}	6.65 ± 1.69 ^{b}

* Average with the same letter means there was not significant difference ($p>0.05$).

5.3. Danish Type Bread

The specific volume for horns ranged between 4.16 and 4.80 cm^3/g, without effect of the employed formulation ($p>0.05$). These values are comparable with those reported for baguettes (4.98 cm^3/g) by Bonnardel and Maitre (1987). For braid shape, the used formulation had effect on the specific volume of Danish type breads, being bigger for formulation 1 (4.35±0.039 cm^3/g) and lower for formulation 3 (3.42±0.61). This fact was in agreement with Gélinas *et al.* (1994), who reported that baking products geometry affects the volume. They calculated values between 5.30 and 5.57 cm^3/g for infinite cylinders and 4.91 to 5.32 cm^3/g for spheres shaped from the same dough formulation.

Dough color was not affected by the used formulation ($p>0.05$). *L, a* and *b* parameters were similar in dough elaborated with the four studied formulations (table 6) and kept without changes during the frozen storage through 6 weeks. However, crumb color of baked bread after the 6-weeks frozen storage was affected by the employed blend of additives ($p<0.05$). Darker (lower values for *L* parameter) and more yellow color (higher *a* values) was measured in Danish type bread from formulations A and B, while lighter appearance was appreciated in the crumb of breads elaborated with the formulations C and D (table 6).

Table 6. Color parameters for frozen dough and baked Danish type breads after 6-weeks of frozen storage at –18°C

Sample	Color parameter	Formulation			
		A	B	C	D
Dough	L	58.9 ± 4.7	58.9 ± 1.7	59.3 ± 1.2	59.3 ± 1.2
	a	-1.3 ± 0.2	-1.7 ± 0.4	-1.2 ± 0.3	-1.2 ± 0.9
	b	12.4 ± 1.4	13.5 ± 0.7	14.7 ± 1.4	13.9 ± 1.4
Baked bread	L	29.9 ± 4.5	29.8 ± 4.2	32.2 ± 3.8	36.4 ±5.0
	a	8.8 ± 1.5	6.6 ± 1.7	5.7 ± 1.1	5.5 ± 1.8
	b	11.1 ± 2.8	10.6 ± 2.2	14.2 ± 1.5	14.9 ± 2.1

For texture and after 6 weeks of storage, the softest formulation was formulation D with firmness of 4.8 N, while the hardest was formulation B with a force of 7.9 N (figure 9). Cryoprotector agent (formulation D) promoted the softness of the bread, even after the frozen storage. However, the manufacturer did not report what kind of cryoprotector was added to their blend.

From the sensory evaluation, the highest qualification was assigned to the formulation D, but without significant difference respect to formulations A and C ($p>0.05$). The lowest qualification was assigned to the formulation B. These results were related to the instrumental analysis, in which higher sensory scores were achieved for the softest bread, and the lowest qualification was given for the hardest bread.

Some defects were observed in horns crust using the formulation C, such as varnish separation (figure 10). This separation was not observed in other formulations. Thus, the addition of emulsifiers only was not enough to preserve the desired quality of frozen bread.

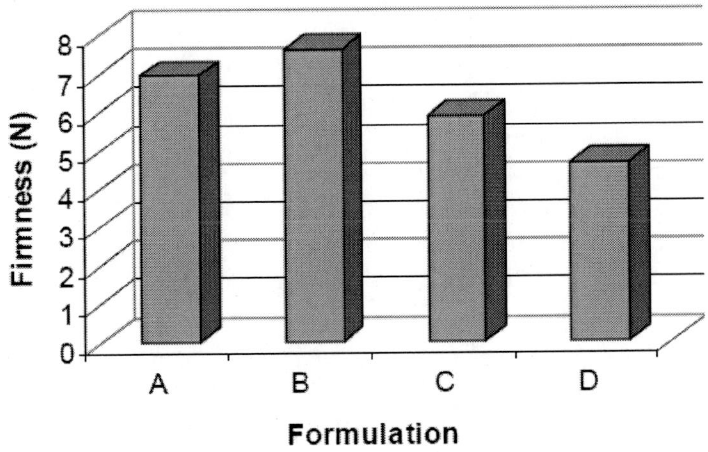

Figure 9. Texture of Danish type bread (horn shape) baked after 6 weeks of frozen storage.

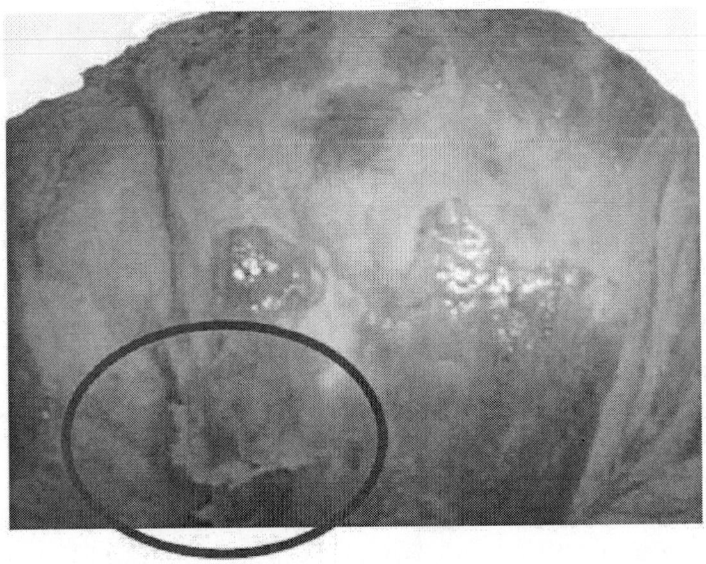

Figure 10. Defect on the crust (varnish separation) in Danish type breads (horn shape) from formulation 3 (DATEM added).

6. CONCLUSIONS

Additives are a convenient way to improve the quality of frozen dough. Commercial blends of additives are useful and easy to get in the market. Furthermore, the advisory of the suppliers is available when the blend is purchased. Commercial blends contain a mixture of emulsifiers, enzymes, oxidant and/or reducer agents, which reach up a synergic effect on the dough. However, the kind of blend would depend of the type of bread formulation, process steps, storage time and expected results. As result of this study, a blend of emulsifiers (soybean flour and DATEM) and oxidant agents (ADA and AA) were enough to keep loaf volume in biscuits, while a blend of emulsifiers (DATEM) and alpha-amylase was good for sponge cake bread. The blend of emulsifiers, proteins, hydrocolloids and a cryoprotector agent was the best to keep the properties of Danish type bread.

Freezing is a good way to preserve dough, and to have bread just baked at anytime. The objective for keeping the good quality of bread even through several weeks of frozen storage was reached up by using commercial blends of additives.

Other important remark is that the collaborative work between food industry and academy developed through this study was fruitful for both parts. Experience in practical and research issues were shared, with feedback for the food plant and the university. Hopefully, more works can be developed in this way, to bring up benefits and learning for the Food Industry, the Faculty and the students of Food Engineering.

REFERENCES

AACC (American Association of Cereal Chemists). 1986. *Approved Methods.* St. Paul, MN, USA.

AIB (American Institute of Baking). 2001. Science of Baking, Frozen Dough. Kansas, USA.

Badui-Dergal, S. 1999. Food Chemistry. Longman de Mexico Editores S.A. de C.V. Naucalpan de Juarez, Mexico. (In Spanish).

Bárcenas M.E., Haros M., Benedito C. and Rosell C.M. 2003. Effect of freezing and frozen storage on the staling of part-baked bread. *Food Research Internationa,.* 36(8): 863-869.

Bhattacharya M., Langstaff T.M. and Berzonsky W.A. 2003. Effect of frozen storage and freeze-thaw cycles on the rheological and baking properties of frozen doughs. *Food Research Internationa,.* 36(4): 365-372.

Bonnardel P. and Maitre, H. 1987. Precooked French bread technology. Conference. Baroeul, France.

Desrosier N.W. 1970. The Technology of Food Preservation. The AVI Publishing Co., Westport, CN, USA. pp. 92–122.

Encinas-Banda D., Gonzalez-Loo H., Garcia M., Velez-Ruiz J.F. and Sosa-Morales M.E. 2005. Effect of different additives on quality characteristics of frozen bread. Poster presentation 36G-2, *IFT Annual Meeting,* July 15-20, New Orleans LO, USA.

Gélinas P. Lagimoniére M. and Rodriguez N. 1994. Performance of cream or compressed yeast in frozen and nonfrozen doughs. Cereal Chemistry, 71(2): 183-186.

Guinoza Matuda T., Chevallier S., de Alcántara P., Filho P., LeBail A. and Tadini C.C. 2008. Impact of guar and xanthan gums on proofing and calorimetric parameters of frozen bread dough. *Journal of Cereal Science.* In press.

Havet M., Mankai M. and Le Bail A. 2000. Influence of the freezing condition on the baking performances of French frozen dough. *Journal of Food Engineering,* 45(3): 139-145.

Heldman D.R. and Singh R.P. 1981. Food Process Engineering, 2nd Ed., AVI Publishing Co., Inc., Westport, CT, USA. pp.127, 135, 160–162.

INEGI, National Institute for Statistics, Geography and Information. 2008. National sales of manufacturer sectors in Mexico. Available in: http://dgcnesyp.inegi.org.mx/cgi-win/bdieintsi.exe/SER89435. Accessed on September 29th, 2008. (In Spanish).

Inoue Y. and Bushuk W. 1991. Studies on frozen doughs. I. Effects of frozen storage and freeze-thaw cycles on baking and rheological properties. *Cereal Chemistry,* 68: 627-631.

Joensson T. and Toernaes H. 2004. The effect of selected surfactants on bread crumb softness and its measurement. *Bulletin.* Danisco A/S, Copenhagen, Denmark.

Kenny S., Wherle K., Auty M. and Arendt E.K. 2001. Influence of sodium caseinate and whey protein on baking properties and rheology of frozen dough. *Cereal Chemistry,* 78(4): 458-463.

Larmond, E. 1976. Laboratory Methods for Sensory Evaluation of Food. Publication 1637. Department of Agriculture of Canada, Ottawa, Canada.

Le Bail A., Grinand C., Le Cleach S., Martinez S. and Quilin E. Influence of storage conditions on frozen French bread dough. *Journal of Food Engineering,* 39(3): 289-291.

Magee T.R.A. and Bransburg T. 1995. Measurement of thermal diffusivity of potato, malt bread and wheat flour. *Journal of Food Engineering,* 25(2): 223-232.

Matz A. S. 1992. Bakery Technology and Engineering. 3rd Ed. Pan-Tech International, Inc.: McAllen TX, USA.

Meric L., Lambert-Guilois S., Neyreneuf O. and Richard-Molard D. 1995. Cryoresistance of baker´s yeast *Saccharomyces cerevisiae* in frozen dough: contribution of cellular trehalose. *Cereal Chemistry,* 72(6): 609-615.

Morales de la Peña, M. 2005. Reformulation of sponge cake dough to increase its shelf-life in frozen storage. Bachelor Thesis, Universidad de las Américas, Puebla, Mexico. (In Spanish)

Murakami Y., Yokoigawa K., Kawai F. and Kawai H. 1996. Lipid composition of commercial baker´s yeasts having different freeze-tolerance in frozen dough. *Bioscience, Biotechnology and Biochemistry,* 60(11): 1874-1876.

Novozymes. 2008. Enzymes at work. Available at: http://www.novozymes.com/NR/rdonlyres/DF82372E-521E-41A3-97CF-71B4A3B6907D/0/ Enzymesat work.pdf. Accessed on August 14th, 2008.

Panozzo J.F., Bekes F., Wrigley C.W. and Gupta R.B. 1994. The effects of bromate (0-30 ppm) on the proteins and lipids of dough. *Cereal Chemistry,* 71(2): 195-199.

Phimolsiripol Y., Siripatrawan U., Tulyathan V. and Cleland D.J. 2008. Effects of freezing and temperature fluctuations during frozen storage on frozen dough and bread quality. *Journal of Food Engineering,* 84(1): 48-56.

Pyler, E.J. 1988. Baking Science and Technology. 3rd Ed. Sosland Publishing Company. Kansas, USA.

Qi Si J. 1997. Synergistic effect of enzymes for breadbaking, *Cereal Foods World,* 42(10): 802-807.

Rahman M.S., Machado-Velasco K.M., Sosa-Morales M.E. and Vélez-Ruiz J.F. 2009. Freezing point: measurement, data and prediction. In *Food Properties Handbook,* 2nd ed., Rahman, M.S. Ed. CRC Press, *Boca Raton,* FL, USA. pp. 153-192.

Rahman S.1995. Food Properties Handbook, CRC Press, *Boca Raton,* FL, USA. pp. 378–380, 385.

Rahman S. and Vélez-Ruiz J.F. 2007. Food preservation by freezing. In Handbook of Food Preservation, 2nd ed., Rahman, M.S. Ed. CRC Press, Boca Raton, FL, USA. pp. 635–665.

Ribotta P.D., León A.E. and Añón M.C. 2003. Effect of freezing and frozen storage on the gelatinization and retrogradation of amylopectin in dough baked in a differential scanning calorimeter. *Food Research International,* 36(4): 357-363.

Ribotta P.D., Pérez G.T., León A.E. and Añón M.C. 2004. Effect of emulsifier and guar gum on micro-structural, rheological and baking performance of frozen bread dough. *Food Hydrocolloids*, 18(2): 305-313.

Sharadanant R. and Khan K. 2003. Effect of hydrophilic gums on the quality of frozen dough: II. Bread characteristics. *Cereal Chemistry,* 80(6): 773-780.

Sosa-Morales M.E., Guerrero-Cruz G., González-Loo H. and Vélez-Ruiz J.F. 2004. Modeling of heat and mass transfer during baking of biscuits. *Journal of Food Processing and Preservation,* 28: 417-432.

Tejero F. 1998. The precooked bread. Editorial Montagud, Barcelona, Spain (In Spanish).

Tramstorf S. 1991. Emulsifiers for bakery and starch products. Bulletin. Grindsted Products, Grindsted, Denmark.

Zghal M.C., Scanlon M.G. and Sapirstein H.D. 1999. Prediction of bread crumb density by digital image analysis. *Cereal Chemistry,* 76(5): 734-742.

In: Food Processing and Engineering Topics
Editors: M. E. Sosa-Morales and J. F. Vélez-Ruiz
ISBN: 978-1-60741-788-0

Chapter 12

NEW FORMULATIONS FOR DEEP-FAT FRIED DOUGHNUTS TO IMPROVE THEIR NUTRITION FACTS

María Elena Sosa-Morales and Jorge F. Vélez-Ruiz
Chemical and Food Engineering Department
Universidad de las Americas, Puebla, Mexico

1. INTRODUCTION

Doughnuts, also called cake donuts or simply donuts, are a kind of bread characterized by a golden-brown exterior color, a crisp crust and an inner core that resembles a baked product more than a fried food (Shih *et al.*, 2001). Doughnuts are the most famous fried bakery product. Their market represented US$3-4 billion business in the US in the last years (Shi and Daigle, 2002).

Based on the origin of volume increase, two types of doughnuts are distinguished: doughnuts prepared with baking powder (sodium carbonate) and doughnuts leavened through a fermentation process carried out by yeasts (*Saccharomyces cerevisiae*). Fermentation may be done in two stages, the first stage involves around 90 min, when only a part of flour, water and the total of yeast are mixed. The resultant fermented dough is called “sponge”. Then, the rest of ingredients (flour, egg, butter, egg, water or milk) are mixed to form the dough and the sponge will be added lately. The final dough is sheeted, and cut in rings, the typical doughnuts shape. Dough rings are submitted to the second fermentation stage for 30 min, where the increasing of volume is high. This procedure is known as the sponge-cake method (Pyler, 1988). When yeast-leavened doughnuts are prepared, the recommended fermentation time is 90 min and the frying temperature is around 190°C (Amendola, 1993).

Being doughnuts a fried product, they absorb substantial amounts of oil/fat during the frying process. The excessive consumption of fat and oils has been related with health concerns, such as obesity, cardiac diseases and digestive troubles. Then, the current trend of consumers is related to reduce the oil content in foods. Food technologists have the challenge to answer that concern and to develop doughnuts without excess of oil content and keeping the desirable properties of the traditional food (Shih *et al.*, 2001).

This chapter is devoted to present the characteristics of the traditional fried doughnuts and the recent advances in the development of formulations to improve nutrition factors of this food, such as to increase the protein and fiber content, reduce the caloric content and, overall, diminish the oil content in this appreciated fried product.

2. Frying Process of Doughnuts

Frying of doughnuts is a unit operation involving simultaneous heat and mass transfer between the frying medium and the food. Heat transfer occurs from the heated fat/oil to the food, with a high driven temperature (more than 150°C). Mass transfer involves two important transferences: the moisture loss from the frying food and the fat/oil uptake within that item. Mathematical aspects and modeling of the process have been reviewed recently by Sosa-Morales and Vélez-Ruiz (2008).

Frying process consists in to introduce the raw ring dough in the frying medium till cook the internal crumb and to get a crispy brown crust. The rings are usually flipped from side to side to have uniform cooking and external color. Even the doughnut floats, because of its lower density in comparison with the frying medium, the process is called deep-fat frying or immersion frying.

Doughnuts may be fried in fats or oils. Common fats are hydrogenated oils where sunflower, cottonseed, soybean, and their mixtures are typical oils used for frying of doughnuts. Olive oil is not commonly used, however, Aggelousis and Lalas (1997) reported that olive oil was more stable to thermal oxidation than cottonseed oil, and a blend of 70% sunflower oil, 25% palm oil and 5% peanut oil were used for frying of doughnuts at 170°C. Hatae *et al.* (2003) reported that there was a preference for doughnuts fried in hydrogenated soybean oil respect to those fried in liquid soybean oil, when both doughnuts were evaluated with a trained sensory panel. Authors related this result to the softer texture got when hydrogenated soybean oil was used.

After the frying process, the excess of fat/oil is removed (by draining) and doughnuts are kept at room temperature for the last step, the application of the cover. The external properties of doughnuts, such as crust color and texture, are affected mainly by the frying conditions (frying temperature and frying time), while the internal characteristics, such as crumb freshness and texture, are affected by time and conditions of the storage (Pyler, 1988).

3. Fat/Oil Content of Fried Doughnuts

3.1. Determination of Fat/Oil Content in Doughnuts

The concern about fats and oils ingestion points out to review the lipids content in foods. The fat/oil content in doughnuts is usually determined following the AACC method, as doughnuts are composite mainly by flour, a cereal derivate. AOAC official method can be followed as well. Basically, the methodology to determine fat/oil content consists in an extraction process through 4-6 hours using a non-polar solvent, such as petroleum ether or diethyl ether. It is common to dry the sample previously to the extraction (AOAC, 1990;

AACC, 2000). Soxhlet or Goldfisch equipments are suitable for the extraction and determination of the fat/oil content in doughnuts.

3.2. Factors that Affect the Fat/Oil Content in Doughnuts

The oil/fat content in doughnuts varies depending of different factors. These are cpmmented next, including some studies related to them.

Factors Related to the Formulation

There are many different doughnut formulations, and the type and amount of ingredients affect the fat content in the final product. The leavening agent affects the fat content in doughnuts as well. In according with Watt and Merril (1963), doughnuts made with baking powder contain lower fat content than those elaborated through a fermentation process using yeast. McComber and Miller (1976) reported that the lipid content of doughnuts increased when tartrate baking powder was used in comparison with those samples leavened with sulfate-phosphate baking powder. They also found that an increase of lecithin level increase the lipid contents in doughnuts.

Presence of polysaccharides, higher protein contents, replacing or substitution of ingredients, use of edible coatings, are facts that affect the final oil content. All these factors are the basis for some of the new doughnuts formulations.

Factors Related to the Frying Process

Frying Temperature

Vélez-Ruiz and Sosa-Morales (2003a) studied the frying of yeast-leavened doughnuts at three temperatures: 180, 190 and 200°C. When higher frying temperatures were employed, lower oil uptake in doughnuts was quantified. This observation was opposite to the observed by Gamble *et al.* (1987) for potato slices fried between 130 and 150°C, who reported that at higher frying temperature, larger oil contents were achieved. The differences could be attributed to the composition and structure of the food item.

Frying Time

During the frying of doughnuts for 120 s, major oil uptake was reported in the first seconds of frying, followed by a diminution in the oil content and then, a small oil gain was quantified at the end of the process (Vélez-Ruiz and Sosa-Morales, 2003b). Tan and Mittal (2006) reported a similar trend, with major oil absorption within the first two minutes of frying and the reduction of the oil uptake at 4 min of frying. That diminution or desorption phenomenon has been previously reported in fried chicken nuggets (Balasubramanian *et al.*, 1995). Desorption was related to a capillarity effect in the crust of the food. Then, the oil content depends of the frying time and the frying stage in which the product is analyzed.

Frying Pressure

The benefits of vacuum frying have been reported for potato chips by Garayo and Moreira (2002). They reported that the oil content for fried potato chips was higher at atmospheric pressure than for those fried under 16.7, 9.9. and 3.1 kPa of vacuum. However, opposite results were found for doughnuts by Tan and Mittal (2006). They used three vacuum levels (3, 6 and 9 kPa vacuum) for frying of doughnuts at 150, 165 and 180°C. Higher oil contents were quantified in doughnuts fried under vacuum conditions, and the oil content increased with the vacuum increase. For example, control doughnuts (fried at atmospheric conditions) had around 13% of oil content (dry basis), while doughnuts fried at 150°C had 24.7, 29.2 and 35.8% when they were fried at 3, 6 and 9 kPa of vacuum, respectively. Those conditions closest, to reach similar characteristics to the control samples, were vacuum frying at 3 kPa and 180°C, in which doughnuts with oil content of 14.5% were obtained.

Factors Related to the Post-Frying Conditions

Draining after the Frying Process

It is well recognized that most of the oil uptake occurs in the post-frying stage, when the food is removed from the fryer (Moreira *et al.*, 1999). Thus, a correct draining of doughnuts after the frying process is required in order to reduce the oil content in the final product. Adsorbent paper or paper towels may be used during cooling of doughnuts, to absorb the excess of oil from the surface of the doughnuts.

Type of Cover

Several covers are used for decorating and imparting sweet complements to the doughnuts. The nature of these covers can affect the total amount of fat in the doughnuts. Doughnuts may be simply sprinkled with sugar, or glazed with sugar syrup, chocolate syrup or other flavoring covers (Moreira *et al.*, 1999). Items covered with chocolate of other kind of fatty ingredients will have higher fat content that those just sprinkled with sugar.

4. New Formulations of Deep-Fat Fried Doughnuts to Improve Nutrition Facts

Different ways to improve the nutritional profile of doughnuts have been explored. The main objective in most of the cases is to reduce the fat/oil uptake. But that reduction must not affect those important expected physical properties of the final product, such as texture, color and sensory acceptance. Next, we present several resources applied to get lower fat/oil contents in doughnuts and their impact on the physical or sensorial attributes of this fried product.

4.1. Addition of Polysaccharides

Curdlan and Cellulose Compounds

Curdlan is a microbial polysaccharide of D-glucosyl residues approved by the USFDA as a food additive since 1996. Addition of curdlan (0 to 1%, w/w) in doughnuts formulation was studied by Funami *et al.* (1999) with the objective to reduce the oil uptake during the frying of doughnuts in cottonseed oil at 170°C. The oil reduction was compared with that got using sodium carboxy methylcellulose (CMC) and methylcellulose (MC). Control samples had a lipid content of 35%, while those samples prepared with curdlan reached 30% of lipids, both contents determined at 4 min of frying. Curdlan added at 0.1% to 0.5% significantly decreased oil uptake of doughnuts and the decrease was even two times greater than that obtained with 1% of CMC. MC had no effect on decreasing the oil uptake. Also, curdlan was more effective to reduce moisture loss during frying. The effects of this additive were related to its thermal gelation during the process, which functions as an oil and moisture barrier. Curdlan gels in doughnuts showed a plasticizing effect similar to shortening in dough, modifying the G_0' storage and G_0" viscous parameters, which increased with the increasing of the curdlan ingredient in the formulation. In puncture tests, doughnuts showed a decrease in the breaking stress when contained curdlan, while this physical property increased using CMC and remained constant with MC, when they were compared with the control doughnuts.

Pineapple Derivates

Pineapple core dietary fiber (PDF) and pineapple core cellulose (PC) were obtained by Prakongpan *et al.* (2002) and were added at 3% (w/w) in a doughnut formulation. Donuts were fried in soybean oil at 190°C. The fat content of fried doughnut with PDF was 14.1%, being 40% less than the control formulation (23.8% of fat content). Doughnuts with PC had 18.3% of fat content. Even addition of PDF resulted in the lowest fat content, it was not the best formulation, because its firmness increased (18.63 N versus 8.86 N of control samples). Doughnuts with PC presented softer texture (9.63 N), similar to the control. Then, PC appeared to be more suitable for reduction of fat content in doughnuts without much augment of texture, and with the consequent caloric decrease in the finished products.

4.2. Partial Substitution of Wheat Flour by Other Ingredients

Rice Derivates

Shih *et al.* (2001) partially replaced the wheat flour from doughnuts formulation by a variety of rice ingredients, such as long-grain rice flour, waxy rice flour, gelatinized rice flour and precooked acetylated rice starch. The replacement was done at 25 or 50%. Doughnuts were prepared with baking powder (sulfate phosphate) and fried at 177°C in Wesson vegetable oil until they were golden brown on the crust and well cooked in the inner core. The replacement of wheat flour for the rice products affected the consistency of the dough. The authors reported 23.8% of oil content in all wheat doughnuts (control), while formulations including long-grain and waxy rice flours had higher oil contents respect to the control. On the other hand, lower oil contents were achieved with formulations including gelatinized rice flour (12.6-14.8%) and precooked-acetylated rice starch (14.85-17.83%). Replacement of wheat flour at 50% resulted in higher firmness values for the doughnuts, but wheat flour at 25% did not affect the texture of the fried product. For example, control doughnut firmness was 801 g_f, while replacement of 25 and 50% of wheat flour by precooked acetylated rice starch resulted in doughnuts firmness of 782 and 6238 g_f, respectively. In conclusion, the formulation with 25% of precooked-acetylated rice starch had a low value of oil content (14.8%) and a comparable firmness with the control.

In other study, Shih and Daigle (2002) used long grain rice flour, waxy rice flour, pregelatinized rice flour and propylene glycol alginate to partially replace flours in the formulation of doughnuts. Control doughnuts resulted with 23.6% of oil uptake while doughnuts with waxy rice flour exhibited the lowest oil uptake, with 11.8%.

Maize and Pigeon Pea Flours

Echendu *et al.*, (2004) formulated blends of maize-pigeon pea flours. These blends were employed to substitute wheat flour in levels of 10, 20, 30, 40, 50, 60, 70, 80 and 90% in formulations of biscuits and doughnuts. In the case of doughnuts, the protein content increased with the content of pigeon pea flour, having the highest content (16.4% of protein) in doughnuts made with the blend of 10:90 maize flour:pigeon pea flour. Authors also commented that products including the tested blends were well accepted in a sensory evaluation.

Integral Wheat, Amaranth or Oat Flours

Wheat (*Triticum aestivum, T. vulgare*) is the principal cereal used for bread making. Integral wheat flour is obtained from the grinding of the grain and the endosperm, its fiber content is from 1.8 to 2.5%. Oat (*Avena sativa*) have higher levels of fiber and protein than wheat, furthermore, its flour contents antioxidants compounds (Kent, 1987). Amaranth (*Amaranthus cruentus*) is a cereal of low production in the world. Its flour is rich in essential aminoacids and its addition into baking products does not affect the leavening of dough (Oke, 1983).

With the objective to increase protein and fiber contents in doughnuts, Treviño-Cuevas *et al.* (2004) partially replaced the refined wheat flour in a yeast-leavened doughnuts formulation by integral wheat, amaranth or oat flour. Replacement was at levels of 30 or 50% (w/w), and doughnuts were fried in soybean oil at 180, 190 and 200°C; the followed procedure is shown in figure 1. Doughnuts composition was affected by the type and proportion of added flour (table 1). The highest protein and fiber content were determined in doughnuts elaborated with 50% of refined wheat and 50% amaranth flour. Also, in this formulation the lowest fat content was determined. However, darker and harder appearance than control items was observed in those doughnuts.

Table 1. Composition (g/100 g) of doughnuts elaborated with different proportions of flours

Formulation	Moisture	Fat	Protein	Fiber	Ashes
100% W	23.13	24.04	5.45	0.29	0.03
70%W–30%I	23.86	25.81	4.36	0.39	0.02
50%W-50%I	24.74	13.81	5.80	0.45	0.04
70%W-30%O	22.66	20.29	5.47	0.48	0.04
50%W-50%O	22.70	14.30	6.87	0.60	0.06
70%W-30%A	22.63	15.05	6.00	1.08	0.10
50%W-50%A	24.36	13.47	7.58	1.41	0.15

Doughnuts were fried at 190°C during 60 s in sunflower oil. *W*=refined wheat flour, *I*=integral wheat flour, *O*= integral oat flour, *A* = integral amaranth flour.

Source: Treviño-Cuevas *et al.*, 2004.

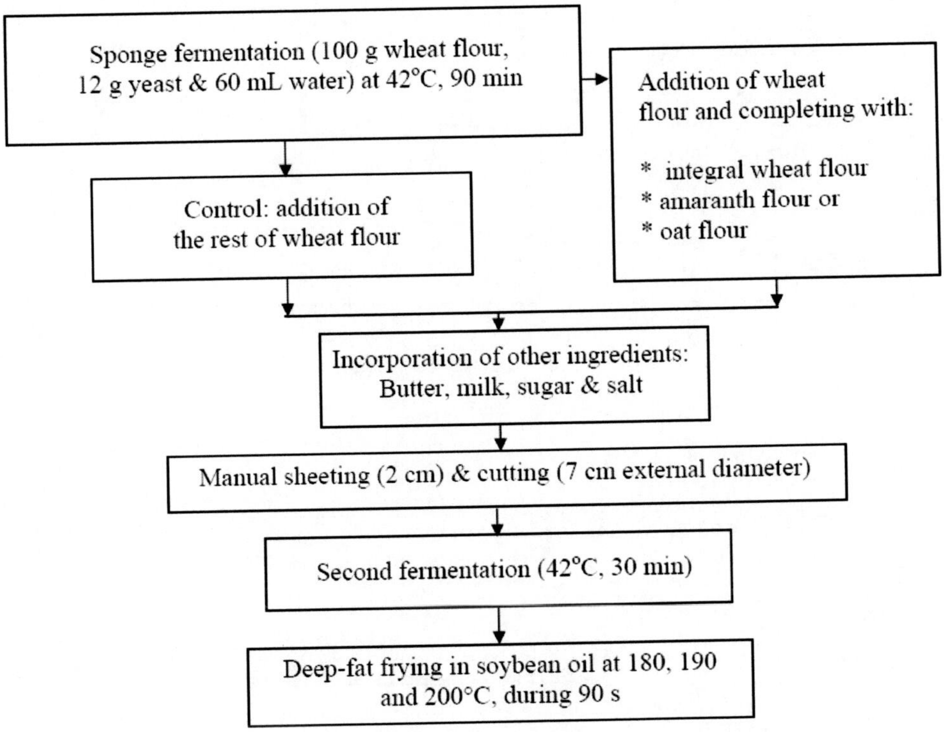

Figure 1. Elaboration process of doughnuts made with mixture of flours (From Treviño-Cuevas *et al.*, 2004).

Oil content increased with frying time (figure 2), and decreased with frying temperature. Highest oil contents were observed in formulations with integral wheat flour, whereas lower oil uptakes were measured in donuts with amaranth flour.

Figure 3 shows the trend for moisture loss during the frying of doughnuts. As expected, moisture decreased with the increasing of frying time and oil temperature, due to a greater evaporation. Final moisture was lower for the control and the highest value was observed for amaranth formulation, at all tested frying temperatures. Firmness of doughnuts was affected by the replacement of wheat flour. Lower firmness was achieved in control (439.3 N) and doughnuts containing integral wheat flour (545.7 N) with 50% of this flour, respect of those containing 50% of oat flour (592 N) and 50% of amaranth flour (657 N). Addition of amaranth flour contributed to a darker dough and without big changes during the frying process. Control, doughnuts with integral wheat flour replaced at 30 or 50% and doughnuts with oat flour at 30% were well accepted by a sensory panel (25 non-trained judges), due to softer texture and lighter color. Eventhough, the amaranth flour at 50% resulted in the highest fiber and protein contents, these doughnuts were hard, dark and evaluated below than 6 in a sensory analysis using a hedonic scale of 9-points. More effort must be conducted in this way to have a more acceptable product with integral amaranth flour.

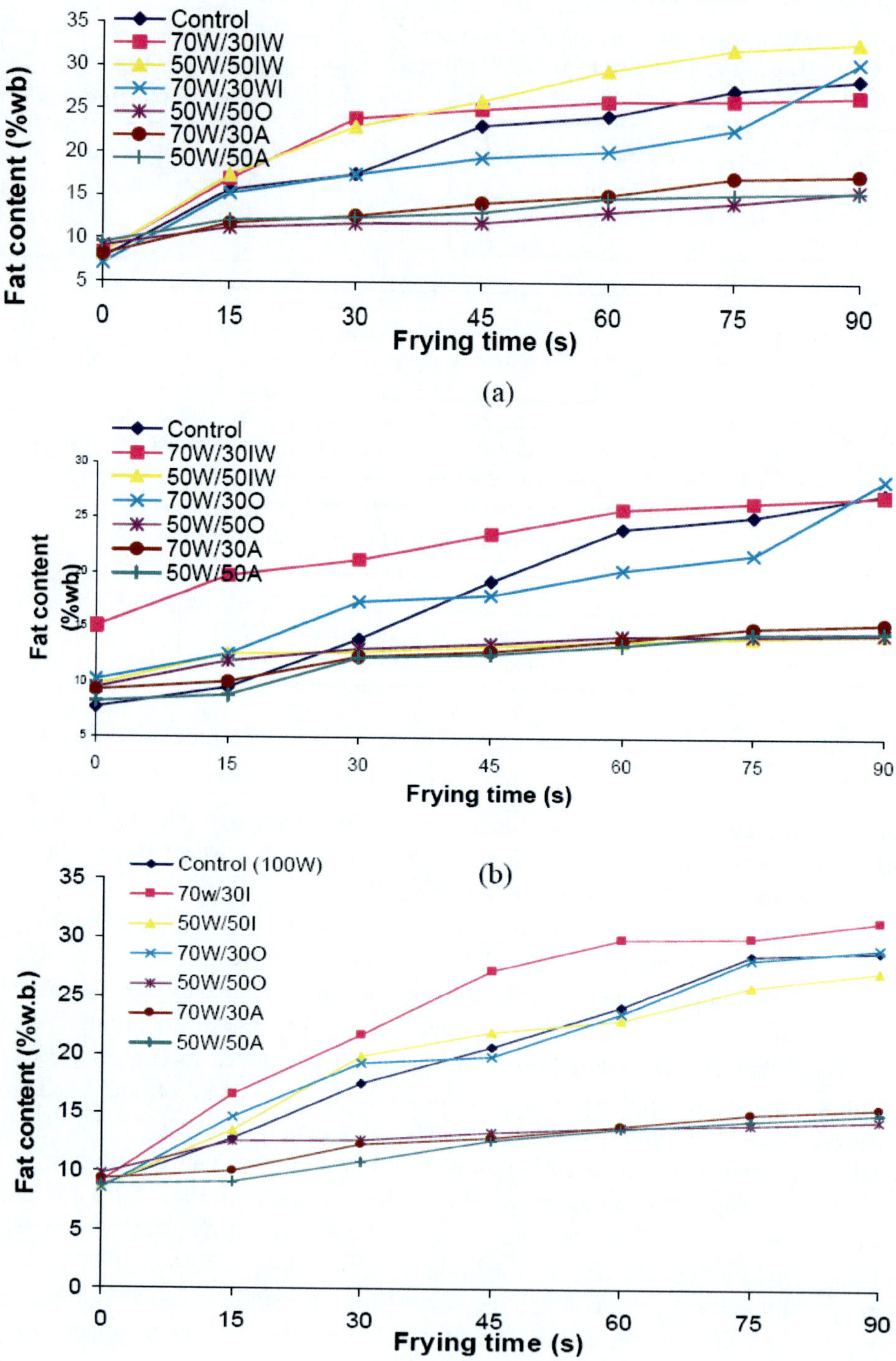

(c)

Figure 2. Fat uptake in donuts elaborated with mixtures of flours during deep-fat frying process at (a) 180, (b) 190, and (c) 200°C (From Treviño-Cuevas *et al.*, 2004).

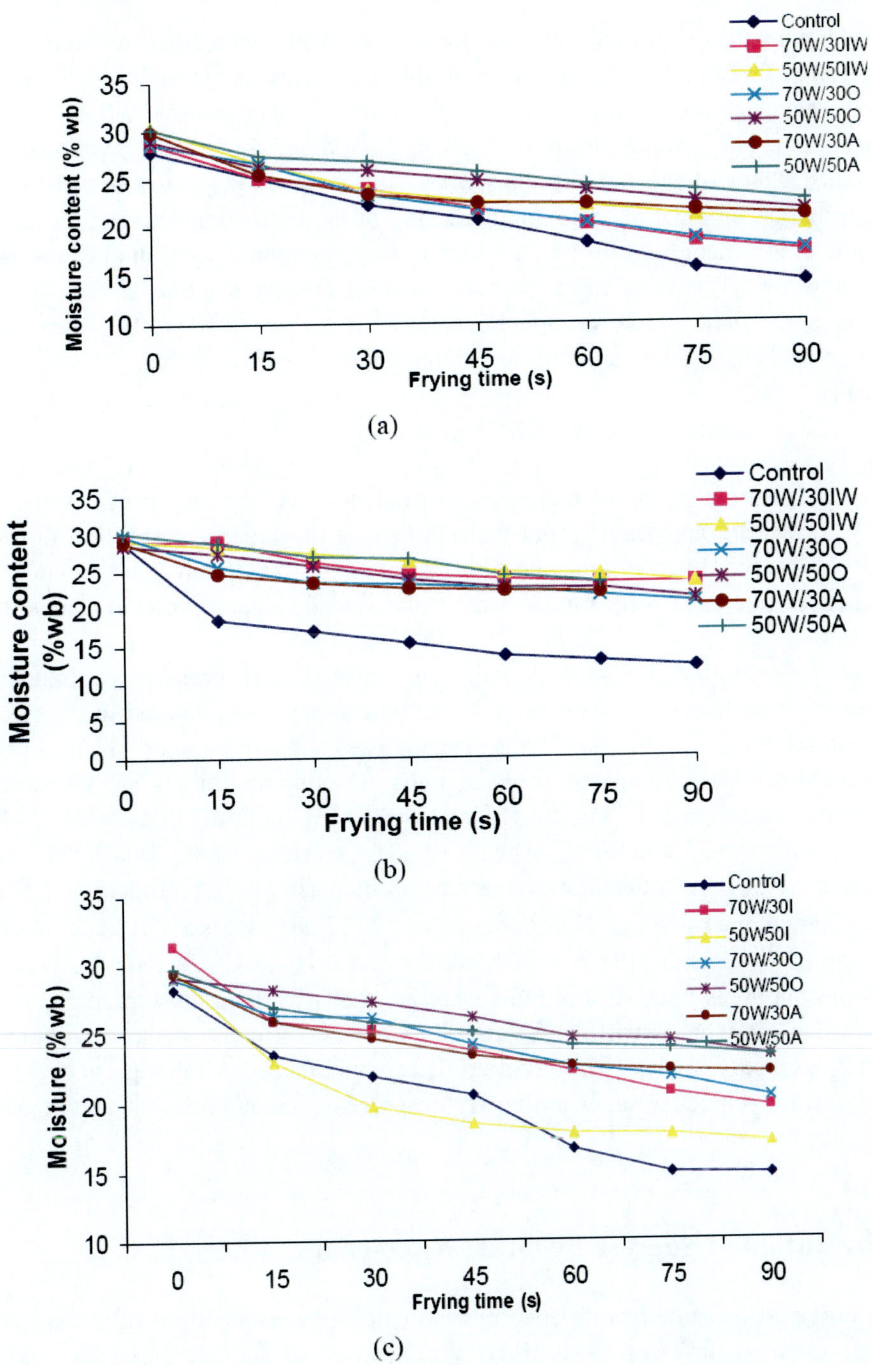

Figure 3. Moisture loss in donuts elaborated with binary mixtures of flours during deep-fat frying process at (a) 180, (b) 190, and (c) 200°C (From Treviño-Cuevas *et al.*, 2004).

Vetch Flour

Vetch is a legume recognized for its low cost in India, Pakistan, Bangladesh and Middle East countries, and for its high protein (28% w.b.) and lysine contents. Rehman *et al.* (2007)

replaced the wheat flour from a doughnut formulation with vetch flour at levels of 5, 10, 15 and 20 g/100 g. Protein, lysine and trypsin inhibitor contents increased in the doughnuts with vetch flour; also, the protein digestibility of these doughnuts increased when in vitro assays were carried out. The vetch flour up to 15 g/100 g improved the protein functional properties of doughnuts. Color of the flours blend (wheat and vetch flour) was affected by the vetch flour presence, became darker with the increasing of the vetch flour proportion; at a level of 5g/100g the color changes were no significant. Also, doughnuts with higher proportions of vetch flour were darker, due to an increase in the Maillard reaction as a consequence of higher content of lysine. However, doughnuts with vetch flour of 10 g/100 g were accepted in the sensory analysis, with qualifications higher than 7 for color, taste, flavor and overall acceptability.

Soybean Derivates

For many years, it is known that soybean flour decrease the fat absorption in doughnuts (Wolf, 1970). Martin and Davis (1986) tried to relate the fat low absorption to the protein dispersibility index, However, the mechanism for less absorption when soybean flour is incorporated is not clear yet, anyway soy flour is added commonly in many doughnuts formulations.

Lee *et al.* (2008) blended soybean hull with wheat flour to prepare doughnuts. Soybean hulls were microparticulated by jet milling and wheat flour was divided in both coarse and fine fraction by using of an air classifying system. The coarse fraction of wheat flour was dry-coated with the microparticulated soybean hulls. Doughnuts containing microparticulated hulls decreased the fat content; in doughnuts from 40.5% (control) to 34.9 and 29.3% when the hull contents were 5 and 10%, respectively. The presence of soybean hulls affected the hardness and crispness of doughnuts, observing an increase of these properties. For example, control doughnuts recorded 312 N as hardness and 1207 N·s as crispness, doughnuts with 5% of hulls had 2149 N and 2489 N·s of hardness and crispness, respectively. Hardness and crispness of doughnuts with 10% of hulls were around twice than those prepared with 5% of hulls. However, doughnuts with soybean hulls were no different to those without soybean hulls after a sensory analysis ($p<0.05$). Authors concluded that microparticulated soybean hulls may form a protective layer during frying process, reducing the oil uptake of the final product.

4.3. Substitution of Sucrose by Other Sweeteners

Low cost and consumer preference by non-caloric sweeteners are important reasons to incorporate them in different foods. However, absence of sucrose can affect the physical properties of bakery products when they are elaborated with other sweeteners. Sucrose replacement affects color of breads because sucrose is the responsible of chemical reactions to promote the crust color in breads, such as caramelization and Maillard's reaction (Badui-Dergal, 1999). Other important objective of sucrose in breads is to give firmness and good texture to the product (Sultan, 1990). Attia *et al.* (1993) employed sweeteners such as acesulfame K, aspartame and binary mixtures of them with polidextrose and fructose to make cakes and Ronda *et al.* (2005) used several polyols and oligosaccharides to prepare sugar-free cakes. The use of sweeteners different to sucrose affected the bread properties in both cases.

The effect of some sweeteners in doughnuts was evaluated by Perez-Perdiguero *et al.* (2007). The employed sweeteners were xylitol, sorbitol, and the commercial products Splenda® (dextrose, maltodextrine and sucralose) and Diabesugar® (Isomaltose, inuline, and acesulfame K). Doughnuts were fried in sunflower oil at 180 and 200°C. Crust color was affected by the sweetener type and frying temperature ($p<0.05$). When frying temperature was 180°C, doughnuts with non-caloric sweetener fried during 60 s showed small values in luminosity, similar to control donuts fried only 30 s. Using higher oil temperature (200°C), luminosity values of doughnuts with replaced sucrose exhibited higher luminosity values, similar to doughnuts fried at 180°C during 45 s without sucrose replacement (table 2). So, high temperatures used in frying were not able to balance the light crust color in doughnuts promoted by the lack of sucrose in the formulation. At 200°C, the presence of commercial blends of sweeteners resulted in lower luminosity values, respect to polyols used individually. This event is in agreement with the reported by Attia *et al.* (1993), who observed that blend of sweeteners work better than the individual presence.

Texture and density of doughnuts were affected by the sugar replacement in doughnuts formulation as well ($p<0.05$). In contrary with the results published by Ronda *et al.* (2005), replacement of sucrose by non-caloric sweeteners resulted in higher firmness values in doughnuts (table 3). In general, substitution of sucrose resulted in doughnuts with lower density. Additionally, blends of sweeteners promoted lower density in doughnuts, in comparison with polyols used individually (table 3).

Besides the effect on physical properties of doughnuts, substitution of sucrose affected the oil content in the doughnuts ($p<0.05$). Doughnuts with xylitol (31.7% w.b.), Diabesugar® (32.2%), and sorbitol (33.2%) showed lower oil contents than the control (37%). A reduction of 15% in the oil uptake employing xylitol was considered an advantage in spite of the differences in color and texture. Highest color and texture qualifications were observed in doughnuts formulated with Splenda®; for flavor, doughnuts with xylitol were the most preferred by a sensory panel. In general, doughnuts with non-caloric sweeteners were evaluated with higher scores than the control. Authors concluded that, nowadays, people have a special taste for sweeteners different to the regular sugar.

Table 2. Luminosity values for crust color in doughnuts elaborated with sucrose (control) and other sweeteners, fried at two temperatures (180 and 200°C)

Formulation/ frying time	Luminosity 180°C	Luminosity 200°C
Control/30 s	45.31 ± 1.21	35.49 ± 0.97
Control/45 s	41.10 ± 1.61	31.62 ± 0.75
Control/60s	37.76 ± 1.07	30.06 ± 1.20
Sorbytol	44.48 ± 0.81	42.30 ± 1.13
Xylitol	45.93 ± 0.87	42.99 ± 1.20
Splenda®	47.86 ± 1.34	45.54 ± 2.20
Diabesugar®	45.84 ± 1.22	44.65 ± 1.37

Source : Pérez-Perdiguero *et al.*, 2007.

Table 3. Texture and density values for doughnuts elaborated with sucrose (control) and other sweeteners, fried at 200°C

Formulation/frying time	Firmness (N)	Density (kg/m^3)
Control/30 s	1.51 ± 0.18	252.21 ± 1.42
Control/45 s	1.61 ± 0.27	275.56 ± 3.99
Control/60s	1.83 ± 0.45	274.81 ± 11.89
Sorbytol	2.44 ± 0.77	261.67 ± 4.42
Xylitol	3.15 ± 0.15	261.95 ± 4.59
Splenda®	2.86 ± 0.22	223.05 ± 1.20
Diabesugar®	2.51 ± 0.30	221.78 ± 0.50

Source : Pérez-Perdiguero *et al.*, 2007.

4.4. Application of Edible Films

To reduce fat uptake during frying, the application of edible films has been studied for different fried foods. For example, hydroxipropylmethylcellulose edible coatings reduced the fat uptake in nuggets fried in peanut oil (Balasubramanian *et al.*, 1995) and methylcellulose films were useful for potatoes and dough discs (Zaritzky *et al.*, 2004). Edible coatings are considered a relatively easy and cheap way to reduce the fat uptake during frying.

Soybean Protein Films

Rayner *et al.* (2000) employed soy protein films with or without plasticizer to observe the effect in the oil uptake of different fried products, among them, dough discs from a doughnut mix.A 10% soy-protein solution without plasticizer, or protein solutions with 3% glycerin or 0.05% gellan gum as plasticizers were held at 80°C for 20 min and then, cooled to provide the film. Films were applied to the foods using the dipping method and films were dried in a convection oven. After this "dehydration", discs of doughnut mix were fried. The use of plasticizer (glycerin or gellan gum) reduced the oil content in fried doughnuts, being 0.179 and 0.176% d.b. in comparison with the determined in the films without plasticizer (0.252%) and in those uncoated samples (0.426%). Additionally, the moisture loss was avoided using gellan gum. The peak force measured in a penetration test was lower for coated items. Uncoated samples resulted with 0.043 N while the coated discs showed 0.027, 0.029 and 0.039 N for doughnuts without plasticizer, with glycerin and with gellan gum soybean films, respectively. From a triangle sensory test, panelists were not able to detect differences between uncoated and coated with soybean films and gellan gum as plasticizier.

Chitosan and Cellulose Films

Martinez-Joffre *et al.* (2007) applied edible films based in chitosan or cellulose films to reduce the oil uptake in yeast-leavened doughnuts. A 1% chitosan solution was prepared in 2% malic acid solution with 0.8% oleic acid as plasticizer. Other 2% cellulose solution was prepared using 1.5% sorbitol as plasticizer. Solutions for films were prepared to get them ready for application after the final fermentation of doughnuts. Chitosan and cellulose solutions were sprayed in the doughnuts surface and the samples were immediately submitted to deep-fat frying at 180 and 200°C during 60 s in fresh sunflower oil. Doughnuts without coating were fried as control.

As it can see in table 4, moisture and oil contents were affected by the application of films ($p<0.05$). Moisture content was higher in doughnuts covered with cellulose films (29.46% w.b.). The highest oil reduction was obtained using chitosan films and frying at 200°C, in this case, doughnuts had 16.9% of oil content, lower than the content of control doughnuts (23.1%). Doughnuts with cellulose coatings had around 20% of oil content at the same frying temperature. Thus, using chitosan films the reduction of oil uptake was twice than that got employing cellulose edible films. Frying temperature affected the moisture an oil contents too. At higher temperature (200°C), lower moisture contents were obtained, while lower oil contents were determined. Also, physical properties of doughnuts were affected by the coating application (table 5). Doughnuts with cellulose coating were softer than the other ones, while chitosan film promoted harder crusts. Densities for control and doughnuts with cellulose film were similar, and denser items were observed in the presence of chitosan coating. Cellulose coating improved lighter appearance, while chitosan coating resulted in darker doughnuts.

In the sensory tests, best qualifications corresponded to doughnuts with cellulose coating. Even a higher reduction in the oil uptake was reached up with chitosan films, the recommended coating is cellulose, due to its physical properties and sensory acceptance.

Table 4. Effect of edible coating application on moisture and oil contents of doughnuts fried at 180 and 200°C for 45 s

Frying temperature (°C)	Control	Chitosan film	Cellulose film
Moisture content (%w.b.)			
180°C	23.66 ± 0.18	24.04 ± 0 .93	29.46 ± 0.30
200°C	21.45 ± 0.47	23.34 ± 0.93	28.46 ± 0.30
Oil content (%w.b.)			
180°C	25.33 ± 0.04	17.70 ± 0.95	23.22 ± 0.25
200°C	23.12 ± 0.15	16.91 ± 0.45	20.05 ± 1.17

Source: Martínez-Joffre *et al.*, 2007.

Table 5. Physical properties of doughnuts with and without edible coating fried at 200°C for 45 s using fresh sunflower oil

Film type	Texture (N)	Density (kg/m^3)	Luminosity
Without film (control)	2.40 ± 0.13	380.39 ± 0.93	29.07 ± 0.35
With chitosan film	3.78 ± 0.30	353.16 ± 0.93	33.45 ± 0.96
With cellulose film	1.99 ± 0.30	733.67 ± 0.93	24.09 ± 0.30

Source: Martínez-Joffre *et al.*, 2007.

4.5. Use of New Frying Media

An indirect choice to modify the formulation of doughnuts is the modification of the frying medium. Due to the fat/oil entrances to the food during the deep-fat frying process, the oil/fat becomes a doughnuts component.

Non-Caloric Frying Medium

In order to reduce the caloric content of doughnuts, Morehouse and Zabik (1989) evaluated the potential use of synthetic polydimethylsiloxane (PDMS) fluids as a non caloric deep-fat frying media. Some foods, including cake doughnuts were fried in the PDMS fluids with different viscosity (35, 100, 350 and 1000 centistokes) and in corn oil as control. Fat contents in PDMS-fried foods were lower than food fried in corn oil (25% less). Most efficient heat transfer and acceptable quality of fried foods were observed when 35 centistoke PDMS fluids was employed. PDMS fluids seemed to have potential to be used as frying media, even they have not been approved by the FDA as a food ingredient.

Trans Fat Free Oil

The concern about ingestion of trans fats has increased in the last years. Trans fats have been related with cancer and heart diseases. Doughnuts show varying amounts of trans fatty acids. van Erp-baart *et al.* (1998) reported till 32% of fatty acids in certain analyzed doughnuts; this value was higher than the measured content of trans fatty acid in pizzas (1-5%) and croissants (15%).

Bordi *et al.* (2007) submitted to a sensory analysis two yeast raised doughnuts fried in a trans fat and trans fat free oil. The sensory panel involved 197 non-trained judges. Based on appearance, taste, texture, sweetness, moisture content and overall liking, no significant differences between the two doughnuts were found out. Only there was difference between the color of doughnuts; doughnuts fried in trans fat free oil ranked with lower qualifications, because they were perceived slightly lighter than those doughnuts fried in trans fat. Authors concluded that trans fat free oils are a good choice for the food industry in order to reduce the trans fats in fried doughnuts.

5. Final Remarks

Most of the new formulations to improve the nutrition facts of doughnuts are related to diminish the oil uptake in them. The reductions in the final fat/oil content of the modified final doughnuts, implementing different ways of formulations are resumed in table 6. As it was commented, there are many different factors involved in the oil content of doughnuts. Factors related to the formulation have been studied more extensively, attending mainly to the addition or replacing of ingredients (polysaccharides, flours with higher protein contents, fiber, other sweeteners) and use of edible coatings, which avoid the oil uptake during the frying. Other way to improve the nutrimental profile is to use no caloric frying media or to use trans fat free oil, but these last choices are in developing yet. The cited proposals of formulations are relatively easy to implement and have potential for commercial application.

The global efforts in this matter are focused to have doughnuts without risks for the consumers health, with the objective to follow enjoying of the crispness, color and special taste and flavor of this unique food.

Table 6. New formulations to improve the nutrition facts of fried doughnuts

Modification in the formulation	Frying conditions	Reduction of the oil content	Reference
Addition of curdlan 0.5-1.0%	Cottonseed oil at 170°C, 4 min	8-9% w.b.	Funami *et al.*, 1999
Addition of dietary fiber and cellulose from pineapple core	Soybean oil at 190°C, 2 min	23.1 - 40.7% w.b.	Prakongpan *et al.*, 2002
Replacement of 25 or 50% of wheat flour: by gelatinized rice flour or	Wesson vegetable oil at 177°C, frying time needed to reach up 90-101°C in the core	37.8 – 47.1% w.b.	Shi *et al.*, 2001
by precooked acetylated rice starch		25.1- 37.5% w.b.	
Replacement up to 30% of wheat flour with pregelatinized rice flour	Wesson vegetable oil at 190°C, frying time till to reach up 90-101°C in the core	54% d.b.	Shi and Daigle, 2002
Replacement of wheat flour at 30 or 50% with integral wheat flour With integral oat flour With integral amaranth flour	Soybean oil at 190°C, 60 s	0.01 - 42.5% w.b. 15.6 - 40.5% w.b. 37.3 – 43.9% w.b.	Treviño-Cuevas *et al.*, 2004
Replacement of wheat flour for 20% of vetch flour	Soybean oil at 185°C, 2 min	11.85% d.b.	Rehman *et al.*, 2007
Replacement of wheat flour for 5 or 10% of soybean hull	Soybean oil at 180°C, 3 min	13.8 – 27.7%w.b.	Lee *et al.*, 2008
Substitution of sucrose for other sweeteners	Sunflower oil at 200°C, 60 s	10.3 -14.3% w.b.	Perez-Perdiguero *et al.*, 2007
Use of soybean protein edible coatings	160°C, 15 min	40.7 – 55.1% d.b.	Rayner *et al.*, 2000
Application of chitosan coatings and cellulose edible films	Sunflower oil at 200°C, 60 s	26.8% w.b. 13.4% w.b.	Martinez-Joffre *et al.*, 2007

REFERENCES

AACC. 2000. Approved methods of American Association of Cereal Chemists. 15th Edition. American Association of Cereal Chemists: Arlington, USA.

AOAC. 1990. Official Methods of Analysis. 15th Edition. Association of Official Analytical Chemists: Washington DC, USA.

Aggelousis G. and Lalas S. 1997. Quality changes of selected vegetable oils during frying of doughnuts. *Rivista Italiana delle Sostanze Grasse*, 74(12): 559-565. (In Italian).

Amendola J. 1993. The Baker's Manual. 4a. Ed. Van Nostrand Reinhold: New York, NY, USA.

Attia E.A., Shehata H.A. and Askar A. 1993. An alternative formula for the sweetening of reduced-calorie cakes. *Food Chemistry,* 48: 169-172.

Badui-Dergal S. 1999. Food Chemistry. Pearson Pub. Mexico City, Mexico. (In Spanish).

Balasubramaniam V.M., Mallikarjunan P. and Chinnan M.S. 1995. Heat and mass transfer during deep-fat frying of chicken nuggets coated with edible film: influence of initial fat content. CoFE. 103-106.

Bordi P.L., Hack D.M., Rager M.D. and Hessert S.W. 2007. Sensory comparison of doughnuts fried in trans fat-free oil to those fried in oil containing trans fats. *Journal of Foodservice,* 18(6): 227-237.

Echendu C.A., Onimawo I.A. and Adieze S. 2004. Production and evaluation of doughnuts and biscuits from maize-pigeon pea flour blends. *Nigerian Food Journal,* 22: 147-153.

Funami T., Funami M., Tawada T. and Nakao Y. 1999. Decreasing oil uptake of doughnuts during deep-fat frying using curdlan. *Journal of Food Science,* 64(5): 883-888.

Gamble, M.H., Rice, P. and Selman, J.D. 1987. Relationship between oil uptake and moisture loss during frying of potato slices from c.v. Record UK tubers. *International Journal of Food Science and Technology,* 22: 233-241.

Garayo J. and Moreira R. 2002. Vacuum frying of potato chips. *Journal of Food Engineering,* 55: 181-191.

Hatae K., Miyamoto T., Shimada Y., Munekata Y., Sawa K., Hasegawa K. and Kasai, M. 2003. Effect of the type of frying oil on the consumer preference for doughnuts. *Journal of Food Science,* 68(3): 1038-1042.

Kent N. L. 1987. Cereal Technology. Acribia Pub. Zaragoza, Spain (In Spanish).

Lee, J.-S., Kim B.-K., Kim K.-H. and Park D.-J. 2008. Preparation of low-fat uptake doughnut by dry particle coating technique. *Journal of Food Science,* 73(3): E137-E142.

Martin M.L. and Davis A.B. 1986. Effect of soy flour on fat absorption by cake doughnuts. *Cereal Chemistry,* 63(3): 252-255.

Martinez-Joffre J, Argaiz-Jamet A., Vélez-Ruiz J.F. and Sosa-Morales M.E. 2007. Doughnuts with reduced fat content based on edible coatings application. *IFT Annual Meeting.* July 29-31, Chicago IL, USA.

McComber D. and Miller E.M. 1976. Differences in total lipid and fatty acid composition of doughnuts as influenced by lecithin, leavening agent and use of frying fat. *Cereal Chemistry,* 53(1): 101-109.

Morehouse S.E. and Zabik M.E. 1989. Evaluation of polydimethylsiloxane fluids as non-caloric frying media. *Journal of Food Science,* 54(4): 1061-1065.

Moreira R.G., Castell-Perez M.E. and Barrufet M.A. 1999. Deep-fat frying. Fundamentals and Applications. Aspen Publishers, Inc.: Gaithersburg, MD, USA.

Oke O. L. 1983. Amaranth. In: *Handbook of Tropical Foods*, H. T. Chan Jr. Ed., Marcel Dekker, Inc.: New York. NY, USA. Pp. 1.

Pérez-Perdiguero L., Vélez-Ruiz J.F and Sosa Morales, M.E. 2007. Effect of addition of non-caloric sweeteners on physical properties of deep-fat fried doughnuts. *Proceedings of the CIGR Section VI 3rd International Symposium*, September 24-26, Napoli, Italy.

Prakongpan T., Nitithamyong A. and Luangpituksa P. 2002. Extraction and application of dietary fiber and cellulose from pineapple cores. *Journal of Food Science,* 67(4): 1308-1313.

Pyler E.J. 1988. Baking Science and Technology Vol. II. 3rd Ed., Sosland Publishing Co.: Kansas City, KA, USA.

Rayner M., Ciolfi V., Maves B., Stedman P. and Mittal G.S. 2000. Development and application of soy-protein films to reduce fat intake in deep-fried foods. *Journal of the Science of Food and Agriculture,* 80: 777-782.

Rehman S.-u., Paterson A., Hussain S., Murtaza M.A. and Mehmood S. 2007. Influence of partial substitution of wheat flour with vetch (*Lathyrus sativus* L) flour on quality characteristics of doughnuts. *LWT-Food Science and Technology*, 40(1): 73-82.

Ronda F., Gómez M., Blanco C. A. and Caballero P. A. 2005. Effects of polyols and nondigestible oligosaccharides on the quality of sugar-free sponge cakes. *Food Chemistry,* 90: 549- 555.

Shih F. and Daigle K. 2002. Preparation and characterization of low oil uptake rice cake donuts. *Cereal Chemistry,* 79(5): 745-748.

Shih F.F., Daigle K.W. and Clawson E.L. 2001. Development of low oil-uptake donuts. *Journal of Food Science.* 66(1): 141-144.

Sosa-Morales M.E. and Vélez-Ruiz J.F. 2008. Heat and mass transfer modeling through deep-fat frying of selected foods. In: *Progress in Food Engineering Research and Development.* J.M. Cantor, Ed. Nova Science Publishers Inc.: New York, NY, USA. Pp: 197-222.

Sultan W. J. 1990. Practical Baking. 5th Ed., Van Nostrand Reinhold: New York, NY, USA.

Tan K.J. and Mittal G.S. 2006. Physicochemical properties changes of donuts during vacuum frying. *International Journal of Food Properties,* 9: 85-98.

Treviño-Cuevas R., Vélez-Ruiz J. F. and Sosa-Morales M.E. 2004. Variation in composition and mass transfer phenomenon in yeast-leavened donuts formulated with binary mixtures of flours during deep fat frying. *Proceedings of the 9th International Congress on Engineering and Food (ICEF9),* March 7-11, Montpellier, France.

van Erp-baart M-A., Couet C., Cuadrado C., Kafatos A., Stanley J. and van Poppel G. 1998. Trans fatty acids in bakery products from 14 European countries: the TRANSFAIR Study. *Journal of Food Composition and Analysis,* 11(2): 161-169.

Vélez-Ruiz J.F. and Sosa-Morales M.E. 2003a. Evaluation of physical properties of dough of donuts during deep-fat frying at different temperatures. *International Journal of Food Properties,* 6(2): 341-353.

Vélez-Ruiz J.F. and Sosa-Morales M.E. 2003b. Heat and mass transfer during the frying Process of donuts. In: *Transport Phenomena in Food Processing.* J. Welti-Chanes, J.F. Vélez-Ruiz & G.V. Barbosa-Cánovas, Eds. CRC Press: Boca Ratón, FL, USA. Pp: 429-441.

Watt B.K. and Merril A.L. 1963. Composition of Foods-Raw, Processed and Prepared. USDA Agricultural Handbook No. 8. Washington, D.C, USA.

Wolf J.W. 1970. Soybean proteins: their functional, chemical and physical properties. *Journal of Agricultural and Food Chemistry.* 6: 969.

Zaritzky N., García M.A., Ferro C., Bortola N. and Martino M. 2004. Methylcellulose coatings applied to reduce oil uptake in fried products. *Food Science and Technology,* 10(5): 339-346.

INDEX

A

B

C

E

F

G

H

I

J

K

L

M

W

X

Y